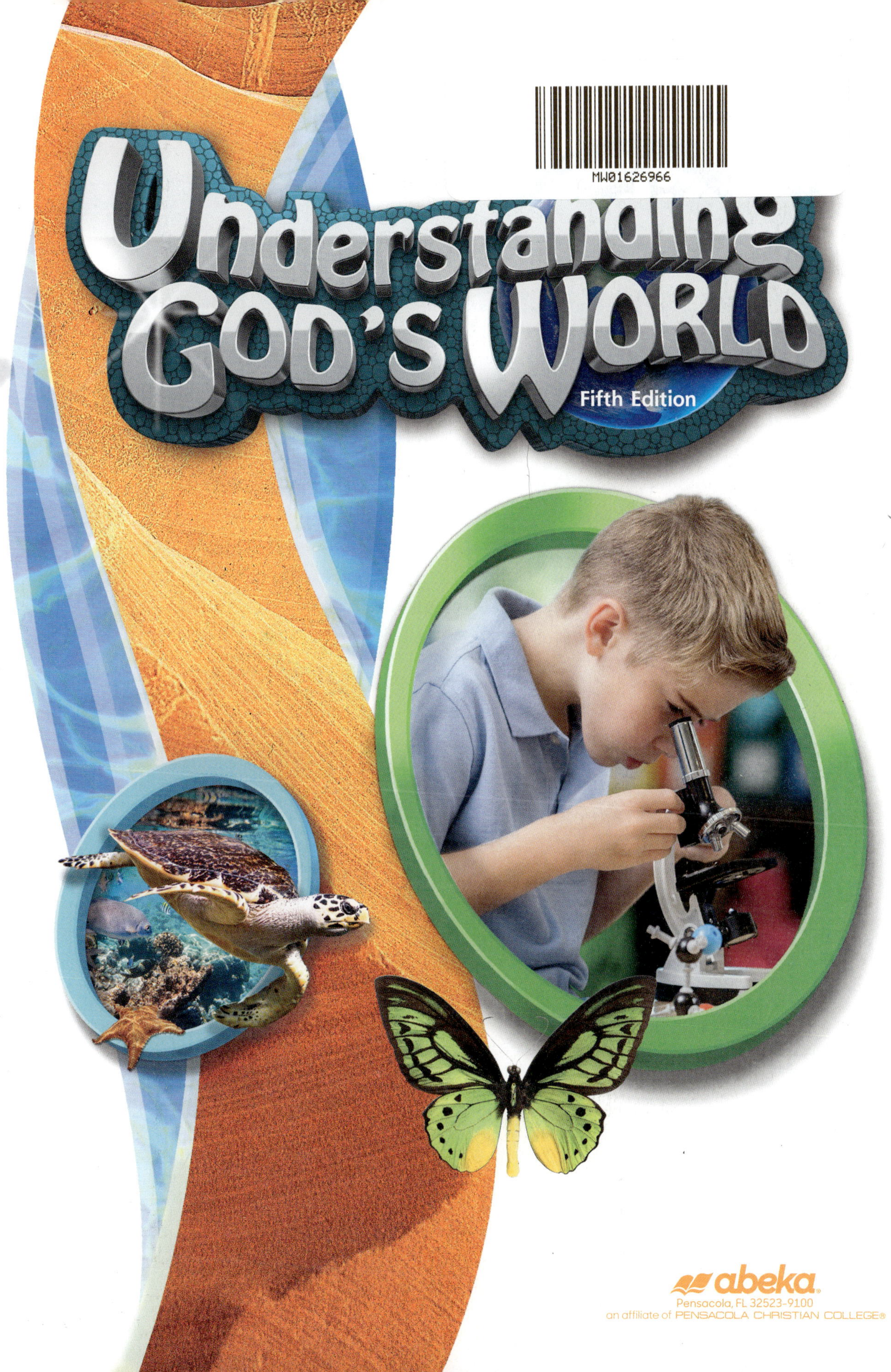
MW01626966
Understanding
God's World
Fifth Edition
abeka
Pensacola, FL 32523-9100
an affiliate of PENSACOLA CHRISTIAN COLLEGE®

Learn to apply scientific principles, such as the scientific method and God's laws in nature, and observe the LIVING parts and NONLIVING parts of the world around you as you explore the world through the pages of Understanding God's World.

To Teachers and Parents

Students will gain a greater interest in science through reading facts about the world around them. By discovering, learning, and understanding science which presents God as the Master Designer of the universe, students will gain a deeper reverence for God while developing a biblical worldview. Students will better understand basic science concepts while building their critical thinking skills through Observe to Understand, Jr. Scientist Lab Activities, and Try This! hands-on activities and demonstrations. Diagrams, Comprehension Checks, Terms boxes, and Chapter Concepts Review sections are designed to prepare students for written evaluation.

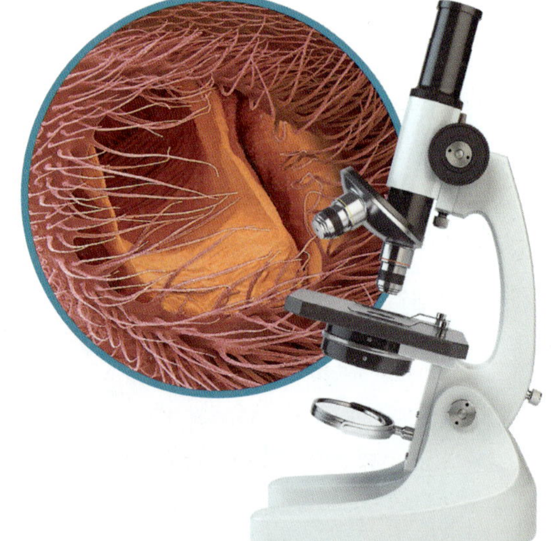

Understanding God's World
Fifth Edition

Staff Credits
Author: Hilary Hasty, Dawn Mereness
Managing Editor: Amy Yohe
Edition Editors: Tanya Harrington, Camilla Kochanowicz, Cheryl Reid, Mary Rhodes
Designers: Todd Hatchett, Grace Larson
Production Artist: Claire Lewis
Illustrators: Margaret Benson, Naomi Ji, Natti Guest, Abeka Staff

Credits are on pages 487–488 which is an extension of this copyright page.

Cataloging Data
Understanding God's world--5th ed.
x, 556 p. : col. ill. ; 26 cm. + teacher's ed.
(abeka science/health series)
For grade 4.
1. Science - Study and teaching (Elementary)
II. Abeka Book, Inc.

Library of Congress: Q161.2 .U62 2020
Dewey System: 500

Contents

Science Foundations

 Try This!

 Scientist Corner

 Jr. Scientist Activity

 Observe to Understand

continued

Science Activities

Chapter 3 Understanding Force and Motion 87

Science Activities

Life Science

Understanding Ecosystems at Work 256

Science Activities

Unit 3 Earth and Space Science

Chapter 7 Understanding the Earth and Its Foundations 321

Science Activities

Chapter 8 Understanding Weather ... 381

Science Activities

Chapter 9 Understanding the Great Expanse of Outer Space ... 436

Science Activities

Pronunciation Key

Symbol	Example	Symbol	Example
ā	āte	ŏ	nŏt
â	dâre	oi	boil
ă	făt	o͞o	fo͞od
ä	fäther	o͝o	bo͝ok
ə	ago (ə•gō′)	ou	out
ē	ēven	th	thin
ĕ	ĕgg	t̶h̶	t̶h̶ere
ẽ (ər)	pondẽr	tū͡	pictū͡re
ī	īce	ū	ūnit
ĭ	ĭt	û	hûrt
ō	ōver	ŭ	ŭp
ô	côrd, taught, saw	zh	measure

Unit 1 Science Foundations

In the beginning was the Word, and the Word was with God, and the Word was God. The same was in the beginning with God. All things were made by Him; and without Him was not any thing made that was made. John 1:1–3

Understanding the Scientific Process

Chapter 1

TERMS

technology: science put to practical use

scientific method: orderly, step-by-step process in which scientists work

hypothesis: a sensible guess; a reasonable prediction

experiment: a procedure designed to answer a question and to test a hypothesis

data: a collection of facts

conclusion: a decision based on evidence

1.1 Using the Scientific Method

No matter where you go in God's world, you will see the evidence of His handiwork.

- ✓ Systems of energy and motion that work exactly as God planned
- ✓ Forests and fields that produce an abundance of plants
- ✓ Thriving plant and animal communities found all over the earth
- ✓ Oceans of life-giving water and an atmosphere that protects life
- ✓ A landscape of mountains, hills, and valleys
- ✓ The heavens that declare God's glory

All of creation shows us that God's world was miraculously designed and created. As we take the time to understand God's world, we are filled with awe and gratitude to our Maker—the Lord of all creation.

God rules and reigns over creation because He designed and created it. He had a special plan for everything He made. Would you like to understand more about God's world? Throughout this science book, you will learn to understand God's world and the way it works.

What Science Is

Science *is the careful study of the wonders of the universe.* A **scientist** *is a person who works to understand more about the universe.* What an enormous job that could be! Could any person understand all of creation? No, even the greatest scientists understand only a small part of God's world.

Our human minds cannot comprehend all of the wisdom of God. Yet God has given people the privilege to observe and study His creation. God created us with curious minds that are able to learn more about Him and His world. He has given us the ability to enjoy His world and use it wisely.

Why We Study Science

One reason we study science is to help others. Many scientists have worked for years trying to find ways to help people. They have thought of ways to prevent or cure disease. Some scientists have designed and invented ways to keep people safe.

We use science in our daily lives. Devices and machines can make hard jobs easier. When *science is put to practical use*, it is called **technology**. Much of technology benefits mankind.

With the use of science through technology, doctors can diagnose patients more accurately. Farmers can produce better crops for food. People can make wise choices to keep their bodies healthy. Designers can make buildings and vehicles that help keep people safe. The best scientific achievements benefit others.

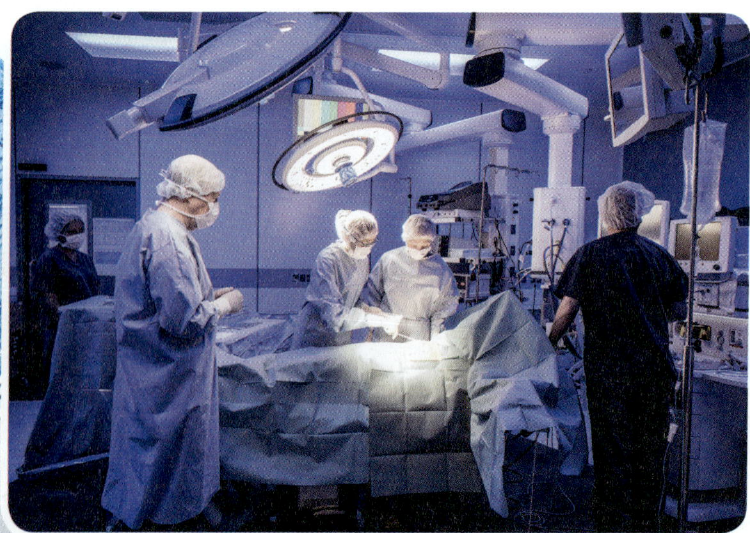

We also study science because it glorifies God. When God created the universe, He did so with wise design and logical order. Understanding more about the order and design God used to create the world shows us some of God's characteristics, such as His perfection, wisdom, power, authority, goodness, creativity, care, mercy, and love.

> *For the invisible things of Him from the creation of the world are clearly seen, being understood by the things that are made.* Romans 1:20

Some people do not believe the universe was created because they do not believe in the Creator. Instead of recognizing the miraculous design of a created universe, they believe that the universe has evolved, or formed gradually, over millions of years. This belief is called evolution.

Believing God's Word is the foundation of a Christian's faith. Over and over again, true science supports what the Bible says—that God created the world. Creation scientists have the special job of carefully investigating God's world to glorify Him.

> *Through faith we understand that the worlds were framed by the word of God, so that things which are seen were not made of things which do appear.* Hebrews 11:3

The Scientific Method

While many people were basing scientific facts on only observation and careful thinking, Sir Francis Bacon introduced a *step-by-step process of finding scientific facts*, called the **scientific method**. This method helps people study each part of nature. When people used the scientific method, they began to understand more about how God's world works. *We remember Sir Francis Bacon as the Father of the Scientific Method.*

Sir Francis Bacon

The scientific method is a step-by-step process scientists use to learn facts and organize information. Scientists from all over the world follow these basic steps:

1. *Scientists observe and ask questions.* Scientists begin the scientific process by **observing**, or *looking carefully* at, a particular part of God's creation or by noticing the way something works. These observations lead to a question that they want to answer.

2. *Scientists form a hypothesis.* Using the information they have, scientists make a reasonable prediction to explain the answer to their question. *This kind of sensible guess, or reasonable prediction, is called a* **hypothesis**.

3. *Scientists experiment and gather data.* A hypothesis is just a reasonable guess, not a scientific fact. Scientists plan an **experiment**, or *procedure to answer their question and test their hypothesis.*

   As scientists complete their experiments, they write down all the facts they are learning. We call this *collection of facts* **data**.

4. *Scientists study data and reach conclusions. A* **scientific conclusion** *is a decision based on evidence.* Sometimes, data supports the hypothesis; in this case, a scientist will conclude his hypothesis is probably correct.

Sometimes, data does not support the hypothesis. *A good scientist follows the evidence rather than his or her idea.* Even if a scientist needs to conclude that a hypothesis is false, new things can be learned from the new data gathered. Scientists are able to share what they have learned with other scientists. Many scientific discoveries and inventions have improved medicine, technology, travel, and the way people live and work. More information helps scientists understand God's world and glorify Him in their work.

Throughout this book, you will be observing God's creation to understand it more. You might ask yourself questions you had not thought of before. You will be studying how the universe works as you experiment and gather your own data. You can be a Junior Scientist, learning how to think like an actual scientist. If you follow the scientific method, you can understand more about God's world, too.

Comprehension Check 1.1

Give the correct answer.

1. Why do we study science? to learn
2. Who is called the Father of the Scientific Method? Sir Francis Bacon

Matching: *Write the letter of the correct answer in the blank.*

D 1. to look carefully at something
C 2. a reasonable prediction or guess
B 3. a procedure designed to test a hypothesis
A 4. a collection of facts

A. data
B. experiment
C. hypothesis
D. observe

1.2 Learning to Observe

Observation is looking carefully at something. Most likely, you have already been doing this! Perhaps you have taken a hike on a nature trail with your family. As you walked, you noticed things. You noticed the color of the leaves on the trees and plants around you. You probably were careful as you walked over dead and decaying leaves, twigs, and branches. You felt the air around you. Perhaps it was hot and humid—or maybe the air felt very cool because the trees were blocking most of the sunlight. As you listened, you could hear insects buzzing or birds chirping. Most of the animals were hiding, but maybe you saw a bird fly from its nest or a squirrel scamper across your path.

To be a good observer, you need your senses. You can learn much about the world around you by using your eyes and ears. You can also observe through your senses of touch, taste, and smell.

As you observe, you are collecting data. During a nature hike, you can collect facts about your surroundings.

- ✓ The weather of the day
- ✓ The kinds of plants around you
- ✓ The kinds of animals living there

As you train yourself to observe more closely, you can collect even more data.

- ✓ The way the forest layers are helping each other
- ✓ What kinds of animal tracks you see
- ✓ How the dead plants are decaying and helping the soil
- ✓ Where mosses and ferns are growing

One of the best ways to develop your observation skills is to carry a notebook and pencil. These observation tools have helped scientists throughout history begin the process of discovering scientific facts. As you make an observation, you could write down your thoughts or sketch a picture.

Observation can lead to research. Perhaps you notice an unusual plant or insect. You might not have the correct field guide to find its name, but later you can compare your sketch with the sources you have available. You can then learn more about the part of God's creation that made you curious.

Observation Ideas

Leaves

Notice the colors and shapes of the leaves you see around you as they change color in autumn. Maple leaves usually turn brilliant red or orange. Leaves from poplars and cottonwoods become bright yellow. Oak leaves become deep red before they turn golden.

Caterpillars

Caterpillars are large and easily seen in the fall. Moth caterpillars are preparing to spin their cocoons,

and butterfly caterpillars are ready to be encased in a chrysalis. Look for them on trees, on buildings, and on the ground.

Shells

If you live near a beach or a pond, you might enjoy collecting shells. Use a field guide to identify the shell; then you could read a book about the animal that once made that shell its home.

Fruits

Hawthorns, chokecherries, apples, wild grapes, and wild cranberries are among the many fruits that ripen in the fall. Be sure not to eat any fruit until you know it can be eaten safely and has been washed thoroughly.

Rocks

If you walk through a stony field, in a gravel pit, beside a pond, or along the seashore, you will enjoy examining the different kinds of rocks to be found.

Feathers

Fall is an especially good time to observe feathers because many birds are replacing their old feathers with new ones. Large feathers come from large birds, such as owls, eagles, and hawks. Smaller feathers come from the smaller songbirds. The colors, patterns, and shape of the feather indicate which bird it came from. An all-red feather probably came from a cardinal.

A bright blue tail feather streaked with black was lost by a blue jay. A feather that seems black at first glance, but sparkles with a rainbow of colors in the sunlight, probably came from a starling.

Animal Tracks

If you take a walk after a rainfall, be alert to animal tracks in the moist soil. Deer tracks are narrow and heart-shaped. Rabbits and squirrels run by putting their hind feet ahead of their forefeet. The prints of the red fox look much like those of a dog. Mice and rats make very tiny tracks; sometimes, you can see the dragging marks made by their tails.

Seeds

A seed is a marvelous package containing a baby plant just waiting to reach the right place for it to grow. Some seeds are carried to their new homes by the wind. Look for milkweed and dandelion seeds being blown by the breeze. Other seeds will find you! Burdocks, cockleburs, beggar's ticks, and tick trefoil will hitch a ride on any person or animal that brushes by. Acorns, the fruit of the oak tree, are plentiful in the fall. If an acorn feels soft, there are probably insects called acorn weevils inside. If you find a small hole in the acorn, the weevils have left the acorn. You might also find some chewed-on acorns that tell you squirrels are nearby.

Observe to Understand

Nature

Here are some things you may need to take along for your observations:

- ✓ notebook
- ✓ pencil
- ✓ hand lens
- ✓ field guides
- ✓ binoculars
- ✓ camera
- ✓ water bottle
- ✓ snack
- ✓ sunscreen
- ✓ insect repellent

1. Decide on a place to observe. It could be your backyard or your neighborhood park. Your family may want to enjoy your observation adventure with you. Together, you can observe state parks, beaches, forests, walking paths, and gardens. Be sure to use insect repellent if you will be exploring near water, plant life, or foliage. If you will be outside a long time, wear sunscreen. You may also need to pack water and a snack.
2. As you explore, find something that you would like to learn more about—perhaps a feather, an old bird's nest, a rock, a shell, a flower, a beetle, a seed, a cone, or a leaf. Ask permission or check for signs before picking up anything.
3. Take the time to photograph or sketch your object of observation. Sketching helps you to observe things that you might not notice at first glance. Write down any questions or thoughts you may have during your observation.
4. Create an exhibit, or display, for your sketch, photograph, or object. You could even make a "What is it?" sign and place your object under the sign. Challenge others to find the answer.
5. Learn more about your object of observation. Field guides are helpful for finding the answer to "What is it?" questions. Libraries provide a variety of good field guides and information books that may have the answer you need. The answer might even be in this book! Check the table of contents, the glossary, and the field guide pages in the back of your science book.

Brainstorm.

1. Write down 5 science topics you would like to learn more about.

   Feather, seed, Plants, people, amleg

2. Make a list of what you will need for your observations.

variable: the parts of an experiment that are controlled

independent variable: the part of an experiment that is changed to test the hypothesis

controlled variables: the parts of an experiment that stay the same

dependent variable: the part of an experiment that is measured to test results

1.3 Making Predictions and Designing Experiments

Observing and Asking Questions

As you learn how to observe God's creation, you will notice things that you want to learn more about. What things make *you* curious? Let's pretend you are curious about growing a garden.

As you observe gardens in your community, you may also want to research gardens or study books from the library. Many of your questions will be answered as you study, but there are still things you may not understand. You may have many questions about growing a garden.

Forming a Hypothesis or Prediction

If you want to think like a scientist, it is *important to start with* **one question**. A good scientific question cannot be answered with a simple *yes* or *no*. It must be tested or proven.

Poor Question: Do carrots grow in all kinds of soil?
Good Question: What kind of soil is best for growing carrots?

A **hypothesis** is a sensible guess, or prediction, based on the data that you have already gathered. Because of your observation and research, you can now form a hypothesis to answer your one question. *Your hypothesis predicts the results of your research.*

Hypothesis: Carrots grow best in sandy soil because most root vegetables need good drainage.

When you form a hypothesis, you need to be able to prove it. How can you show that sandy soil is best for growing carrots? It is time to plan an experiment to test your hypothesis.

Designing a Fair Experiment

Scientists plan a fair experiment to test a hypothesis. Sometimes, the data supports the hypothesis, but other times it does not. Scientists should be willing to *follow the evidence rather than their own ideas.*

As scientists design experiments they are aware of **variables**, *the parts of an experiment a scientist controls.* These variables may or may not change. First, scientists must choose one independent variable. *An* **independent variable** *is the part of an experiment a scientist changes.* Soil type may be the independent variable tested in the experiment. One planter could have a sandy-soil mixture while the other planter will have a clay-soil mixture.

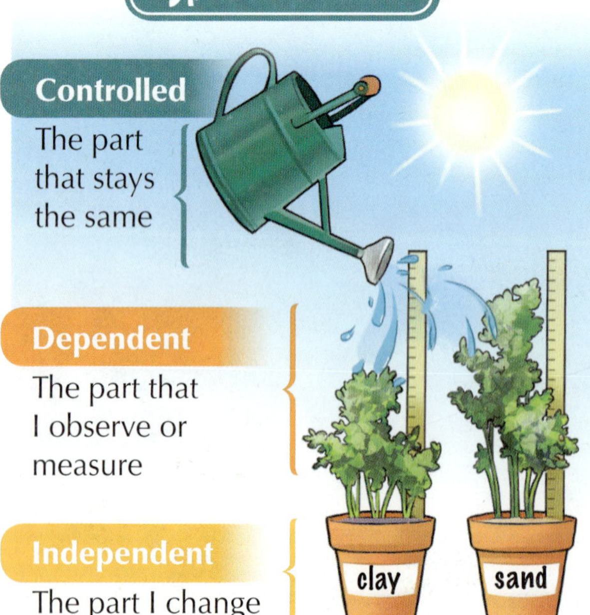

Second, a scientist lists *all the other variables that will stay the same.* These are called **controlled variables**. If you are designing an experiment to answer a question about soil type, then all the other variables in your experiment stay the same. The seeds, planter, water, temperature, air, and sunlight are all controlled variables—they never change. The independent variable—soil type—is the only thing that can change.

Next, scientists determine a dependent variable to measure for data. *The* **dependent variable** *is the part of an experiment a scientist measures to test the results.* In the carrot experiment, all the carrots could be measured for color, length, and size. If the carrots in the sandy soil are healthier, better formed, and larger than the carrots in the clay soil, you can conclude that sandy soil *does* help carrots grow better. Your data supports your hypothesis.

As you can see, science is the *careful* study of the wonders of the universe. After hypothesizing, planning, and experimenting, a careful scientist wants to ensure that his or her results are accurate. Accuracy is why a scientist will repeat the same experiment several times. Consistent results help to support or disprove a hypothesis.

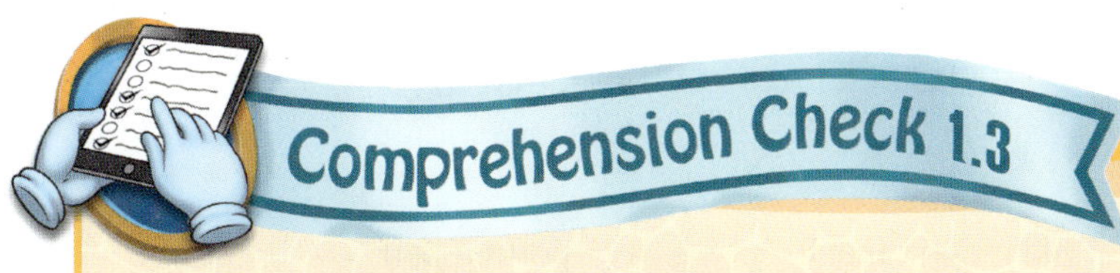

Comprehension Check 1.3

Matching: *Write the letter of the correct answer in the blank.*

____ 1. a procedure scientists design to test a hypothesis	A. controlled variables
____ 2. the part of an experiment that is changed to test the hypothesis	B. dependent variable
____ 3. the parts of an experiment that stay the same	C. experiment
____ 4. the part of an experiment measured to test the results	D. hypothesis
	E. independent variable

True/False: *Write the word* true *or* false *in each blank.*

____ 5. Forming a hypothesis starts with one statement.

____ 6. A hypothesis predicts the results of your research.

____ 7. A wise scientist follows his own ideas.

____ 8. Use only one independent variable in an experiment.

Discuss: *Explain why the false answers are incorrect statements.*

Grow plants from seeds for observation.

Materials needed:

- ✓ several plant containers
- ✓ large plastic bag for each planter
- ✓ tray to set the planters on
- ✓ water
- ✓ potting soil
- ✓ seeds
- ✓ *plant light
- ✓ pebbles

1. Poke several small drainage holes in the bottom of each plant container. Place a pebble over each hole to allow extra water to drain out while keeping the soil in. (Plants need water, but if they get too much water, they will rot.) Fill the planters with potting soil. Sprinkle the seeds on the soil and mix them into the top layer of dirt with your hands. Sprinkle the soil with water; then place each planter inside a plastic bag, tying the end of each bag shut.
2. Check the moisture of the soil occasionally. Do not let it become dry and crumbly or wet and soggy. When the seeds sprout, remove the planters from the bags. Water the plants when necessary, and be sure the plants get enough light.
3. Observe the plants as they continue to grow. For example:

- Plants grown from kidney beans and lima beans may become mature enough to produce seeds if the plants receive enough light.
- Scarlet runner beans grow flowering vines.
- Vines will also grow from pumpkin, acorn squash, and cucumber seeds.
- Cucumber vines often produce little cucumbers.
- If you are patient, you can get little tomatoes from your tomato seeds.
- Marigolds and zinnias will flower if given enough light.
- Morning glory plants grown from seeds will produce new seeds in about three months.

*optional

Try This! experiments with this symbol have an additional worksheet in *Understanding God's World STEM Activities* to help you apply the scientific method.

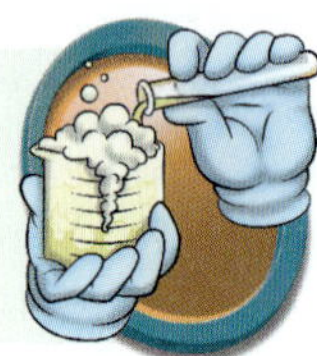

graph: a diagram that shows data

sector: a part of a circle graph that represents a certain fraction or part of the whole

interval: the amount from one line to the next on a graph

point: a location on a line graph that represents data

trend: data on a graph that can help predict future data

1.4 Recording and Graphing Data

As scientists gather data, they write down measurements, document details using computer programs, and sketch or take pictures.

Tools for Recording Data

Paper and pencil are the tools you will use most often because most experiments require them. Use them to write down or sketch your observations and results. Write down details, such as measurements. You may choose to use a special notebook or folder for your scientific observations.

A camera is also a helpful tool to use for observation. Use it to take pictures of things you observe and want to study later. You can also use a camera to record the stages of an experiment or its results.

Graphing Data

Imagine a scientist, surrounded by piles and piles of data. How easy would it be to access data or share it? It would be difficult, wouldn't it? A group of random, or mixed up, facts is not helpful, but a collection of organized facts can help scientists learn new things. It is very important that data is organized accurately. Scientists organize data by creating charts and graphs. **Graphs** *are diagrams that show data.* Charts and graphs can clearly show patterns and answer questions.

A circle, or pie graph, shows us parts of a whole. Each part of a circle graph is called a sector. *A* **sector** *is a part of a circle graph that represents a certain fraction or part of the whole.*

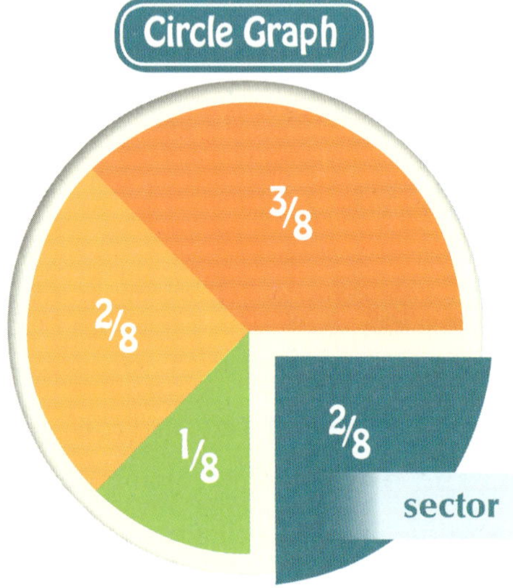

A bar graph compares different amounts. Each bar represents a specific part of the data. You can read a bar graph by looking at the numbered lines. *The amount from one line to the next on a graph is called an* **interval**. Interval counting makes it easier to count and read data. In this graph, we can easily count by tens.

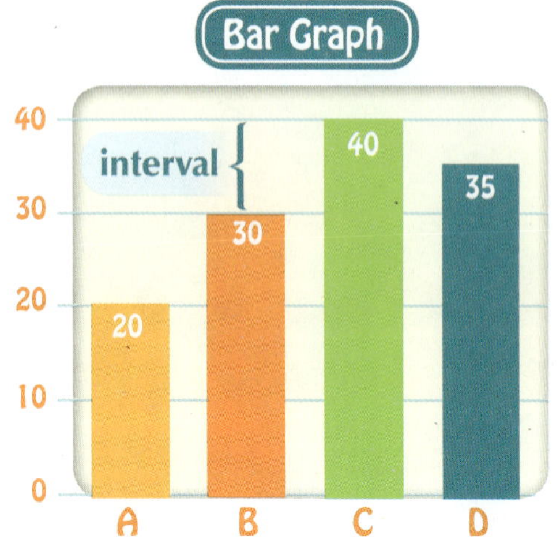

A line graph shows how things change over time. It shows intervals both vertically and horizontally. To read data on a line graph, look across, then count up. Each **point** on a line graph *represents your data* and may fall where the lines intersect, or cross. After connecting the points, study the line to help you see if there is a trend.

A **trend** shows a clear direction of the *data on a graph that can help predict future data.* A consistent upward direction of data is called an upward trend. A downward trend shows a downward direction of data. If there is no change in your data or your data is not consistent, there is no trend.

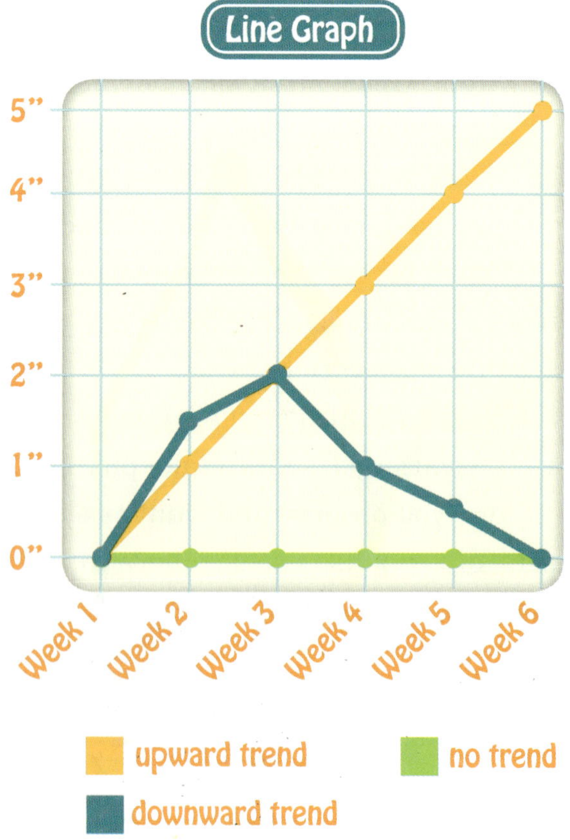

Tools for Graphing Data

By using the right tools, you can learn to graph data. You can use graphing paper and colored pencils to create several kinds of graphs. Computer programs can also help you create charts and graphs.

Comprehension Check 1.4

Give the correct answer.

1. What kind of graph shows the parts of a whole? pie graph
2. What kind of graph can compare different amounts? Bar graph
3. What kind of graph shows how things change over time? line graph
4. What kind of graph can show trends? line graph

Think and Apply: *Choose the best graph for the given situation.*

5. You need to record the data for rainfall each month over a one-year period. line graph
6. You need to show the fraction of time you spend each day for a week on schoolwork. circle graph
7. You need to show the number of each kind of plant in your garden. Bar graph

laboratory: a place where scientists work

metric system: a universal decimal measuring system based on the meter and kilogram

1.5 Learning Laboratory Procedures

A **laboratory** *is a place where scientists work.* When entering a laboratory, or lab, scientists follow certain procedures for safety and accuracy. As a Junior Scientist, you can begin to learn lab procedures as you use this book.

In a lab, you can find a variety of tools for observations and experiments. Each tool has a specific purpose. You will not use every tool for every experiment. It is important that you learn to match the correct tool with its use.

Tools for Safe Science

Scientists use many tools. A few of these tools are specifically used to protect the scientist.

Goggles *protect your eyes.* Some materials could be harmful if they come in contact with your eyes.

Scientists wear **gloves** to protect their skin from certain materials and from harmful germs.

A **mask** protects your nose, mouth, and lungs. It keeps you from inhaling materials that could irritate your lungs or from breathing in harmful germs.

Safe Science Lab Tips

During science lab activities, some materials can be unpredictable or harmful if improperly handled. Here are some laboratory safety tips to remember:

- ✓ Before you start an experiment or handle materials, wash your hands. Wear gloves if your experiment requires them.
- ✓ Use safety goggles to protect your eyes when handling materials that could come into contact with them.
- ✓ Use a safety mask to protect you from inhaling materials that could hurt your lungs or give you germs.
- ✓ Do not taste science materials or place them near your nose or mouth.
- ✓ Do not run near science tools or experiments. Move carefully while you handle science materials.
- ✓ Clean up after each experiment. Put away each material and tool.
- ✓ Wash your hands after you finish an experiment or handle materials.

Tools for Measuring Time

Scientists often use **clocks** to record specific times that experiments start or stop. They also use clocks to record how much time passes before any changes occur.

A **timer** can count *down* time and remind you to check an experiment for observation or to gather data.

A **stopwatch** counts *up* time. It can measure time passed and can stop at a fraction of a second. It helps *record amounts of time very precisely.*

Tools for Measuring, Mixing, and Pouring

Scientists use metric measures when measuring and recording data. *The* **metric system** *is a universal decimal measuring system based on the meter and the kilogram.* It is used all over the world. A scientist in Europe can use the data an American scientist has gathered. The metric system helps scientists work together.

A **thermometer** can measure the temperature of a substance.

A **scale** can measure weight. If you are using a scale to measure a substance in a container, be sure to weigh the container first. Subtract that amount from the measured amount to find the accurate weight. Some scales can automatically do this; know how your scale works first.

A **ruler** is used to measure length, width, or height. You may also use a meterstick or tape measure for longer measurements.

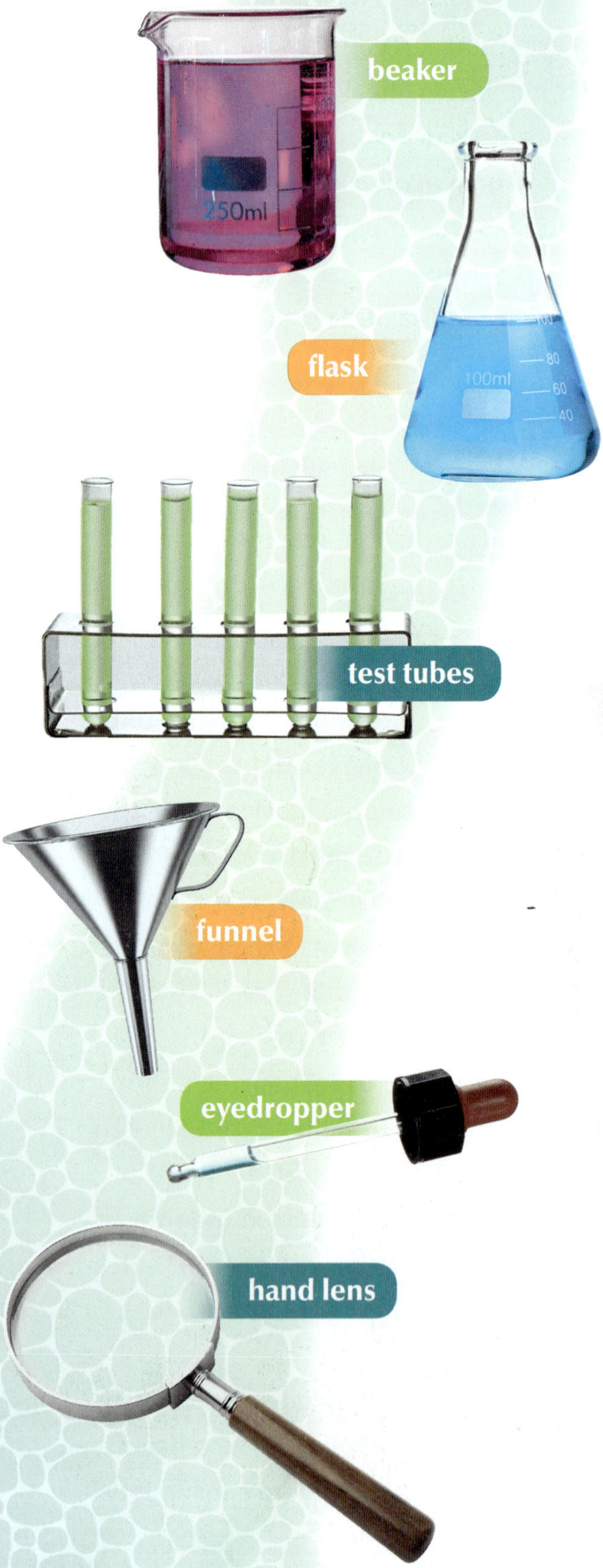

beaker

flask

test tubes

funnel

eyedropper

hand lens

A **beaker** is a cylinder-shaped container used to measure liquid. Allowing the liquid to settle before taking a measurement will help you to be accurate when you use a beaker.

A **flask** is similar to a beaker, but it has a narrow neck and opening. This prevents gases from escaping too quickly. A flask is a good tool to use when mixing two materials because it can also prevent spills.

A **test tube** is used to hold, mix, or heat small amounts of liquid. Because test tubes are narrow, they are usually held by test-tube holders to prevent spilling. Sometimes, a scientist needs to pour a liquid into a flask or test tube. A **funnel** can be used to prevent spilling.

An **eyedropper** is a good tool to use when *measuring out or using very small amounts of liquid.*

Many tools can be used to measure, but picking the best kind can help you to be accurate. Remember, use metric measures when possible. It will help you work like a scientist!

Tools for Observing Near and Far

A **hand lens** is a common tool scientists use to see something more closely. Its lens is designed to **magnify**, or make a small thing look larger. A hand lens is

a good tool for observing insects, plant textures, or details in soil and rock.

A **microscope** also uses lenses, but it works with a light and can also focus on small details. That is why scientists use microscopes to observe tiny things. A microscope is an excellent tool for examining fibers, germs, bacteria, small creatures, and cells.

As scientists began using microscopes, the miracle of God's created designs became more evident. The closer scientists look at the tiny parts that make up things, the more they can see their miraculous complexity.

Binoculars are an excellent tool for *observing things that are far away,* such as animals in their habitats. You can even use binoculars to get a better look at the stars in the night sky.

The best tool for *observing things in outer space* is the **telescope**. It uses lenses and mirrors to help scientists see things that are very far away, such as stars, planets, or other heavenly bodies. Some telescopes are more powerful than others.

Before using a type of viewing tool, be sure its lenses are clean. Follow the instructions for the viewing tool you are using. You will need to learn how to focus its lenses or prepare microscope slides.

Even if you may not have a scientific laboratory, you can practice what you know about choosing the best science tools for your experiments. Be prepared with the right materials for safety and accuracy. As a curious Junior Scientist, you will enjoy many lab activities that will help you better understand God's world.

Whatsoever thy hand findeth to do, do it with thy might.

Ecclesiastes 9:10

Comprehension Check 1.5

Multiple Choice: *Circle the correct answer to choose the best science tool.*

1. You are conducting an experiment using chemicals. What do you need to wear to protect your eyes?

   a. gloves b. goggles c. mask

2. You need to measure how long it takes your friend to run up the stairs. What time-measuring tool will help you be most accurate?

   a. clock b. stopwatch c. timer

3. You are preparing microscope slides to examine pond water. Which measuring tool would help place a drop of water onto a slide?

   a. beaker b. eyedropper c. scale

4. You are observing a deer in its natural habitat. What viewing tool would best help you make your observations without disturbing the deer?

   a. binoculars b. hand lens c. telescope

5. You are observing the stars. What viewing tool would help you best see heavenly bodies?

   a. binoculars b. microscope c. telescope

Practice the scientific method using the carrot experiment, pp. 12–14. Record your data on Worksheet 5.

Materials needed:

- ✓ carrot seeds
- ✓ 2 large, deep planters
- ✓ sandy-soil mixture
- ✓ clay-soil mixture
- ✓ water and beaker (or measuring cup)
- ✓ warm, sunny place
- ✓ permanent marker

What to do:

1. Observe and ask questions.
   - What kind of vegetable is a carrot?
   - Do root vegetables need more or less drainage?
   - What kind of soil gives plants better drainage?

   Question to answer:
   What kind of soil is better for growing carrots?

2. Form a hypothesis.
   - Your **hypothesis**: Carrots grow better in (clay soil, sandy soil) because most root vegetables need good drainage.

3. Experiment and gather data.
   - Decide on your **independent variable** (what will change).
   - Decide on your **controlled variables** (what stays the same).
   - Determine your **dependent variable** (what is observed/measured).
   - Complete the experiment.
     - Fill one planter with sandy soil and one planter with clay soil; label each.
     - Plant 4–5 carrot seeds in each planter. Be sure the seeds have enough space to grow.
     - Keep both planters watered. Use a beaker to keep water amounts consistent.
     - Observe both planters every week. After 10–12 weeks, pull the carrots and observe differences. Record your data.

4. Study your data and reach conclusions.
   - Compare your results and data.
   - Discuss the results.
   - Did you support your hypothesis?

Chapter 2

Understanding Matter and Energy

TERMS

physical world: the world we live in and interact with, made of matter

matter: a physical substance that has weight and takes up space

molecule: a tiny particle of matter

atom: a particle that makes up a molecule

energy: the ability to do work

thermal energy: heat energy

2.1 Matter and Energy in the World God Made

Imagine the world before God created it. Formless. Empty. Dark. Then out of nothing, God created all that we see. The great and powerful oceans, the green grasses, the tall trees, each creature, and every molecule was designed and created by God. The **physical world**, *the world we live in and interact with*, is here because God said, "Let there be...."

Matter

Matter *is the substance of the physical world. It is real, has weight, and takes up space*. Wherever you go in this world, all that you will ever see is made of matter. Even things you do not see are made of matter. What about the smell of a fresh apple pie? That scent is matter. What about the taste of a juicy strawberry? That is matter, too. If you can see, touch, taste, or smell it, it is matter. Even the air you breathe is made of matter.

Though all matter is real, has weight, and takes up space, not everything real is classified as matter. Love, wisdom, and truth are real things that are not matter. None of these real things take up space or have weight. Your brain is matter, but your mind is not. Our thoughts, emotions, and feelings are real, but they are not of the physical world. For example, love is real; but it is real because you can experience it and give it to others, not because it has weight and takes up space.

All physical objects and substances are made of matter. Because there is so much matter in the universe, we could never learn about it all in one book. But we can learn much about matter by studying a very common, yet unique substance—water.

Matter Molecules

God designed each molecule to make up the kind of matter He created it to be. How can all matter—even the matter we cannot see—have weight and take up space? *Matter is made up of tiny particles called* **molecules**. That means the smallest particle of water that is still water is a water molecule. Molecules are so small that you cannot see them without using a powerful microscope.

Atoms

If a molecule of water is broken down into even smaller particles, the particles are no longer water. That is because *a molecule is made of even tinier particles called* **atoms**.

Hydrogen and oxygen are gases found in the air. When two atoms of hydrogen and one atom of oxygen join together, they make a molecule of water. Other kinds of molecules are made of other combinations of atoms.

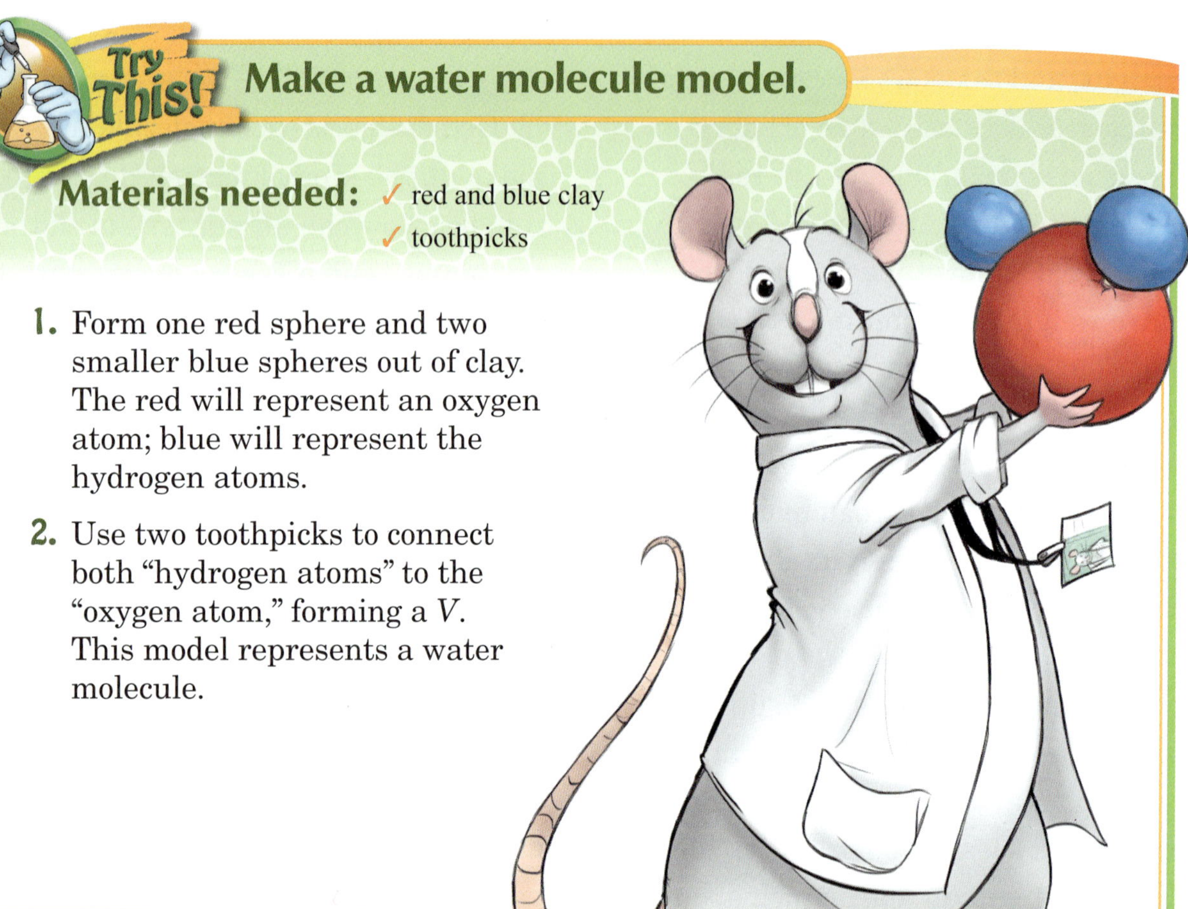

Make a water molecule model.

Materials needed:
- ✓ red and blue clay
- ✓ toothpicks

1. Form one red sphere and two smaller blue spheres out of clay. The red will represent an oxygen atom; blue will represent the hydrogen atoms.
2. Use two toothpicks to connect both "hydrogen atoms" to the "oxygen atom," forming a *V*. This model represents a water molecule.

Show that matter takes up space.

Materials needed:
- ✓ glass jar (filled halfway with water)
- ✓ tape
- ✓ marbles

1. Mark the water level in the jar with tape; add several marbles, one at a time to the jar.
2. Observe the water level. Why did it change?

The water level changed because both the marbles and the water are matter. They both need to take up their own space. The water level change is equal to the volume of marbles. Volume is the space that matter takes up.

Energy

Matter can be moved or changed. To move or change, matter needs energy. **Energy** *is the ability to do work.* Light, heat, and sound are all types of energy. They all can make things work. Light moves as its waves travel from one place to another. Heat energy makes molecules move. Another name for *heat energy* is **thermal energy**. Sound energy is sound waves, or vibrations, traveling through air.

Moving air is called wind. Wind energy can cause other things to move. Its energy can cause a sailboat on an ocean to move from shore to shore. It can help wet things dry more quickly. It can also make electricity.

Could there be matter and energy without a Creator Designer? With faith as our foundation and a scientific observation of the physical world, we know that matter and energy came from a powerful and creative Designer. As we understand how matter and energy work, we can give praise to God for the miraculous world that He made.

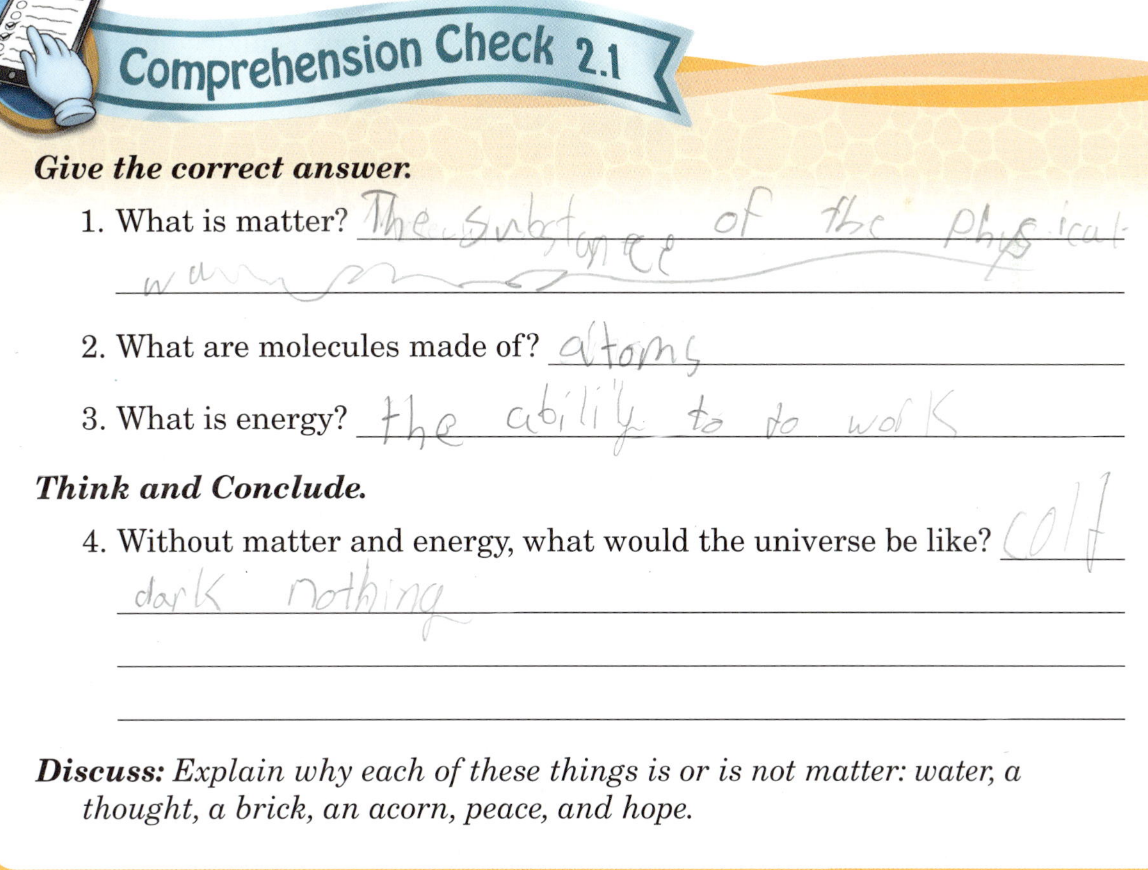

Comprehension Check 2.1

Give the correct answer.

1. What is matter? ______________________________

2. What are molecules made of? ______________________________

3. What is energy? ______________________________

Think and Conclude.

4. Without matter and energy, what would the universe be like? ______________________________

Discuss: *Explain why each of these things is or is not matter: water, a thought, a brick, an acorn, peace, and hope.*

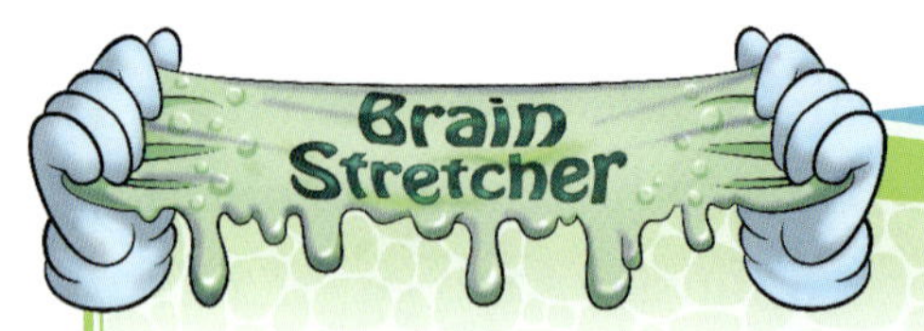

Other Types of Energy: Electrical and mechanical energy also give things the ability to do work. Electrical energy can power your home and be seen on display during a thunderstorm. Mechanical energy is in all objects that are moving. It is also the kind of energy that helps you run and play.

Study this chart to understand how different types of energy work. Try to identify what moves in each type of energy.

Type of energy	Example	What it does
light		makes colorful images move to your eyes in waves
thermal (heat)		moves molecules in matter
sound		causes vibrations to move through the air in waves
wind		moves air to make electricity
electrical		causes energy particles (electrons) to flow through a wire as a current
mechanical		allows an object to move

property: a way to describe or sort matter according to its characteristics

solid: matter that can keep its shape

dense: packed closely together

liquid: matter that cannot keep its shape, but can fill the shape of a container

gas: matter that can fill a space but escapes easily

2.2 Three States of Matter

We will learn about three forms, or **states**, of matter: *solid*, *liquid*, and *gas*. Water is the only substance that exists naturally in all three states. When water is a solid, it is called ice. Water in its liquid state is called water. Water is a gas when it is water vapor. Whether water is ice, liquid water, or water vapor, it is still water. The substance of the water has not changed. In each of its states, water is always made of the same type of molecule: two hydrogen atoms and one oxygen atom.

Solid Matter

We can describe each state of matter by its own unique properties. *A* **property** *of an object is a way to describe or sort it according to its characteristics*. By knowing the properties of each state of matter, you can easily classify them.

A property of all **solid** *matter is that it can keep its shape*. Many kinds of matter are found naturally as solids. Rocks and metals are usually solids. When water is frozen, it is in its solid state. Ice keeps its shape until it melts.

Solids keep their shapes because the molecules in them are very **dense**, or *packed closely together*. If you were packed closely together with other people in a small space, such as an elevator, would it be easy for you to move? Because the molecules in a solid are very dense, they cannot move quickly; however, they do vibrate. When you look at a brick, your desk, or some other solid, you do not see it vibrate. The molecules are so small and vibrate so weakly that you cannot see any movement.

Liquid Matter

When you think of water, you probably picture it in its liquid state. There are other liquids found in nature, too, such as oils and juices. Your blood can flow throughout your body because it is a liquid. In fact, all liquids have the property of being able to flow. **Liquids** *cannot hold their shape but can fill the shape of a container instead.* Liquids spread out to fill something because the molecules in them are not packed tightly together. They are *loosely connected, making them less dense than solids.* Molecules in liquids vibrate far more rapidly than the molecules of solids do. They vibrate within the liquid.

Gaseous Matter

Gases *can fill up space, but escape the area they are contained in very easily.* This is because gas molecules have too much energy and move too quickly to stay together. They are only weakly attracted to each other.

Gaseous matter is less dense than solid or liquid matter. Its molecules are far apart and are the fastest moving molecules. They whiz around and bump into each other all the time. Like all matter, gases have weight and take up space.

Many gases are colorless. Air is a mixture of gases, such as oxygen and nitrogen. Carbon dioxide is the gas we breathe out. Water in its gaseous form is called water vapor. Though water vapor is invisible, it can always be found in the air. You have probably seen clouds of steam above boiling water. You did not actually see the water vapor itself. You saw extremely tiny drops of liquid water pushed higher and higher into the air by the water vapor.

Matter

Throughout your study of this chapter, use your senses to observe matter. Think about the molecules that make the things you smell, feel, see, and taste. Ask yourself: What space is this matter taking up? Is this a solid, liquid, or gas? What properties do I observe?

- ✓ Things you use at school: pencils, books, and art supplies
- ✓ Things you play with: sports equipment and toys
- ✓ Things that are part of the weather: clouds and rain
- ✓ Things you eat and drink: apples and juice
- ✓ Things you use at home: furniture, dishes, and blankets
- ✓ Things you see at parties: balloons

Comprehension Check 2.2

Fill in the Blank: *Choose the correct word from the box below and write it in the blank.*

1. Liquid matter cannot hold its shape, but can fill the shape of a container.
2. Solid matter can keep its shape.
3. Solid matter is very dense.
4. Gaseous matter molecules can escape very easily.
5. Liquid matter molecules are loosely connected.
6. Gaseou matter is less dense than any other state of matter.

Gaseous	Liquid	Solid

physical property: a property that can be observed or measured

volume: the amount of space matter takes up

mass: the amount of matter in an object

weight: how heavy an object is because of the amount of gravity pulling it

density: how tightly packed molecules are in an object

2.3 Properties of Matter

Matter has some properties that are physical. **Physical properties** *can be observed and measured.* We can observe these kinds of properties with our senses. A material's state, color, size, shape, odor, taste, texture, strength, hardness, and flexibility are all examples of its physical properties.

We can also identify the physical properties of matter by measuring. An object's size, weight, mass, volume, and density are also physical properties.

Volume and Mass

All matter has volume. **Volume** *is the amount of space matter takes up.* Some types of matter take up more space than others. Imagine your mother comes home from grocery shopping and sets a gallon of milk and a two-liter bottle of soda on the counter. The soda bottle may be taller than the gallon of milk, but the gallon is a greater measure of volume than two-liters. It takes up more space.

Mass *is the amount of matter an object has.* A baseball has greater mass than a golf ball because it is made of more matter. Sometimes, it is easy to tell which objects have more mass just by looking at them; but remember that mass and volume are not the same property. A bowling ball and a beach ball of equal size have the same volume but do not have the same mass. The bowling ball is made of dense, solid matter; the beach ball is filled with gaseous matter, which is not as dense. Which object do you think has greater mass? The bowling ball has a greater mass because it has more matter.

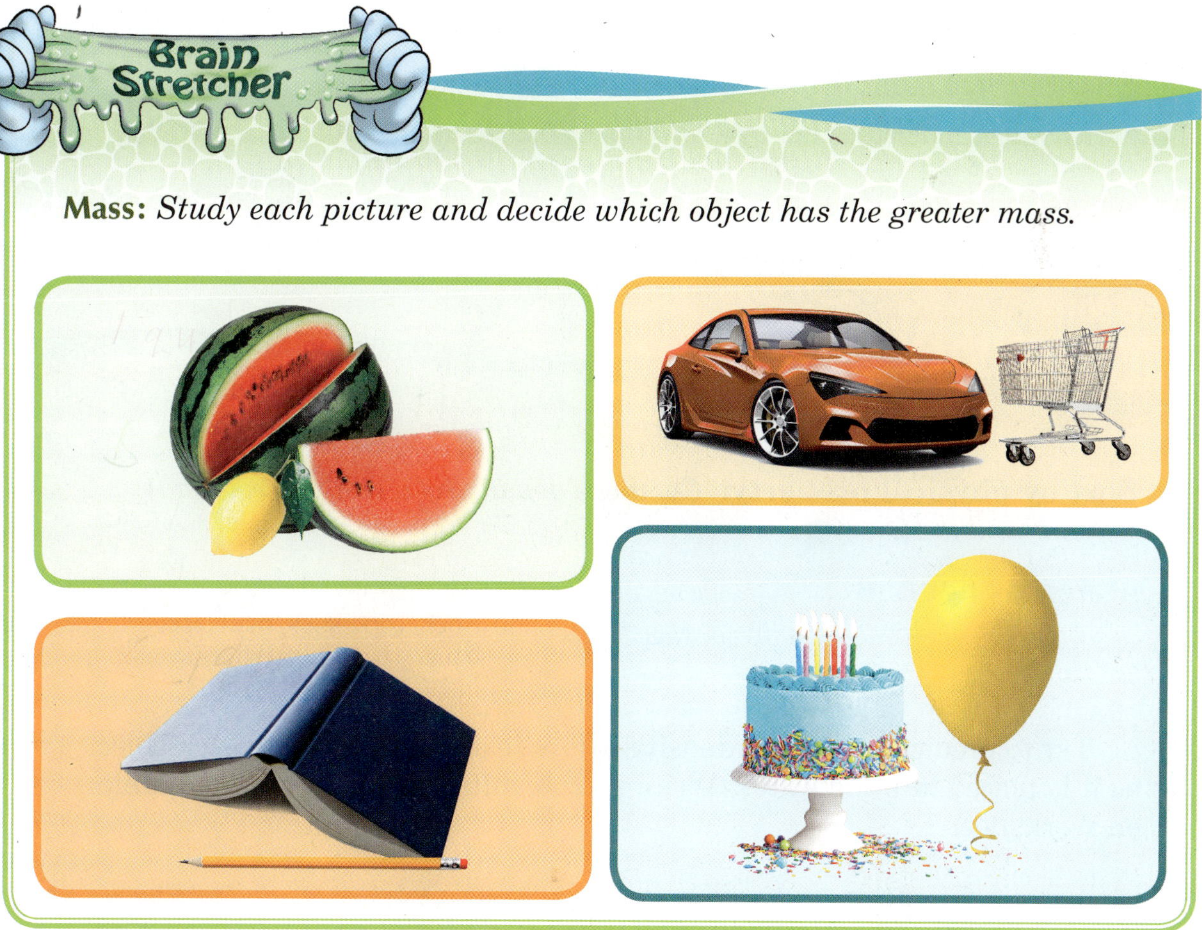

Mass: *Study each picture and decide which object has the greater mass.*

Weight and Density

Weight *is how heavy an object is because of the amount of gravity pulling its mass.* Though your mass remains the same, your weight would change if you were away from the gravity of Earth. We can measure the property of weight using a scale.

Some types of matter can float on water, while other types sink. Do things sink because they are heavy? No, a ship is very heavy, but it was designed to stay afloat. Ship designers and builders use what they know about water, gravity, and other forces to make ships. Other things can float because of the molecules they are made of or because of the way water molecules behave.

All solid matter is made of molecules that are packed tightly together. Yet, some solid matter has molecules that are more tightly packed than others. **Density** *is how tightly packed molecules are in matter.*

A rock will sink right away because its molecules are packed *very* closely together. It is very dense. A piece of cork, however, is made of molecules that are not as close together. Because cork has a lower density, it floats.

Sort by physical property: *Compare density by testing objects that sink and float.*

Materials needed:

- ✓ plastic tub of water
- ✓ Worksheet 7
- ✓ materials to test (rock, cork, stick, marble, egg, bouncy ball, metal spoon, small inflated balloon, toy, etc.)

Try to predict an object's density before testing it. Observe each object as it is placed into the water. Write your results on the information table on Worksheet 7.

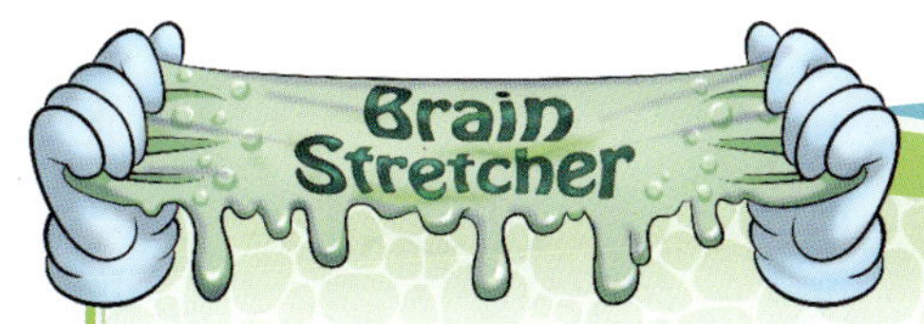

Surface Tension: Because of the way molecules of water fit together, water is able to do some amazing things. The molecules in liquid water stick together better than the molecules of many other liquids. This property is called surface tension. Because water has such a high degree of surface tension, objects that are heavier than water will float on the surface unless the surface film of the water is broken.

Comprehension Check 2.3

Matching: *Write the letter of the correct answer in the blank.*

_____ 1. the amount of space matter takes up

_____ 2. how much matter an object has

_____ 3. how heavy an object is because of the amount of gravity pulling it

_____ 4. how tightly packed molecules are in an object

A. density
B. mass
C. volume
D. weight

Try This! Design a buoyant object.

Materials needed:
- ✓ rubber-like putty
- ✓ large bowl of water

1. Form the putty into a compact, solid ball; drop it into the water. Observe what happens.

   Since the ball of putty does not have a large surface area, it does not displace, or move, much water. It sinks because of its density and because its design is not buoyant, or able to float.

2. Stretch the putty out, shaping it into a wide boat. Place your boat on the water, observing what happens.

   The boat design has a larger surface area and displaces more water. This design gives it buoyancy. Buoyancy is the upward force from the water to push an object to the surface.

Observe surface tension.

Materials needed:
- ✓ bowl of water
- ✓ eyedropper
- ✓ paper clip
- ✓ liquid detergent
- ✓ paper towel

1. Carefully place the paper clip flat on the surface of the water. Does it float on top of the water, or does it sink to the bottom of the bowl? If it sinks, you may need to increase the surface area of the paper clip.

2. Repeat step 1, but place a paper-towel piece under the paper clip to increase the surface area. Surface tension will keep the paper clip afloat, even when the paper towel eventually sinks to the bottom of the bowl.

3. Now that the paper clip is floating, use the eyedropper to add a drop of liquid detergent next to it. What happens? Write down your observations.

melting point: the temperature at which a solid changes into a liquid

boiling point: the temperature at which a liquid rapidly changes to a gas

expand: to get larger

contract: to get smaller

freezing point: the point at which a liquid changes to a solid

2.4 Heating and Cooling Matter

Molecules of solids are packed tightly together and move slowly. In most solids, the molecules have a strong attraction for each other. This means that they will stay close to each other and hold together. When heat energy is added to a solid, the molecules begin to move more rapidly. If they move fast enough, the attraction between molecules is not strong enough to hold them together as a solid. The solid can melt and become a liquid. The molecules have not changed, but their motion has overcome the attraction that holds them together.

Water's Melting Point

Not all solids melt. Aren't you glad that you never need to worry about your science book melting before you finish your lesson? Solids that melt do not all melt at the same temperature. *The temperature at which a solid changes to a liquid is its* **melting point**. The melting point of water is **32°F**.

What happens when you put ice into a glass of water? Does the ice melt, or does the water freeze? The ice melts because the warmer water has more energy. *Heat energy always moves from the warmer space to the cooler space.* This is how God designed thermal energy to work. Because heat has energy, it is able to move to a cooler space; coldness does not have energy and cannot move into a warmer space. As the ice cube increases in temperature, it melts.

Water's Boiling Point

Just as heat energy can cause a solid to change into a liquid, it can also cause a liquid to change into a gas. Increased heat energy causes the molecules to move faster. The attraction between the molecules is not enough to hold them together as a liquid. The molecules break apart, and the liquid becomes a gas. In water, the gas that begins to escape is water vapor. *The temperature at which a liquid rapidly changes to a gas is its* **boiling point**. The boiling point of water is **212°F**.

Water's Freezing Point

The temperature at which liquid water changes to a solid is **32°F**. *It is called the* **freezing point** *of water*. Did you notice that water's freezing point and melting point are the same? When water is at 32°F, some of its molecules are solid, and some are liquid.

Water Is Unique

Because of the way its molecules were created, water does not behave as other liquids do when it changes state. Most substances **expand**, or *get larger*, when they are heated; they **contract**, or *get smaller*, when they become colder. This change is true of solids, liquids, and gases.

Just before water gets cold enough to freeze, a strange thing happens. *At 39°F, water begins to expand and become less dense as it gets colder.* Water that is colder than 32°F has completely changed to a solid—ice. Ice has a greater volume and less density than the same amount of liquid water. This is what causes ice to float. If water behaved as other liquids do, ice would be smaller and denser than an equal amount of liquid water, causing it to sink.

Imagine a very cold winter day at a pond. The water on top of the pond has begun to freeze. If water acted like other liquids, the ice would contract, become denser, and sink to the bottom of the pond. More and more ice would form and sink. Soon the pond would be

solid ice. What would become of the fish and plants in the pond? What if the same thing happened in lakes, rivers, and oceans?

Instead of contracting and sinking, ice expands and floats. Ice on top of a pond or lake helps protect water plants and fish during very cold weather. They live securely beneath the ice until it melts in warmer weather. Because water expands when it reaches its freezing point, water life continues.

What do water's properties tell you about its Creator? God's unique design for water protects every water habitat. He is a wise and caring Creator.

O Lord, how manifold are Thy works! in wisdom hast Thou made them all: the earth is full of Thy riches. So is this great and wide sea, wherein are things creeping innumerable, both small and great beasts.

Psalm 104:24–25

Try This! Observe a heat energy transfer.

Materials needed:

- ✓ mug of steaming water
- ✓ beaker of cold water
- ✓ small metal object
- ✓ tongs
- ✓ timer

1. Carefully place the metal object in the steaming water. After two minutes, remove it with tongs. Be careful; it may feel very hot.
2. Now place the hot metal object in the cold water for two minutes. What do you predict will happen? Will the water get hot, or will the metal object get cooler?

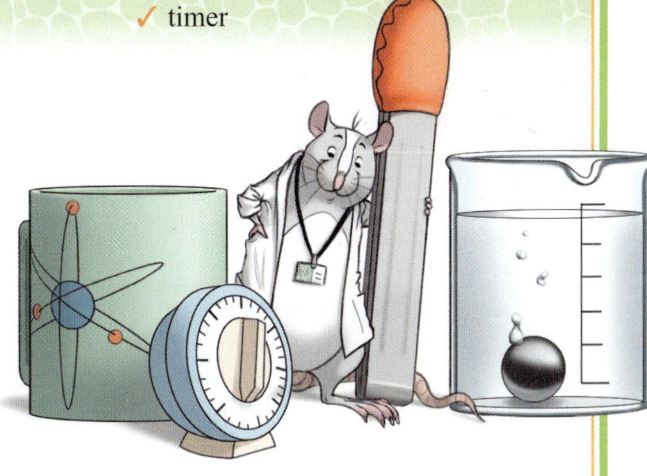

Because heat energy always moves from the warmer space to the cooler space, it will transfer from the metal object to the cooler water. There is not enough heat energy to make the water hotter than the metal. Both the metal and the water will probably be about the same temperature after two minutes.

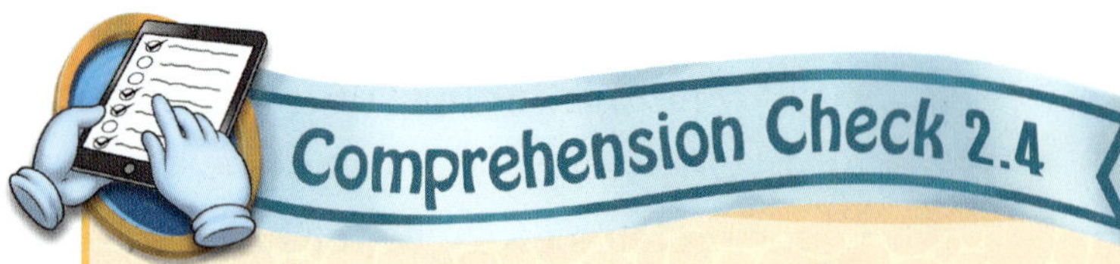

Comprehension Check 2.4

True/False: *Write the word* true *or* false *in each blank.*

True 1. Not all solids have a melting point.

False 2. Heat energy always moves from a cool place to a warm place.

True 3. A liquid can change into a gas when its temperature reaches its boiling point.

False 4. Water contracts when it freezes.

Discuss: *Explain why the false answers are incorrect statements.*

TERMS

mixture: two or more different kinds of matter combined together

solution: a mixture in which one kind of substance is dissolved into another

dissolve: to break down into smaller pieces until it can no longer be seen

solvent: the part of a solution that dissolves another substance

solute: the part of a solution that is dissolved

2.5 Mixtures

Not everything you see around you is made of only one kind of matter. Think about vegetable soup. Many kinds of matter, or ingredients, are mixed together to make the soup. When you *combine different kinds of matter together*, it is called a **mixture**.

With a mixture, you can separate the kinds of matter again. In the vegetable soup, you can separate each ingredient that was combined to make that soup. Sand and water are two different kinds of matter. If we mix them together, they also form a mixture. The sand is still sand, and the water is still water, but together they can form a mixture. We can use a strainer, lined with a filter, to separate the sand from the water again. The water drains out while the sand remains in the strainer.

Parts of this mixture can be separated again because there has been no physical change in each kind of matter. The pieces of sand are still hard and granular. The water is still a liquid.

Solutions: A Type of Mixture

A **solution** is a mixture in which *one substance is dissolved into another substance.* When a substance is **dissolved**, *it is broken down into smaller pieces until it can no longer be seen.* There are *two main ingredients* in a *solution*: a *solvent* and a *solute.*

Salt water is a solution. The **solvent**, water, is *the part of the solution that dissolves another substance.* The **solute**, salt, is *the part of the solution that is dissolved.* Salt water tastes salty even though you cannot see the salt. It has dissolved into the water, making it more difficult to separate each type of matter again. Water is a solvent for many kinds of substances. God gave water this special property so that it can carry nutrients that plants, animals, and people need. God knows our needs—even before we do. He always takes care of us.

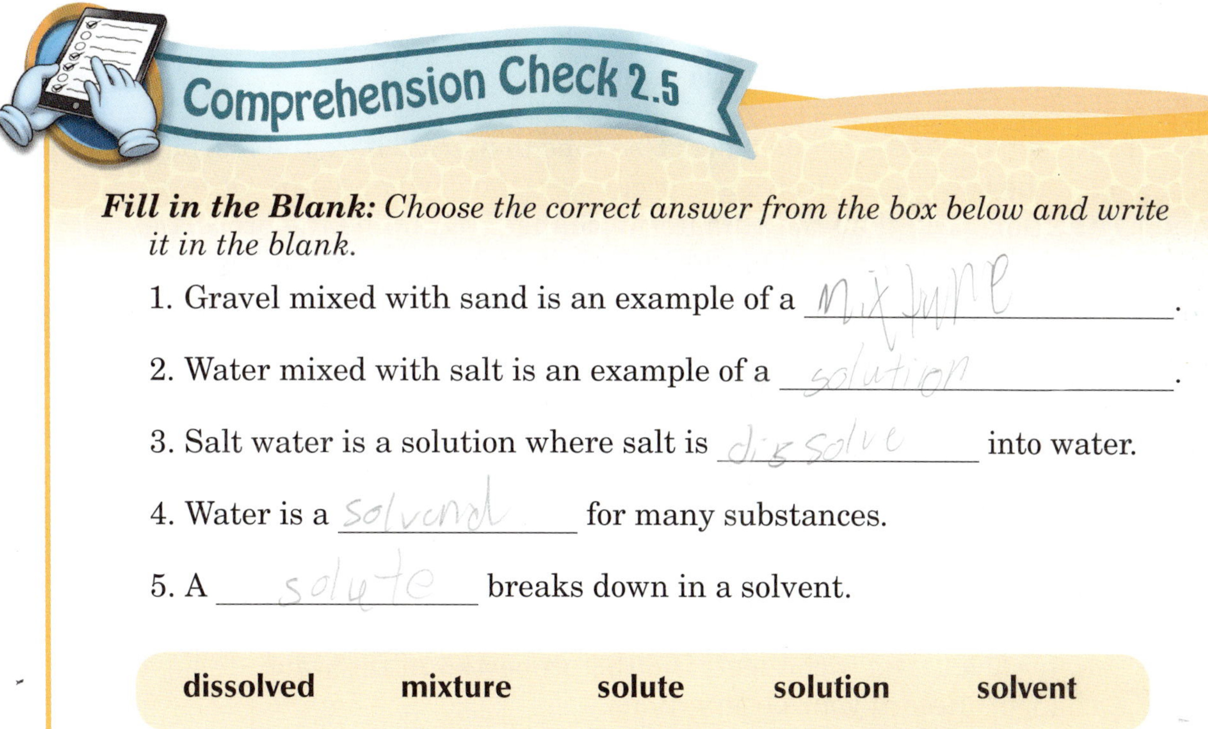

Comprehension Check 2.5

Fill in the Blank: *Choose the correct answer from the box below and write it in the blank.*

1. Gravel mixed with sand is an example of a ____________.
2. Water mixed with salt is an example of a ____________.
3. Salt water is a solution where salt is ____________ into water.
4. Water is a ____________ for many substances.
5. A ____________ breaks down in a solvent.

dissolved	mixture	solute	solution	solvent

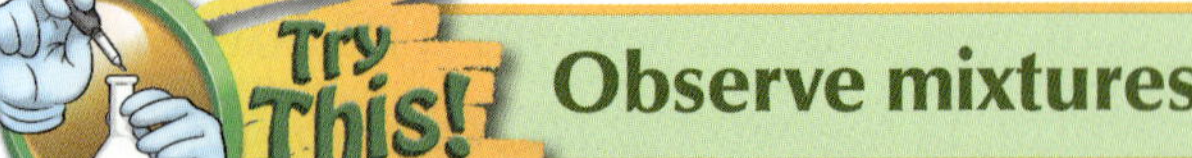

Observe mixtures.

Materials needed:
- ✓ 2 glass bowls
- ✓ warm water
- ✓ 12 marbles
- ✓ 24 sugar cubes
- ✓ spoon
- ✓ tape

1. Examine twelve sugar cubes, the marbles, and the water. Write down the physical properties you observe in this information table.

matter type	sugar cubes	marbles	water
hardness			
texture			
color			
shape			
taste			

Remember to place only food items in your mouth.

2. Add the marbles to one bowl. Use tape to mark where the marbles fill the bowl. Add twelve sugar cubes to the bowl and stir. Observe the tape. Did the level of material change? Examine the properties of the sugar cubes and marbles. Did the properties change?

The level of matter in the bowl did change because there are still two substances in the bowl, each taking its own space. The properties did not change. We could sort the marbles and sugar cubes out again.

3. Partially fill one bowl with water. Use tape to mark the water level. Add twelve sugar cubes, one at a time, stirring the water after adding each cube. When all of the cubes have dissolved, observe the water level. Did it change? Did the properties of the water or sugar change? Record your observations.

The level will be the same, even though you added several cubes of sugar because the water dissolved the sugar completely. There are empty spaces too small for you to see between the molecules of water. When the water dissolved the sugar, or broke it down into molecules, the molecules of sugar fit into the empty spaces between the molecules of water. The water now tastes sweet.

Write what you have observed about the marbles and sugar cubes, and the water and sugar cubes.

Marbles and Sugar Cubes	**Water and Sugar Cubes**
______________________	______________________
______________________	______________________
______________________	______________________
______________________	______________________
______________________	______________________
______________________	______________________
______________________	______________________
______________________	______________________
______________________	______________________
______________________	______________________
______________________	______________________
______________________	______________________
______________________	______________________

List three other mixtures.

____

____

____

____

energy: the ability to do work

kinetic energy: moving, working energy

potential energy: stored, waiting energy

converted energy: one kind of energy changed to another kind of energy

fuel: a material that is burned to release energy

transferred energy: energy moved from one place to another

Two Forms of Energy

Potential and Kinetic Energy

Energy *is the ability to do work.* It can move or change matter. What type of energy can cause water to change its state? Adding heat energy to water can cause its molecules to move more quickly. Water can change from a solid to a liquid then from a liquid to a gas because of thermal energy.

Water itself has energy. Have you ever observed a stream? If you wade in a stream, you can feel the energy of the water as it pushes against your feet. But what if the moving water is stopped by a dam or large boulder? Here it forms a pool. If you put your feet in this pool, you will not feel a moving current. The water is still. Would it surprise you to discover that the swiftly flowing stream and the still, quiet pool both have energy?

The water in the stream that rushes over your feet is moving. *All moving things have working energy, which is called* **kinetic energy**. A baseball zooming out to center field is full of energy because it is *moving*. The motorcycle that passes your family car is also full of kinetic energy. What about your little sister? You can probably see her kinetic energy in action when she eats too much sugar!

Some objects which are not moving have the *potential to move*. They have energy stored in them, waiting to be used. *This kind of stored, waiting energy is called* **potential energy**. If you stretch a rubber band, it is full of potential energy. However, as soon as you let the rubber band go into the air, its potential energy becomes moving, kinetic energy.

The water stored up behind the dam also contains potential energy. If the dam breaks or water is allowed to spill over, the water's energy will change to moving, kinetic energy. If it is used to power a water wheel, its kinetic energy can be used to make mechanical energy.

If you sit at the top of a water slide, you are storing energy by being in a high place. Because of gravity, you are full of potential energy. But as you begin to slide down, that potential energy turns into kinetic energy.

Energy Can Be Converted

Potential energy is constantly being changed, or converted, into kinetic energy and back again. *To* **convert energy** *means to change it from one kind of energy to another.* Coal, oil, and natural gas are sources of **fuel energy**. *Fuel sources contain potential energy that release kinetic energy as they are burned.* As fuel energy converts to thermal energy, it can now give something else the power to work. Many vehicles depend on fuel to burn for energy.

God provided us with plenty of natural energy sources. Manmade structures, such as windmills, water dams, and solar panels can all harness natural energy and convert it into electrical energy at a power station.

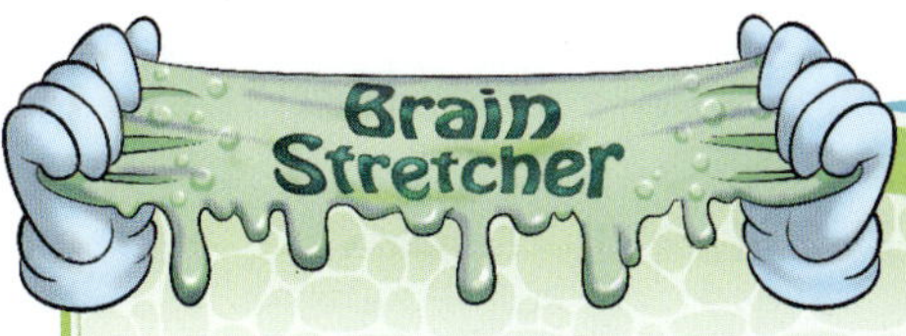

Energy Types: Did you know this apple is also an example of potential energy? Your body uses the food you eat as chemical energy. As molecules of food are broken down by your digestive system, chemical reactions take place that change food into energy. Chemical energy from food becomes either thermal energy (keeping your body at the right temperature) or mechanical energy (giving your body the ability to move). Mechanical energy is a kind of kinetic energy. Your body has kinetic energy because of the foods you eat.

Energy Can Be Transferred

Not only can energy be converted, but it can also be transferred. *An* **energy transfer** *happens when energy is moved from one place to another.* Do you remember how heat energy always moves from a warmer space to a cooler space? That is how God designed thermal energy to transfer. Heat, light, sound, and electrical energy are all kinetic energy that can be transferred. God designed unique ways for each of them to move from one place to another. We will learn more about these special designs throughout this chapter.

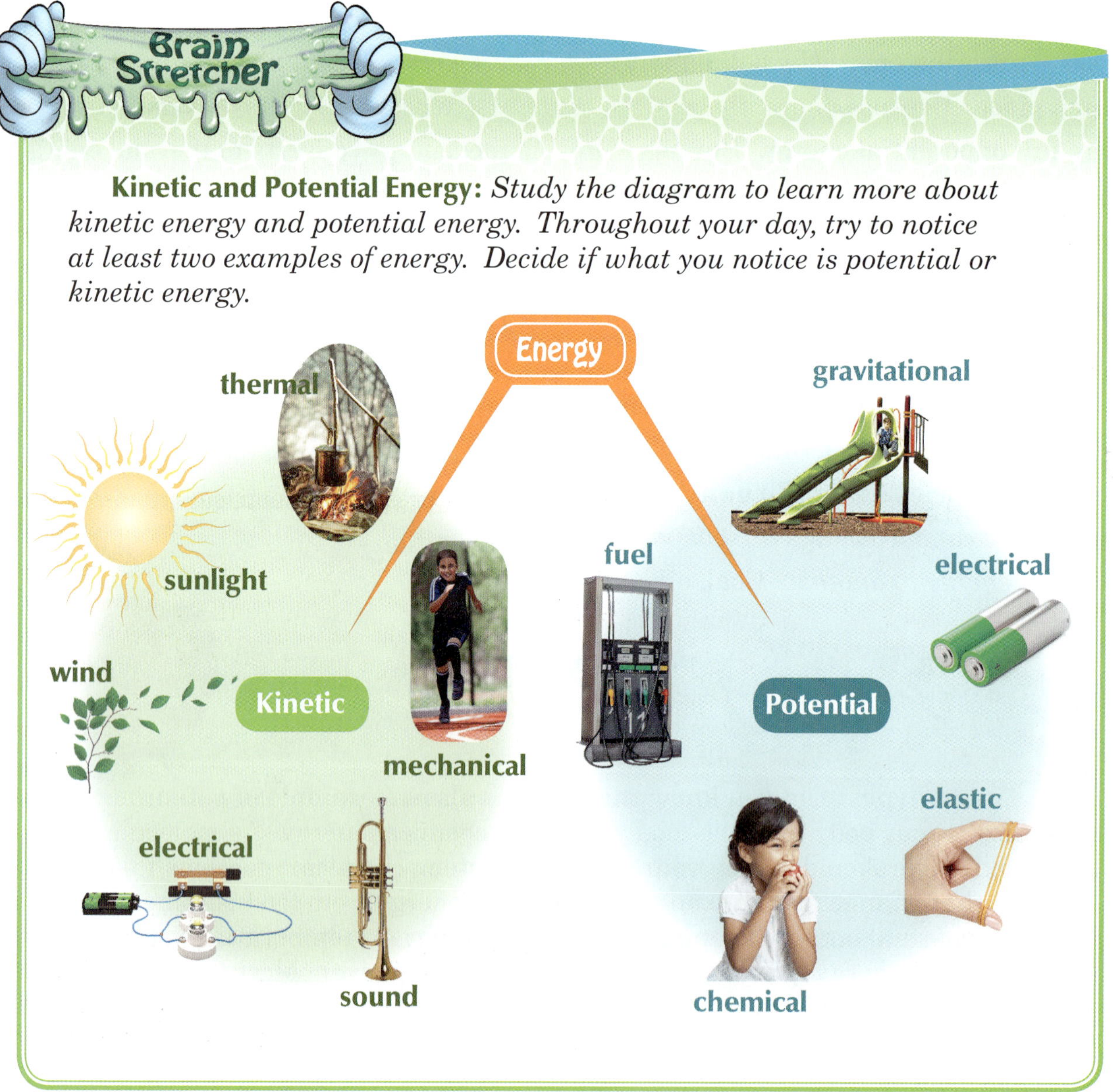

Kinetic and Potential Energy: *Study the diagram to learn more about kinetic energy and potential energy. Throughout your day, try to notice at least two examples of energy. Decide if what you notice is potential or kinetic energy.*

Observe to Understand — Energy

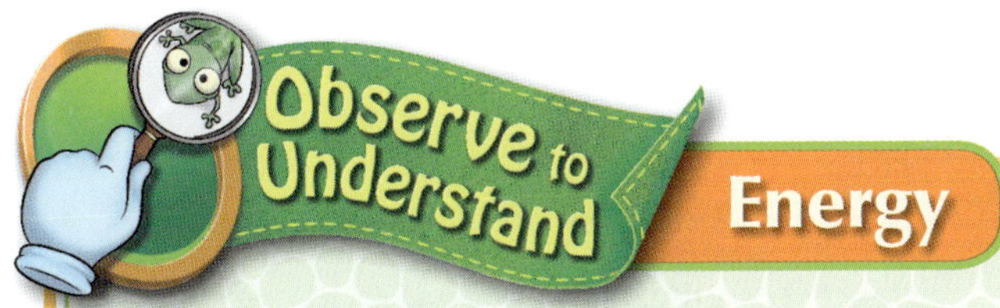

Throughout your study of this chapter, look for things that have energy. As you observe, ask yourself, "What gave that object the energy to do work, or move?" You may choose to keep a journal of your observations.

Comprehension Check 2.6

Matching: *Write the letter of the correct answer in the blank.*

B 1. moving or working energy

C 2. stored energy

D 3. energy changed from one kind to another

A 4. energy moved from one place to another

A. converted energy
B. kinetic energy
C. potential energy
D. transferred energy

Think and Classify: *Differentiate potential energy from kinetic energy by writing each answer in the correct column.*

battery	**light turned on**	**pool of water**
boiling kettle	**moving car**	**running dog**
fuel	**plate of food**	

Kinetic	**Potential**
light turned on	batter
moving car	fuel
running	plate of food
boiling kettle	pool of water

static electricity: a buildup of electrical charge

nucleus: the positive center of an atom

electron: a negative particle of an atom

attract: to pull together

repel: to push apart

2.7 Static Electricity

You have seen the power of electrical energy in a thunderstorm. Benjamin Franklin saw the similarity between lightning and electricity. He set out to prove that lightning is electricity with his famous kite and key experiment. As other scientists learned more about electricity, people were able to use it to power their homes and businesses.

Lightning is a different kind of electricity from the kind we use to give power to things. It is a form of static electricity. **Static electricity** *is a buildup of electrical charge.* Perhaps you have rubbed your feet on the carpet and then touched a doorknob. Did you feel a little spark? That was

static electricity. What happened to make that spark?

You might remember that all matter is made of **molecules**, and that molecules are made of even smaller particles called **atoms**. If we were able to look very closely at an atom, we would see that it has two main parts:

- a positive **nucleus**, or center
- negative **electrons** whizzing around the nucleus

Because the atom has both a negative and a positive part, it is balanced. But, if a negative electron gets separated from the atom's positive nucleus, the atom is no longer balanced.

Rubbing two objects together can cause atoms to lose or gain electrons, creating this imbalance. If you rub your feet on the carpet, static electricity builds up on your body. It is building up to create a charge. In static electricity, opposite charges **attract**, or pull each other together, but like charges **repel**, or push each other away. The spark you felt touching a doorknob is a sudden discharge of static electricity.

Try This! Observe the push and pull of static electricity.

Materials needed:
- ✓ 2 balloons
- ✓ string
- ✓ wool or clean, dry hair
- ✓ permanent marker
- ✓ tape
- ✓ scissors

1. Tie a long piece of string to each inflated balloon. Mark one balloon with "A," and the other with "B."
2. Tape the string ends to an open doorway keeping the balloons 8–10 inches apart.
3. Charge balloon A by rubbing it on a piece of wool or your hair, and let it go.
4. Observe what happens.
5. Repeat step 3 with balloon B and let it go.
6. Observe what happens.

Do you see how electrical charges can either push or pull? Their positive or negative charges can attract and repel.

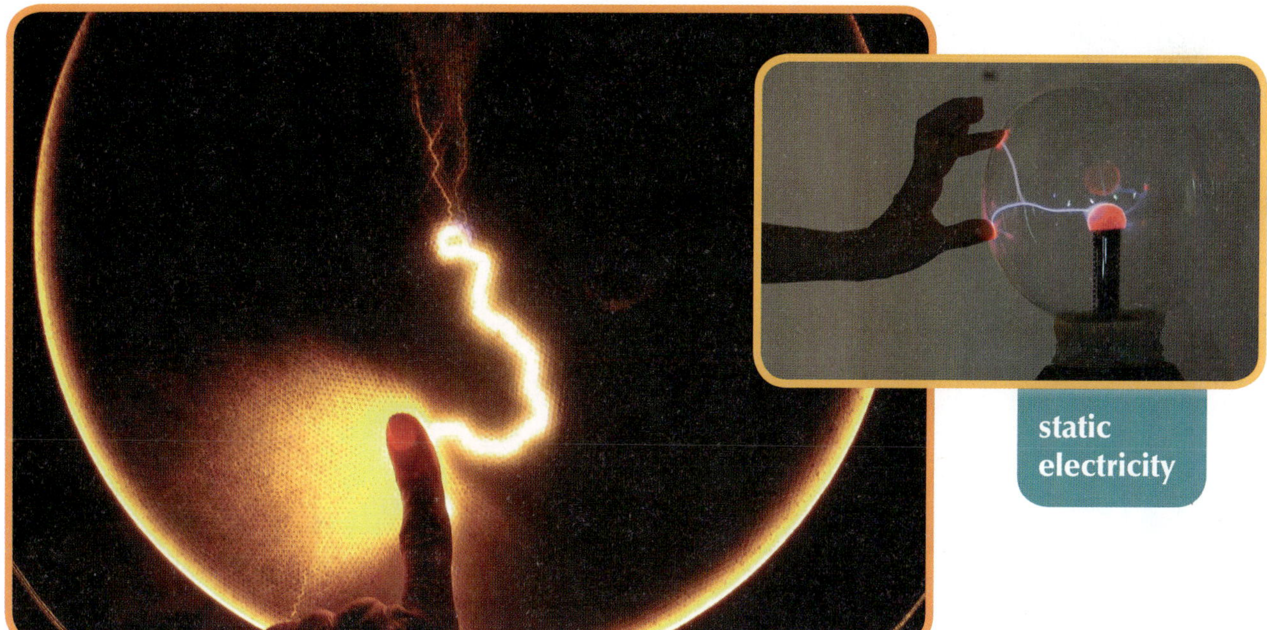

static electricity

Fill in the Blank: *Choose the correct answer from the box below and write it in the blank.*

1. The spark you sometimes feel when you touch a doorknob or other object is ______________.
2. An atom has a ________ nucleus and ________ electrons.
3. Rubbing two objects together can cause atoms to lose or gain ______________.

electrons **negative** **positive** **static electricity**

Think and Conclude: *Write these terms in order from largest to smallest.*

atom **electron** **molecule**

4. ________ 5. ________ 6. ________

TERMS

current electricity: the flow of electrons along a path

generator: a machine that converts mechanical energy into electrical energy

conductor: a material that can easily transfer (electrical) energy

cable: bundled wires, covered in plastic or another insulator

insulator: a material that cannot easily transfer energy

closed circuit: an uninterrupted path of electricity; allows an electrical current to flow

open circuit: an interrupted path of electricity; does not allow electrical current to flow

2.8 Current Electricity

Lightning and other forms of static electricity are examples of naturally occurring electricity. We cannot use static electricity to power our homes, however, because it cannot be controlled. That is why another kind of electricity has to be made.

Fuel, water, wind, and solar energy can all be converted into current electricity at a power station. Unlike static electricity, which can *discharge suddenly*, **current electricity** *flows along a path*. To make current electricity, you need a generator. *A* **generator** *is a machine that converts mechanical energy into electrical energy*.

A generator has metal coils and magnets. The coil, like any type of matter, is filled with atoms. Metals such as copper are very good electrical conductors. *A* **conductor** *can easily transfer energy*. Electrical energy can easily flow through copper. The magnet can pull electrons out of their paths around an atom to flow in one direction along the coil. This happens when the coils or the magnets spin. The power to make them spin often comes from fuel, wind, water, or solar energy.

Current electricity is different from static electricity; instead of sudden bursts of electrical charges, current electricity flows continuously. *As electrons are flowing, they transfer electrical energy from one place to another.* When the electricity is flowing, we call this its **current**. Is current electricity potential or kinetic energy? Because the electrons are moving from one place to another, it is kinetic energy.

How does current electricity transfer from a generator and into other objects, such as traffic lights or our homes? Electrical current flows from

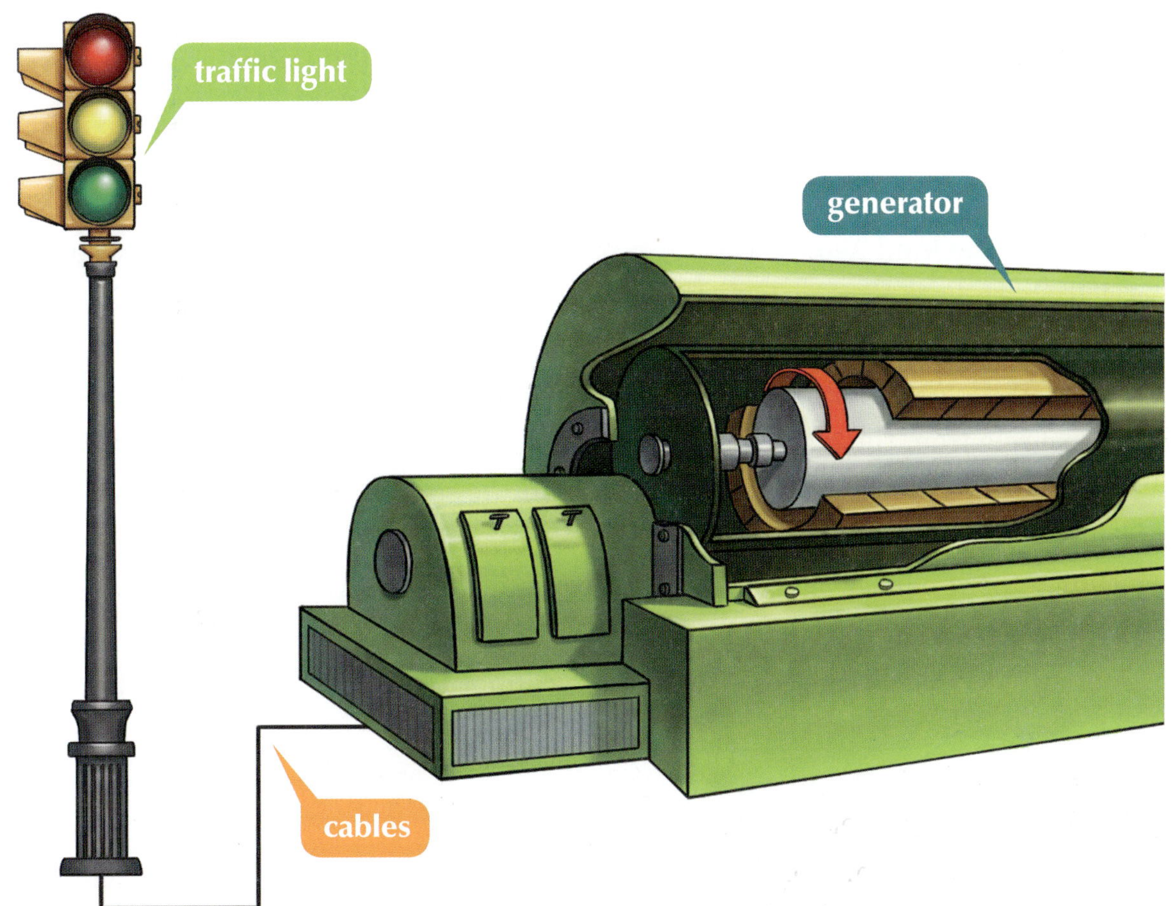

the generator and through a series of wires called **cables**. An electrical cable contains a bundle of copper wires. Why do you think these wires are made of copper? Copper is a good conductor; electricity can easily flow through these wires.

The bundle of copper wires is covered in plastic for a very important reason—electricity can be dangerous. The electricity flowing through a wire is very powerful, and touching the wire could be deadly. These wires need to be covered in something to block the electricity from flowing out to you. Plastic is a good electrical **insulator**. *An insulator does not easily transfer energy*. A plastic wall switch protects you from electricity as you turn your light on.

There are two kinds of wire in each cable. One kind brings electricity into your home, while the other kind takes electricity back to the generator.

Why does one kind need to take electricity back to the generator? Electricity needs a path to travel on called a **circuit**. If the electricity makes a complete circle from the generator and back to the generator, the circuit is complete, or **closed**. *A* **closed circuit** *is an uninterrupted path of electricity that allows an electrical current to flow. An* **open circuit** *is an interrupted path of electricity that does not allow electrical current to flow.*

As electricity comes into your home, it branches off into many circuits. Each of these circuits has a switch that can be opened or closed. When you use a light switch at home, you can safely open or close an electrical circuit. *An open switch turns off the electrical current; a closed switch turns on the electrical current.* Remember, if a switch is open, the circuit is incomplete and electricity cannot flow in a complete circle.

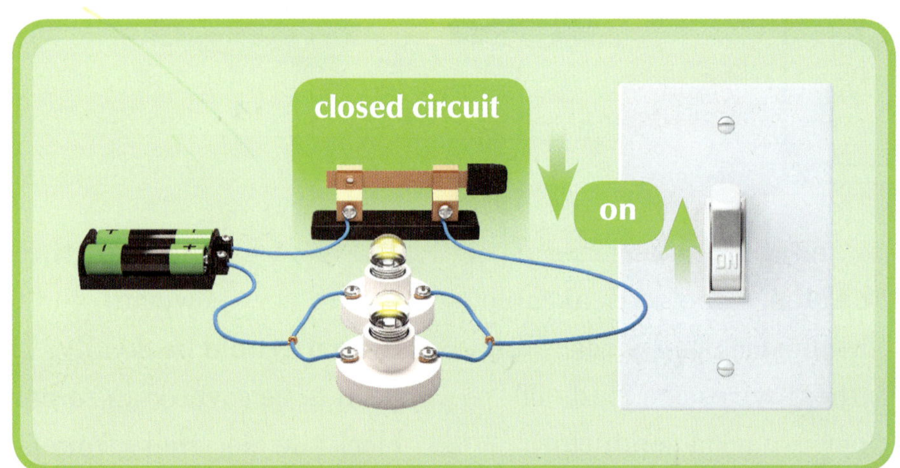

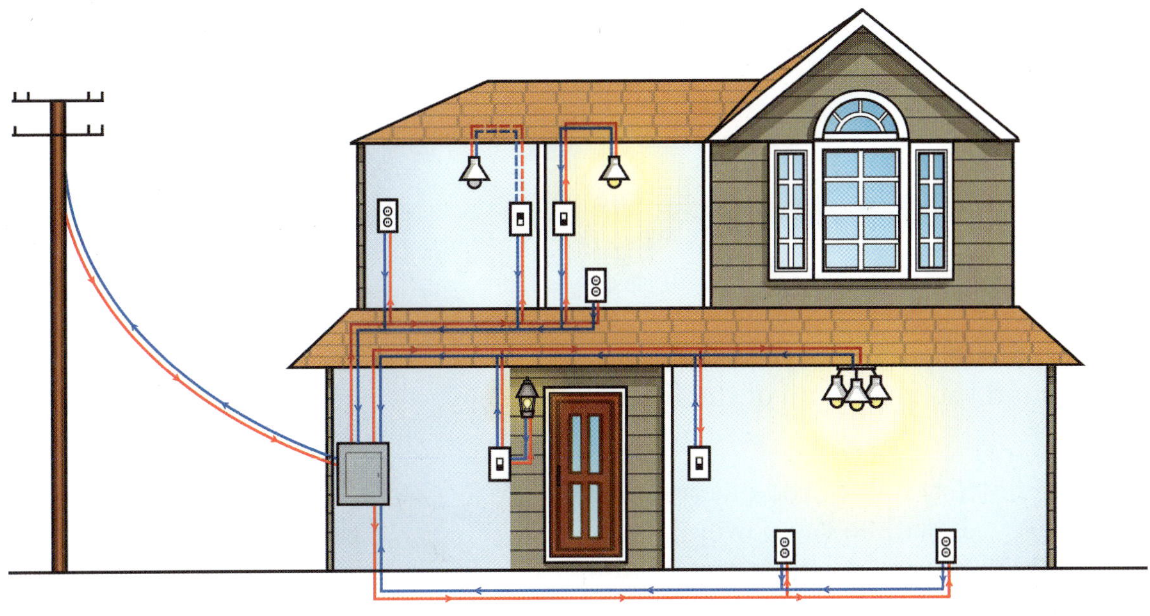

Try This! Make a miniature generator.

Materials needed: ✓ magnet wire ✓ strong bar magnet ✓ compass ✓ heavy cardboard

1. Make a coil by wrapping one end of the wire around your hand ten times; then pull your hand out. Wrap the other end of the wire around the cardboard-enclosed compass five times.
2. Twist the metal wire ends together to make a complete circuit for electrons to flow. Insert a bar magnet inside the empty coil and move it back and forth.
3. Observe the compass.

Why is the needle moving? The needle is showing us that electricity is flowing inside the wire. The magnet pulled the electrons into a path. Do you remember what we call electrons flowing on a path? We just generated current electricity! Current electricity can transfer from one place to another, giving something else energy.

Convert stored electrical energy into light energy.

Materials needed:
- ✓ 1 ft. bell wire
- ✓ 1½ volt battery (AA battery)
- ✓ 1½ volt flashlight bulb

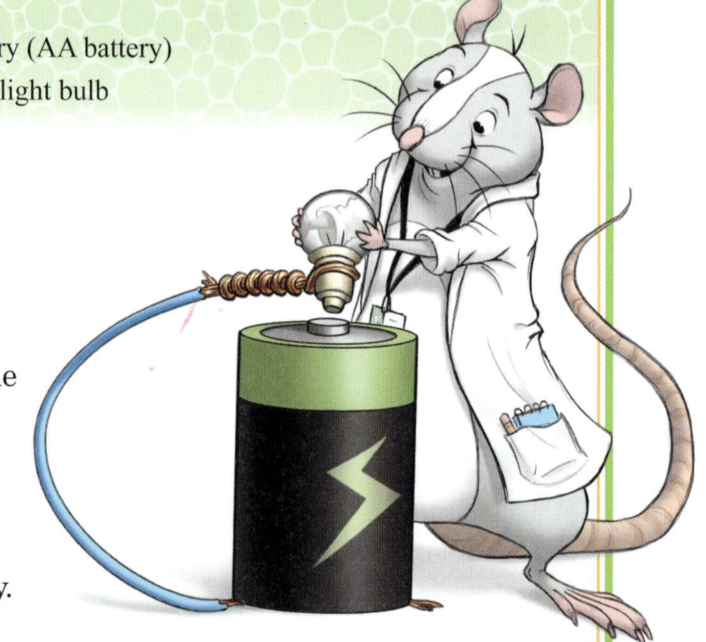

1. Twist one end of the exposed wire around the metal end of the light bulb.
2. The battery has stored electrical energy. It will be the power source. Place the battery on a table with the negative (–) end down. Slide the other side of the wire underneath the battery.
3. Touch the bottom of the light bulb to the positive (+) tip of the battery. Write down your observations.

When the metal bottom of the light bulb touched the battery's positive end, electricity could flow in a complete circle all the way to the negative end. With a closed circuit, electrons are flowing, giving the light bulb power to work. You have just converted (potential) electrical energy into (kinetic) light energy!

Electrical Storm Safety

Did you know that water can conduct electricity? That is why it is dangerous to be in a pool, lake, or other body of water during a thunderstorm. Lightning can strike the water, and its electricity can travel to you.

Electricity can be deadly. After a storm, stay away from downed power lines, cables, or wires.

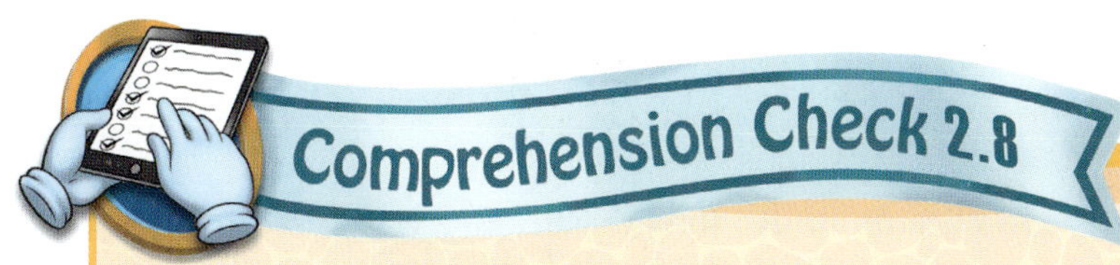

True/False: *Write the word* true *or* false *in each blank.*

_______ 1. A generator can make current electricity.

_______ 2. Electricity can transfer from one place to another because of flowing electrons.

_______ 3. Plastic is a good conductor of electricity.

_______ 4. When you turn an electric light on, you are opening the circuit.

_______ 5. A closed circuit can turn off a light.

Discuss: *Explain why the false answers are incorrect statements.*

Think and Classify: *Can you sort these materials by conductivity?*

Some materials have the property of conductivity. Metals such as copper can conduct electricity. Water also has this property. Do you remember what we call materials that do not have the property of conductivity? Any material that cannot transfer energy is called an insulator.

copper

rubber

plastic

wood

water

conductor of electricity	insulator of electricity

luminous: giving off light energy

crest: the highest point of a wave

trough: the lowest point of a wave

amplitude: the measure of a wave's strength, measured from its height to its resting position

2.9 Light-Wave Energy

Properties of Light

A long time ago, people believed light came from our eyes. They thought it traveled from the eye to objects that were seen. Though we need light to see, light does not come from our eyes. It is energy moving from its source, to its destination, and then to our eyes. Light energy helps us to see.

God created light energy. *When an object can give off light, we say that it is* **luminous**. Can you think of some objects that can give off light energy? The sun, like all stars, is luminous. It gives off light. Light bulbs are used to light our homes because they have been made to be luminous. Fire is also luminous. Before the light bulb was invented, people used oil lamps, candles, lanterns, and torches for light.

Stars, light bulbs, fire, certain sea creatures, and fireflies are all luminous. Even certain metals can be luminous if they are heated to very high temperatures.

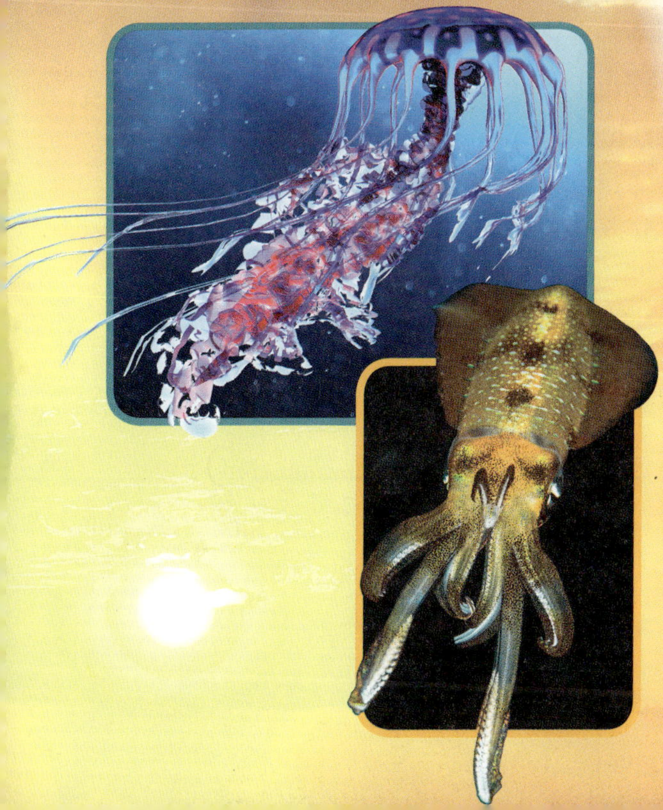

Light energy goes out in every direction as it shines. Sunlight rays shine in every direction as they leave the sun and travel to their destinations. When it is daytime where you live, rays of sunshine give light to the whole continent. A light bulb works in the same way, glowing in all directions to give light to an entire room.

God created light to always travel in a certain way. There are two things you need to remember about the way light travels. *First, it always travels in a straight line*—from its source to its destination. *Second, light travels faster than anything else*—about 186,000 miles per second. That means it takes light under three seconds to travel from the earth to the moon and back.

> *And God said, Let there be light: and there was light. And God saw the light, that it was good.*
>
> *Genesis 1:3–4*

Wave Energy

Light energy can be transferred from one place to another because it travels in waves. Have you ever watched the waves of a big lake or ocean? A wave is powerful because it is moving energy. Waves can change water's potential energy into kinetic energy.

If you imagine what an ocean wave looks like, you can imagine what a light wave looks like. In a wave, we see both a high point and a low point. *The highest point of a wave is called a* **crest**; *the lowest point is called a* **trough**.

Some waves are stronger than others because they have more energy. We can measure the *strength*, or **amplitude**, of a wave by looking at the distance between the height of the wave and its resting position.

If you have a jump rope, you can make a model of a light wave. Tie one end of the rope to a doorknob, hold the other end, and take several steps away from the door. Shake your hand up and down, watching for crests and troughs. This movement is how a light wave is able to travel from its source to its destination.

crest
amplitude
trough

Try to give your light-wave model more energy, or amplitude, by moving your arm higher and lower. Do you see the difference in the wave? More energy makes larger, stronger waves.

Is light energy potential or kinetic energy? Remember, light can be transferred from one place to another because of its wave energy. It always travels in a straight line from its source and travels more quickly than anything else. Because of its movement, light is kinetic energy.

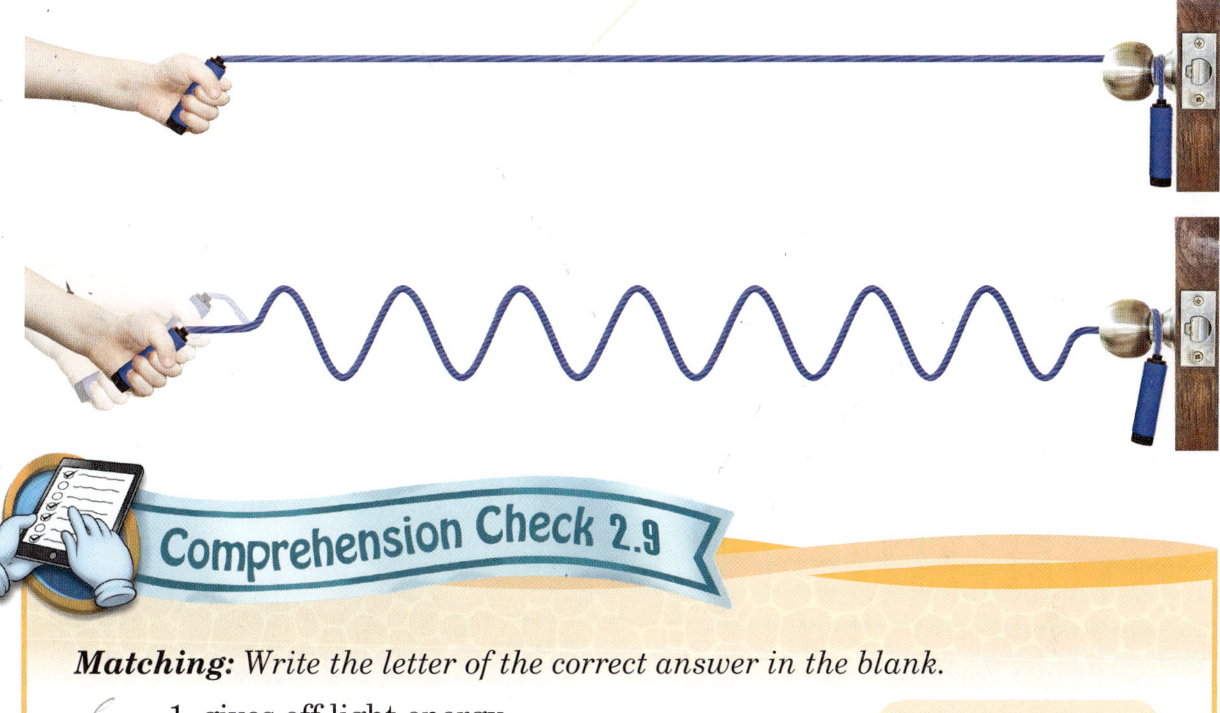

Comprehension Check 2.9

Matching: *Write the letter of the correct answer in the blank.*

___C___ 1. gives off light energy

___B___ 2. the highest point in a wave

___t___ 3. the lowest point in a wave

___A___ 4. the measure of a wave's strength

A. amplitude

B. crest

C. luminous

D. trough

Think and Conclude.

5. Give an example of a luminous object. ______________________

6. What travels faster than light? ______________________

7. What gives light energy the ability to be transferred from one place to another? ______________________

transmit: to allow waves to pass through

opaque: allowing no light to pass through

shadow: a dark area or shape cast by an opaque object

reflect: to throw back light, changing its direction

transparent: allowing light to pass straight through

translucent: allowing light through but scattering it so objects cannot be seen clearly

2.10 Transmitting Light

Because light travels in a straight line, it cannot turn corners or go through barriers. This means some objects can block light, while other objects **transmit** light, or *allow light to pass through.*

An **opaque** material blocks light. The walls in your home are opaque. You cannot look through the walls of your house to see your backyard. Your backyard is still on the other side of your walls, but because they are made from an opaque material, *no light can pass through.*

A **shadow** *is a dark area or shape cast by an opaque object.* At certain times of the day, your house may cast a shadow on your backyard. Because your house is an opaque object and sunlight rays travel in a straight line, some rays are blocked. Usually a shadow has the same shape as the object it is cast from.

Do you play basketball? A bounce pass to your teammate changes the direction of the ball. The ball bounces *down* on the court and bounces *up* to your friend. Light can bounce, too. We call bouncing light **reflection**. *To* **reflect** *is to throw back light, changing its direction*. Light still travels in a straight line, but the direction is now changed. Some materials are better reflectors than others. A mirror is a very good reflector because its surface is shiny.

A **transparent** material transmits light because *it allows light to pass straight through it*. The clear, glass windows in your home are transparent. If you look out your window, you can see everything in your backyard. Transparent materials, like clear glass, can transmit light.

A **translucent** *material allows light through but scatters it so objects cannot be seen clearly*. It can both block and transmit light. The sun can "hide" behind a cloud on a cloudy day. It is still in the sky, and enough of its light passes through the clouds for it to be daytime, but the sun itself is hidden behind a cloud. *Light is scattered when it passes through a translucent material*.

Try This! Classify materials according to light transmission.

Materials needed:
- ✓ plastic wrap
- ✓ aluminum foil
- ✓ cardboard
- ✓ waxed paper
- ✓ parchment paper
- ✓ mirror
- ✓ flashlight

1. Test each material by holding each one up about five feet from a wall.

2. Shine the flashlight on each material, observing the effects of the light on the wall. Write down your results.

Comprehension Check 2.10

Fill in the Blank: *Choose the correct word from the box below and write it in the blank.*

1. Opaque materials block light.
2. Transparent materials transmit light, allowing it to pass straight through.
3. Translucent materials both block and transmit light, scattering it.

Opaque	**Translucent**	**Transparent**

Give the correct answer.

4. What is a dark area cast by an opaque object? Shadow
5. What is thrown-back or bouncing light called? refelt

Think and Predict.

6. What objects are good reflectors? Morrits

refraction: the bending of light rays as they enter a new material

wavelength: the distance from one crest to the next crest

prism: a transparent, solid figure used to separate white light into colors

color spectrum: the series of colors created when white light is separated

Bending Light

Refraction

We know that light must travel in a straight line. Sometimes, it can "bounce" off the surface of an object. Do you remember what we call light when it is thrown back in the opposite direction? Bouncing light is reflected light. Light can also change direction if it bends.

Light can bend and change direction because of refraction. **Refraction** *happens when light is bent while traveling through one kind of matter into another kind of matter.* Light energy traveling through air moves very, very quickly; but it slows down a little when it travels through water. It is still traveling in a straight line, though it appears to have changed direction just a little.

Place a pencil in a clear drinking glass. Now fill the glass halfway with water. Do you see how the pencil looks bent? The pencil itself did not bend; bent light rays make it appear bent. What two materials did the light travel through to cause this refraction? Light traveled very quickly through the air, but it slowed down a little when it traveled through water.

What happens to its waves when light energy slows down? Use your jump-rope model to see how light waves change when they slow down or speed up. Try shaking your hand up and down very quickly. Now try it very slowly. Do you see the difference in the waves? They look farther apart, don't they? *The distance from one crest to the next is called a* **wavelength**.

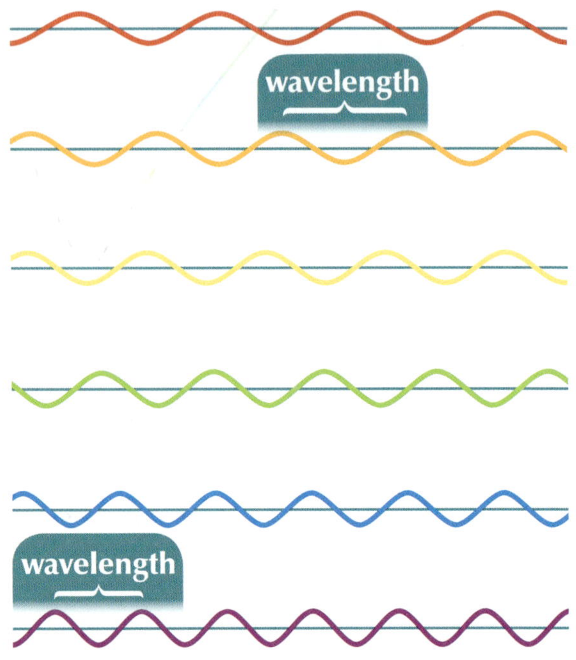

The Color Spectrum

Did you know that light is actually all the colors of the rainbow? *White light is made of colors—red, orange, yellow, green, blue, and violet.* The initials of these colors spell ROY G. BV.

A **prism** helps us *to separate white light into its different colors.* Remember, light can change speed when it travels from one kind of matter to another. Light can travel through air and into a prism, changing its speed. Do you remember what happens to light then? It bends. A prism makes each color of light bend in a different way, spreading out a *series of colors* we call the **color spectrum**. *Each color of the spectrum has a different wavelength. Red has the longest wavelength,* while *violet has the shortest.*

Can you imagine how beautiful the rainbow looked to Noah the very first time God placed it in the sky? Now that you understand the color spectrum, you can see that God uses white light to make each rainbow. Little droplets of water in the air after a rain act like tiny prisms, separating the light into all the colors that spread across the sky. Light works exactly as God created it to work, making His world colorful.

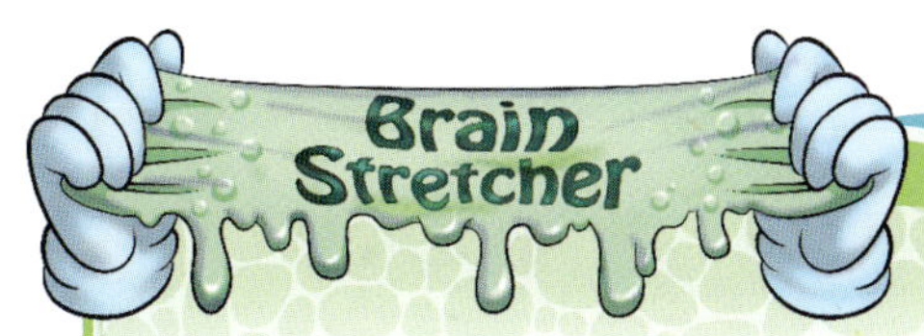

Light Energy and Your Eyes: Light energy helps us to see color. Your eyes work like a camera; when they are open, they are constantly "taking pictures." When there is plenty of light, your eyes see in bright color. If there is only a little light, you can see shadows and figures in black and white.

In the middle of your eye is a black circle called the pupil. The pupil is really a round hole that lets the right amount of light into your eye. Your pupil gets smaller or larger depending on how much light is available. The size of your pupil is controlled by your iris, or colored part of the eye. The iris is a ring of muscles that controls the amount of light that enters your eye.

With the right amount of light, the lens of your eye can focus what you are seeing onto the retina. The retina helps your brain interpret color. The optic nerve in the back of your eye carries the message of sight to your brain. Your brain will then tell your body how to respond to what you see.

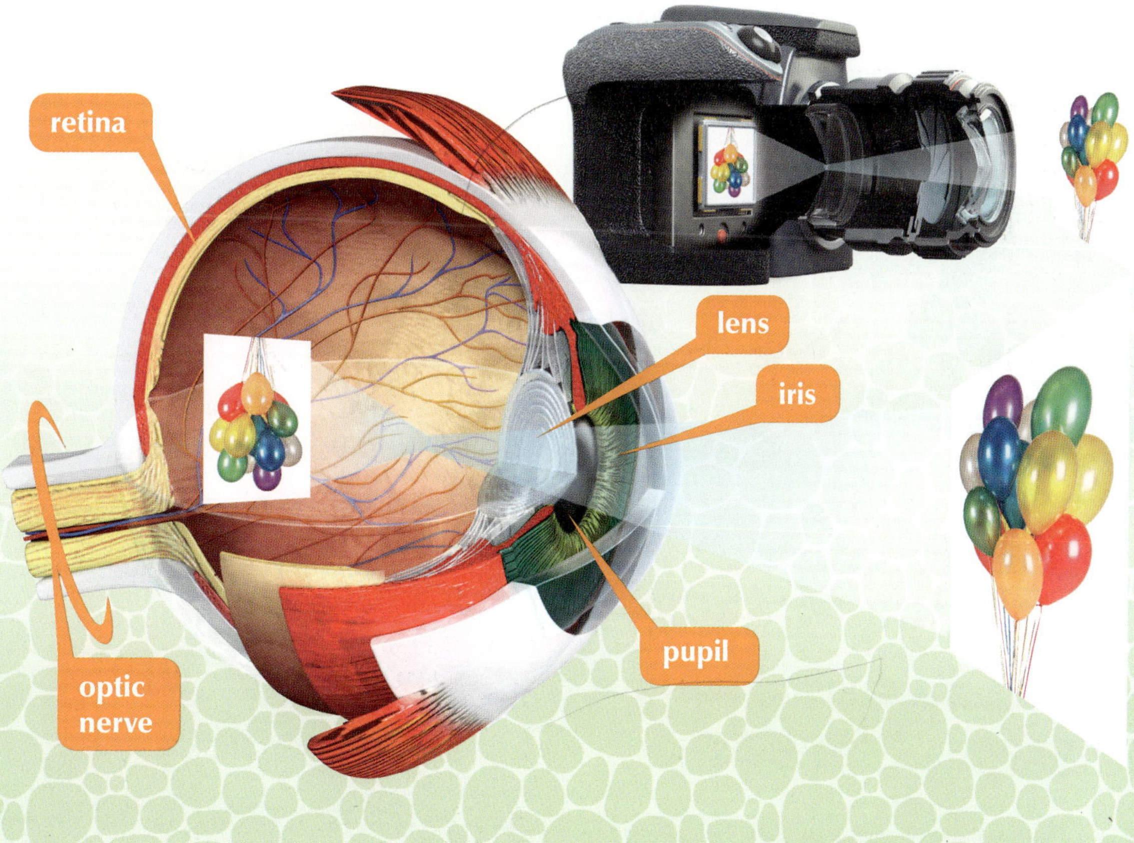

retina
lens
iris
pupil
optic nerve

Comprehension Check 2.11

Multiple Choice: *Circle the correct answer.*

1. Bent light is called __?__.

   a. reflected light b. refracted light c. a shadow

2. A __?__ is the distance between two crests in a wave.

   a. prism b. spectrum c. wavelength

3. A prism can separate __?__ light into colors.

   a. black b. rainbow c. white

4. The color of the spectrum with the longest wavelength is __?__.

   a. blue b. orange c. red

Label: *Write the colors of the spectrum in order. Use your colored pencils to trace each wavelength.*

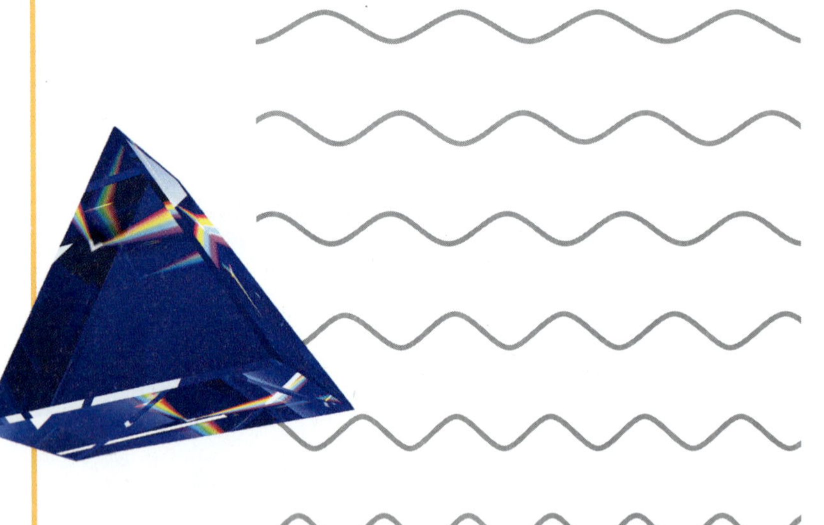

5. ______________

6. ______________

7. ______________

8. ______________

9. ______________

10. ______________

vibrate: to move rapidly back and forth

2.12 Sound-Wave Energy

Can you imagine a world without sound? If you could travel to every planet in the solar system, you could not hear the sounds you can enjoy on Earth. The voice of a friend, a bird's song, the rustle of leaves on a windy day, ocean waves splashing on a sandy shore—many things can be recognized by the sounds they make.

How is sound produced? How does it travel? God designed a unique way for sound energy to travel from one place to another.

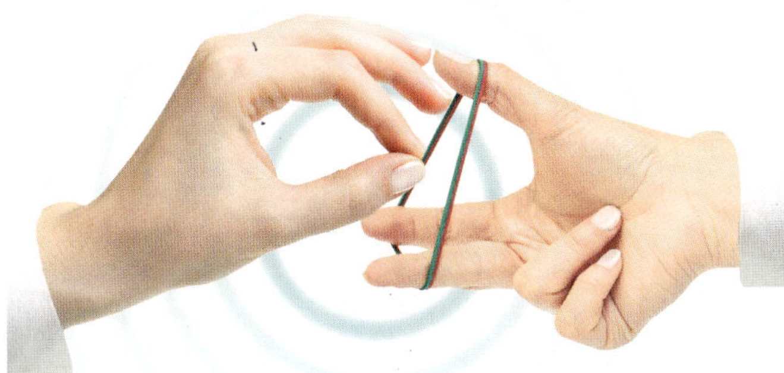

Sounds Are Caused by Vibrations

Sound is a form of energy that is released when an object **vibrates**, or *moves rapidly back and forth*. To see how vibrations produce sounds, loop a rubber band around a doorknob and stretch it as far as you can without breaking it. Now pluck it with your finger. What happens? The rubber band moves rapidly back and forth, producing a hum. What happens to the hum when the rubber band stops vibrating? The hum stops because the vibrations stop.

When the rubber band vibrated, it made the air molecules around it move. The moving energy of the rubber band was converted into sound energy when it caused the molecules of air to bump into one another.

This energy creates a wave of air molecules bumping into other air molecules as the sound spreads out. We hear sound when sound waves enter our ears. These vibrations continue as they travel through the parts of our ear to the auditory nerve.

Try This! Observe the effects of sound vibrations.

Materials needed:
- ✓ bowl
- ✓ plastic wrap
- ✓ metal pan
- ✓ large metal spoon
- ✓ *large rubber band
- ✓ salt

1. Place plastic wrap over the top of the bowl and pull it tightly to create a smooth surface. If you cannot pull tightly enough, try placing a large rubber band around the bowl.
2. Sprinkle a little salt on top of the plastic.
3. Place the metal pan on its side near the bowl and, without touching the plastic or the bowl, hit the metal pan with the spoon several times. Observe the salt. Did you see any of the salt crystals move?

The sound vibrations you made by striking the metal pan traveled through the air, bumping into air molecules. The air disturbance traveled until it reached the air near the salt crystals, causing them to vibrate.

*(optional)

Sound Travels through Matter

Like light energy, sound energy can also be transferred from one place to another because of its waves. Sound waves and light waves work differently; *sound waves can travel only when there are molecules of matter to carry the vibrations.* Sound waves cannot travel through empty space.

In 1660, English scientist Robert Boyle proved that this was true. He placed a watch with a dependable alarm inside a sealed glass jar. Using a pump, he removed the air that was inside the jar. When it was time for the alarm to ring, Boyle heard nothing. When he allowed air back into the jar, he was then able to hear the alarm ringing. Boyle repeated his experiment many times, sometimes using a bell instead of the watch. When air was removed from the jar, Boyle could see the bell vibrating but could hear nothing. Molecules were not there to transfer sound energy in waves. No sound could be heard.

Sound waves can travel not only through air, but also through liquids, such as water, and solids, such as metal. Sound travels through solids the fastest because the molecules in a solid are closer together. Musical instruments depend on solid material to help them make sound.

Sound waves travel slowest through gases, faster through liquids, and fastest through solids. Is sound potential or kinetic energy? Because of its movement, sound energy is kinetic energy.

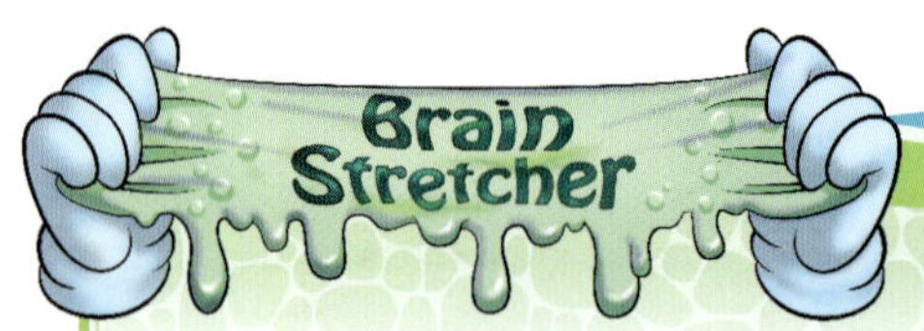

Sound Waves and Your Ears: The ear is designed to catch sound waves and direct them to the brain. We can see only a small part of the ear. Most of the ear is actually inside the skull.

The ear has three main parts. They are the outer ear, the middle ear, and the inner ear.

The outer ear begins with the pinna. This is the part of the ear you can see, and it acts like a cup. It is designed to catch sound waves.

The sound waves travel through your outer ear into a little tube called the auditory canal. You can see the beginning of this canal by looking into someone's ear.

The eardrum is at the end of this canal. The eardrum is a thin piece of membrane that is stretched tightly, like a drum. It separates the outer ear from the middle ear. The sound waves strike the eardrum, making it vibrate. These vibrations can then travel to your middle ear.

In the middle ear, there are three tiny bones: the hammer, the anvil, and the stirrup. Each little bone amplifies the sound waves as they are transmitted to the inner ear.

The cochlea is in the inner ear. It is filled with a liquid. The three little bones from the middle ear vibrate this liquid, which causes the auditory nerve to send a message to your brain. It is then that you hear the sound.

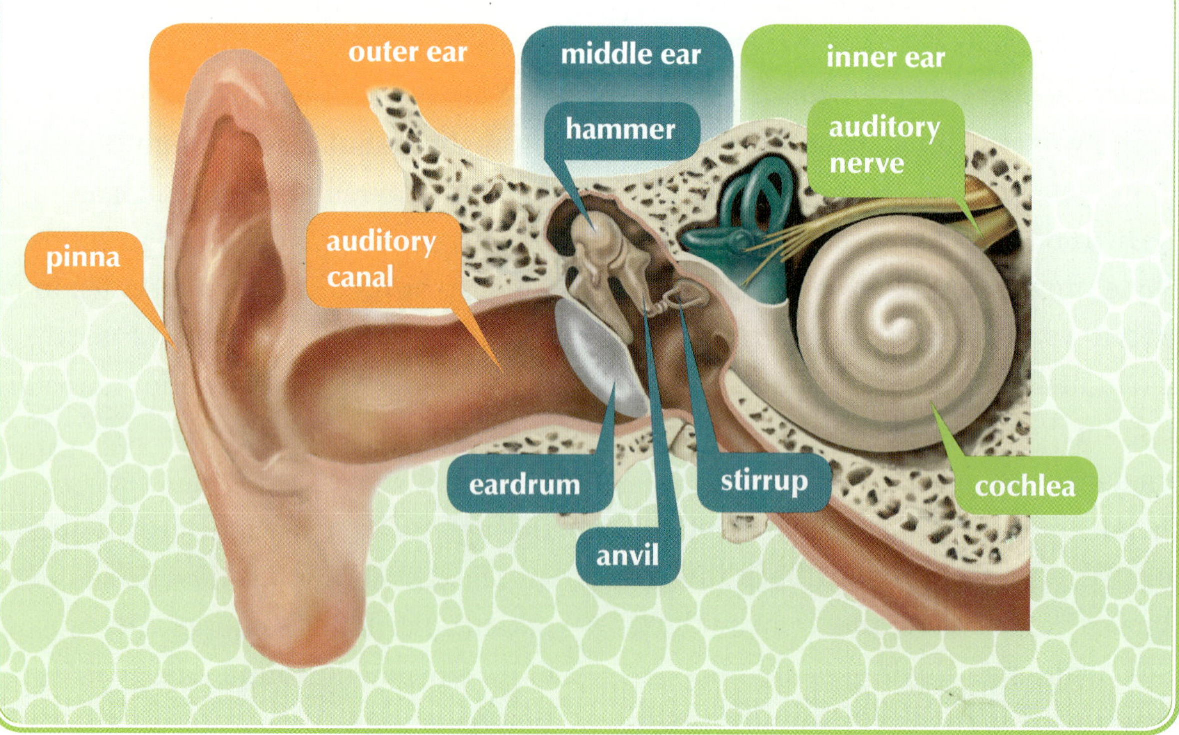

Comprehension Check 2.12

Fill in the Blank: *Choose the correct answer from the box below and write it in the blank.*

1. __________ occurs when an object vibrates.
2. Sound energy can travel because of __________.
3. Sound needs __________ through which to travel.

matter	**sound**	**sound waves**

Think and Predict.

4. Would sound waves travel more quickly through the earth or through the air? Why? ______________________________
5. Could sound waves travel in outer space? Why or why not? ______________________________

TERMS

volume: the loudness or softness of sound

pitch: the highness or lowness of sound

2.13 Volume and Pitch

Sound is a gift from God. He gave Earth an atmosphere of matter so that sound waves could travel. He also designed sound waves to produce a variety of sounds. Can you imagine what life would be like if everything sounded the same? God gave nature the beautiful and interesting sounds we hear as we work and play outside.

Different sound waves make different sounds. Some sound waves are big; others are smaller. Some travel quickly while others travel more slowly. The differences in sound waves make volume and pitch.

Volume: Loud or Soft

As sound travels from its source, it spreads out in all directions. Sound-wave energy becomes weaker as it travels until finally the sound cannot be heard at all. This is why it is easier to hear someone talking when they are close to you and harder to hear someone far away.

Some sounds are loud while other sounds are soft. You can use your voice to shout loudly or whisper quietly.

Volume *is the loudness or softness of a sound.* Larger, more powerful sound waves make loud sounds. Softer sounds are made by smaller sound waves with less energy.

More energy makes a bigger sound wave; less energy makes a smaller sound wave. We can measure the *strength of a sound wave* by its **amplitude**.

Below is the sound wave from a loud sound. Notice how the amplitude decreases as the sound moves away from its source.

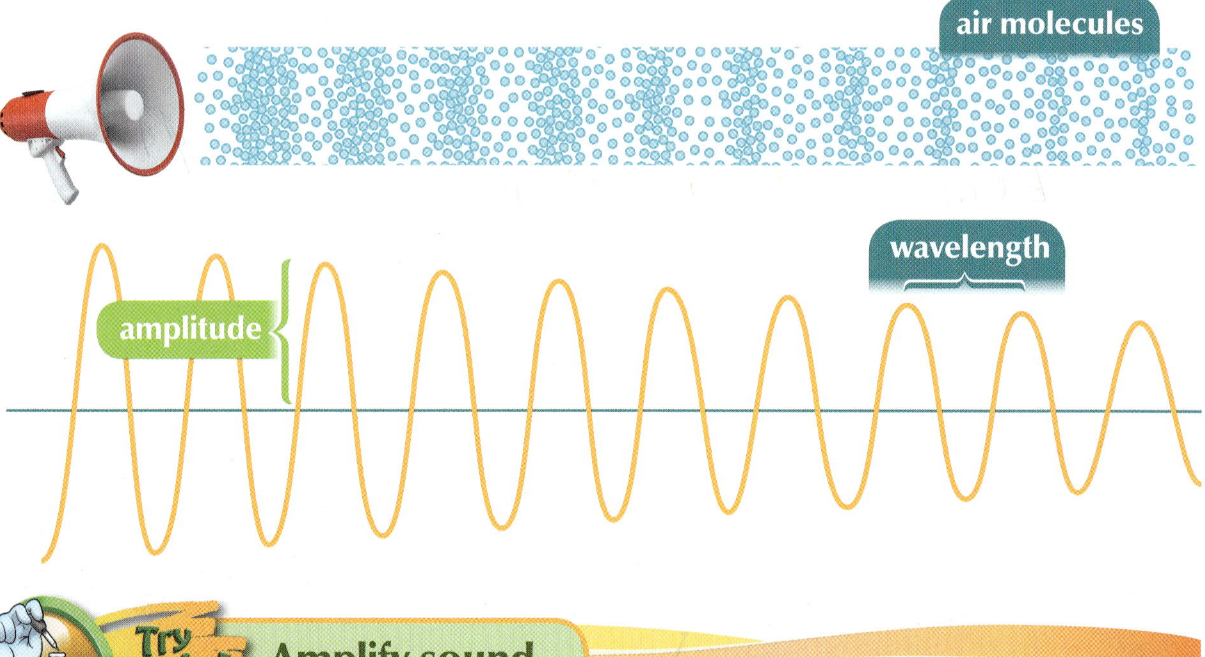

Try This! Amplify sound.

Materials needed: ✓ paper ✓ tape

Make a megaphone with a piece of paper, taping it securely. Try talking with and without it, observing the difference. Why did you sound louder using your megaphone? By using a megaphone, you *are giving its waves more amplitude*.

Pitch: High or Low

If you listen to a piece of music, you can hear a variety of sounds. There are high sounds and low sounds. **Pitch** *is the highness or lowness of a sound*. Pitch is different from volume. It does not describe how loud or soft a sound is. Can you think of some high and low sounds in nature? A bird singing is a high sound; the roar of a bear is a low sound. Sounds are high or low because of the speed of the vibrations making them.

Attach a rubber band to a doorknob. Loop the other end around a pencil. Stretch the band very tightly and pluck it. The vibrations of the rubber band are so fast that all you can see is a blur. *Faster vibrations make higher pitches.* The "twang" you hear as you pluck the rubber band has a high-pitched sound.

Now, allow the rubber band to become fairly loose. Pluck it again. Can you hear the difference? The sound of the rubber band is much lower. Now pluck it again and watch.

You can now see the vibrations, because they are much slower. *Slower vibrations make lower pitches.*

The design of the violin helps it to make both high- and low-pitched sounds. Thicker strings vibrate more slowly and make lower-pitched sounds. Thinner strings vibrate more rapidly, making higher-pitched sounds.

Heat, light, sound, and electricity can all be transferred because of God's unique design for each kind of energy. What do these special designs tell you about their Designer? Our powerful God gave His creation energy. His perfect design for each energy transfer system also shows us that He is wise beyond measure. As you understand how matter and energy work, give all glory and praise to God.

> *And God saw every thing that He had made, and, behold, it was very good.* Genesis 1:31

Try This! Observe high and low pitch.

Materials needed: ✓ funnel ✓ balloon ✓ scissors

Cut the balloon so that it can be loosely stretched over the large end of the funnel. Pat the large end of the funnel and listen to the sound. Now stretch the balloon so that it is tighter and pat it again. Is the sound higher or lower than the first sound? Why? Write down your observations.

Comprehension Check 2.13

Give the correct answer.

1. How do we describe the loudness or softness of a sound? volume

2. How do we describe the highness or lowness of a sound? pitches

Alexander Graham Bell

Inventor of the Telephone

Whenever you answer the telephone, think of its inventor: a Scottish-American scientist who wanted only to be remembered as a teacher of the hearing-impaired. Alexander Graham Bell sought to understand the wonders of God's world to help others, especially the hearing-impaired people whom he loved so much.

Aleck Bell, as his family called him, was born in Edinburgh, Scotland, in 1847. Bell's father and grandfather were important public speakers. Aleck's father also taught people who could neither hear nor speak, how to speak. To help them, he invented a system known as "Visible Speech," a code in which written symbols represent sounds by indicating the position of the throat, the tongue, and the lips.

Aleck's mother, who was partially deaf, was an accomplished artist and musician. She taught Aleck and his two brothers the Bible as well as reading, writing, and arithmetic. Their father taught them how to speak correctly. As teenagers, they began going with their father on lecture tours to demonstrate the usefulness of Visible Speech in teaching the hearing-impaired to speak and in learning foreign languages.

Aleck also began teaching music and speech to people who could neither hear nor speak. When his family moved from Scotland to Canada, Aleck went to Boston, giving his own lectures on Visible Speech and opening a school for training teachers of the hearing-impaired. The next year he became a professor at Boston University where he taught how the organs of speech function.

Bell was very interested in the transmission of sound. At that time, people could send Morse Code messages over the electric telegraph. Morse Code is a combination of dots (short sounds) and dashes (long sounds). Bell began experiments to find a way to send several telegraph messages over a single wire at the same time and to transmit sounds of any sort over a wire.

A young telegraph mechanic, Thomas Watson, began working with Bell. Together, they worked to make both the "harmonic telegraph" and the telephone. Bell and Watson worked long into the night almost every evening until they were successful. At first, the new machine could transmit only the sound of a plucked reed. After many experiments, they were able to transmit human vocal sounds, but not words, over the telephone.

On March 10, 1876, Bell and Watson were in different rooms in a Boston boarding house trying out a new transmitter. Bell clumsily spilled battery acid all over his clothes and called for his assistant. Watson heard the first words to be transmitted by telephone: “Mr. Watson, come here. I want you!”

That summer, Bell demonstrated the telephone to the emperor of Brazil at the Centennial Exposition in Philadelphia. Bell’s invention soon became a worldwide sensation. After that, Bell set up an experimental sound laboratory and established the American Association to Promote the Teaching of Speech to the Deaf, later called the Alexander Graham Bell Association for the Deaf and Hard of Hearing. Many important inventions in sound transmission happened at Bell’s laboratory. Alexander Graham Bell became an American citizen and spent the remainder of his seventy-five years on Earth promoting the causes of science, invention, and education for the hearing-impaired.

Today, we can use satellite transmission and radio waves to transmit sound. Radio waves do not need an atmosphere of matter to carry them. This is how astronauts are able to communicate with each other in space. We also depend on this kind of transmission to use cellphones and wireless technology.

Chapter 2 Concepts Review

Matter Concepts 2.1–2.5

A. Remember

True/False: *Look at the underlined word to determine if the statement is true or false. If the statement is true, write* true *in the blank. If the statement is false, write* false *in the blank. Be prepared to explain why the false answers are incorrect.*

______ 1. Gases can keep their shape.

______ 2. The molecules in a solid are very dense.

______ 3. Liquids can fill the shape of a container.

Fill in the Blank: *Write the correct number in the blank.*

4. The melting point and freezing point of water is ______ degrees Fahrenheit.

5. The boiling point of water is ______ degrees Fahrenheit.

B. Think Like a Scientist

Label: *Write* S *for solid,* L *for liquid,* G *for gas, and* N *for not matter. Be prepared to explain your answers.*

______ 1. oxygen

______ 2. love

______ 3. happiness

______ 4. milk

______ 5. rock

______ 6. metal

Short Answer: *Write the correct answer in the blank.*

7. Give an example of a physical property of matter. ______

8. What happens to water when it freezes? ______

Identify: *Circle the correct answer.*

9. Which object has molecules of greater density?

10. Circle each atom in this molecule.

C. Fun with Terms

Matching: *Write the letter of the correct answer in the blank.*

_____ 1. a physical substance that has weight and takes up space

_____ 2. the substance in a solution that dissolves into the solvent

_____ 3. the substance in a solution that is able to dissolve another substance

_____ 4. how tightly packed molecules are in an object

_____ 5. two or more different kinds of matter mixed together

_____ 6. a mixture in which one kind of substance is dissolved into another

_____ 7. a tiny particle of matter

_____ 8. a way you can describe or sort matter

A. density
B. matter
C. mixture
D. molecule
E. property
F. solute
G. solution
H. solvent

Word Search: *Find and circle the words listed in the box above.*

2 Energy Concepts 2.6

A. Remember

Short Answer: *Write the correct answer in the blank.*

1. What is energy? ______________________________
2. What is kinetic energy? ______________________________
3. What is potential energy? ______________________________

B. Think Like a Scientist

Sort: *Write each energy type from the box in the correct column below.*

cellphone battery	**fuel**	**running**
firewood	**moving water**	**sliding**
food	**musical sounds**	

Kinetic	Potential
____________	____________
____________	____________
____________	____________
____________	____________

C. Fun with Terms

Crack the Code: *Write the letter that matches each number in the blank to discover the missing term.*

1. ___ ___ ___ ___ ___ ___ : the ability to do work
   5 14 5 18 7 25
2. ___ ___ ___ ___ : material that is burned to release energy
   6 21 5 12
3. ___ ___ ___ ___ ___ ___ ___ : moving, working energy
   11 9 14 5 20 9 3
4. ___ ___ ___ ___ ___ ___ ___ ___ ___ : stored, waiting energy
   16 15 20 5 14 20 9 1 12
5. ___ ___ ___ ___ ___ ___ ___ : to change
   3 15 14 22 5 18 20
6. ___ ___ ___ ___ ___ ___ ___ ___ : to move
   20 18 1 14 19 6 5 18

1	a	14	n
2	b	15	o
3	c	16	p
4	d	17	q
5	e	18	r
6	f	19	s
7	g	20	t
8	h	21	u
9	i	22	v
10	j	23	w
11	k	24	x
12	l	25	y
13	m	26	z

3 Electricity Concepts 2.7–2.8

A. Remember

Identify: *Circle the electrons in this atom.*

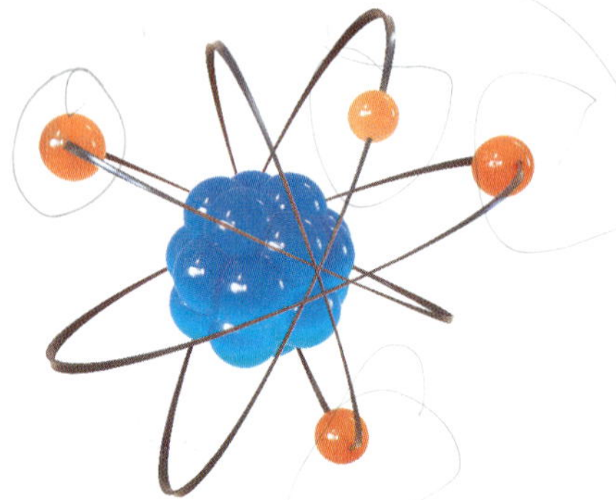

Short Answer: *Write the correct answer in the blank.*

1. Which part of an atom is positive? ____________________

2. Which part of an atom is negative? ____________________

B. Think Like a Scientist

Label: *Write* S *for static electricity and* C *for current electricity next to each clue. Be prepared to explain your answers.*

_____ 1. able to power our homes

_____ 2. buildup of electrical charge

_____ 3. can be controlled

_____ 4. caused by rubbing two objects together

_____ 5. flow of electrons in a complete circuit

_____ 6. gives sudden discharge

_____ 7. causes lightning

_____ 8. made with a generator

Identify: *Label the closed circuit and the open circuit. Be prepared to explain which circuit can turn on a light and why.*

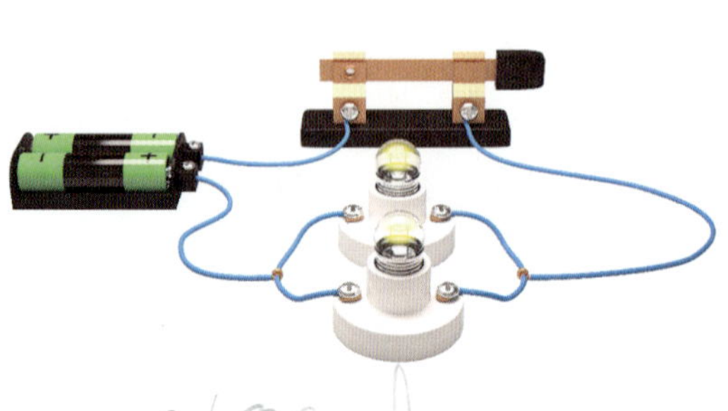

9. ____________________

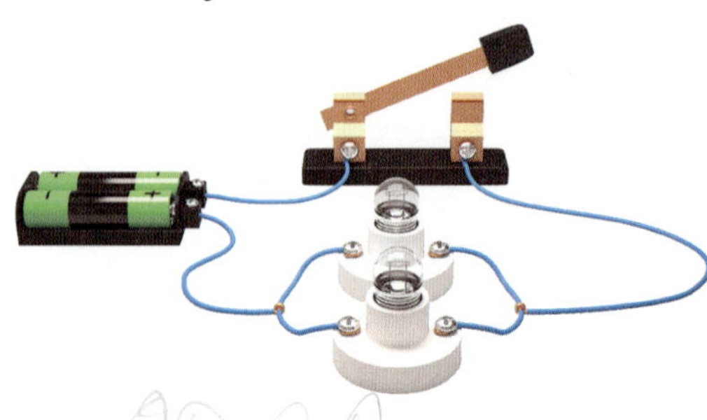

10. ____________________

C. Fun with Terms

Matching: *Write the letter of the correct answer in the blank.*

____ 1. bundled wires, covered in plastic or other insulator	A. atom
____ 2. a particle that makes up a molecule	B. attract
____ 3. a material that can easily transfer (electrical) energy	C. cable
____ 4. a machine that converts mechanical energy into electrical energy	D. conductor
____ 5. a negative particle of an atom	E. electron
____ 6. a material that cannot easily transfer (electrical) energy	F. insulator
____ 7. to pull together	G. generator
____ 8. a tiny particle of matter	H. molecule
____ 9. to push apart	I. repel

4 Wave-Energy Concepts 2.9–2.13

A. Remember

1. Trace the light wave. Label *crest*, *trough*, *wavelength*, and *amplitude*.

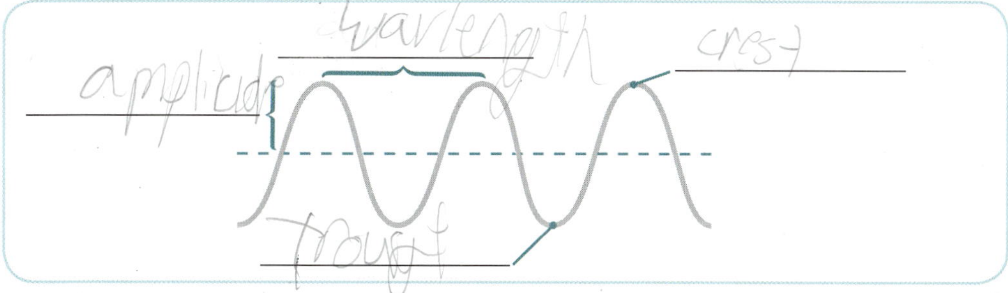

2. List the colors of the spectrum in order.

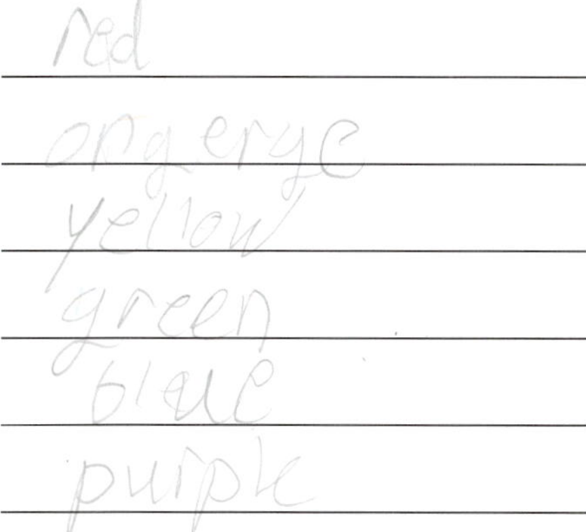

B. Think Like a Scientist

1. Circle the wave that shows a greater amplitude.

2. Label the materials correctly according to light transmission.

______ ______ ______

Matching: *Write the letter of the correct answer in the blank.*

______ 3. Which picture shows refraction?

______ 4. Which picture shows reflection?

______ 5. Which picture shows a luminous object?

A. B.

C.

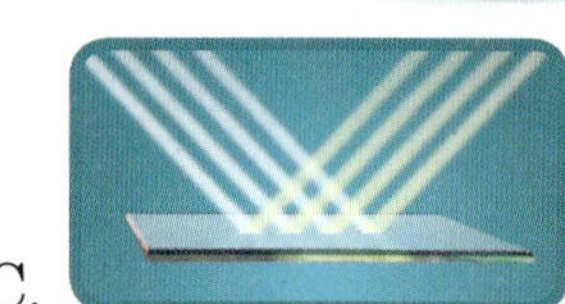

C. Fun with Terms

Crack the Code: *Find the letter of the correct term and write it on the numbered line below.*

______ 1. the measure of a wave's strength

______ 2. to pass through

______ 3. a dark area or shape cast by an opaque object

______ 4. to move rapidly back and forth

______ 5. the loudness or softness of sound

______ 6. the highness or lowness of sound

E. vibrate
G. pitch
N. volume
R. shadow
V. transmit
W. amplitude

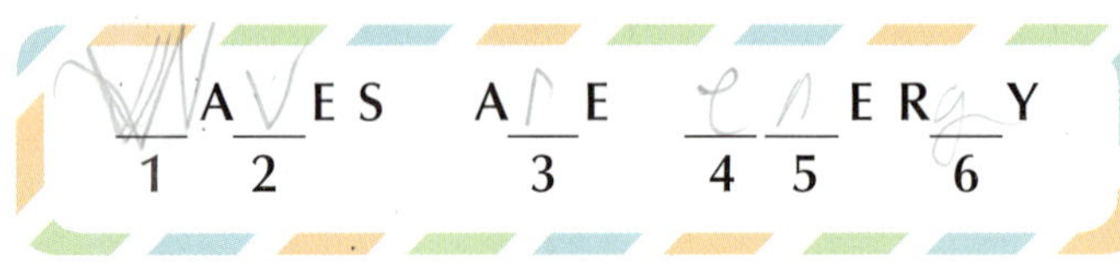

Understanding Force and Motion

Chapter 3

TERMS

motion: a change in position

force: the push or pull on an object

work: movement resulting from force

friction: a force that works against motion along a surface

balanced forces: opposing forces that prevent work from being done

unbalanced forces: a greater force that overcomes opposing forces, resulting in work

3.1 Forces That Affect Motion

What makes it possible for a snowboarder to make a flip on a halfpipe? How does a hockey player change direction quickly enough to defend his goal? What can a swimmer do to increase her speed? What causes a football to travel down the field so quickly? Forces and motion are what make sports possible. Athletes who understand how to use forces and motion can become excellent in their sports.

We can understand how force and motion work because God created them to work in a certain way. We can discover the order He planned for things that move.

What Is Motion?

Objects do not always remain in one position. *As objects change position, we say they are in* **motion**. An object needs a force to stop or start movement. A force can speed something up or slow something down. A force can change the direction of a moving object or change an object's shape. Motion is impossible without force.

What Is Force?

Force *is the push or pull on an object.* Have you ever tried to lift a very heavy box? As you pushed and pulled and struggled, you probably thought you were working very hard. But if you were unable to move the box, you did not do any work—not in the scientific sense. *When something has been moved, we say that* **work** *has happened.*

Are you doing any work right now? If you are sitting at your desk, holding your book, and reading silently, you are not doing any work in the way scientists use the term. Your mind is very busy, but you are not moving. If you hold your book up long enough, your arms may get tired, but you are not doing any work unless you move the book or turn the page. If you close your book and put it away, then you have done work. *All work is the result of force.*

What force did the work of pushing this sailboat out to sea? The wind's force can push things along and move them.

What force did the work of pulling this raft down the river? River currents are filled with energy. What force pushes seashells to the shore and pulls the sand out from under your feet? Ocean waves and currents exert force, giving the ocean the ability to pull and move things.

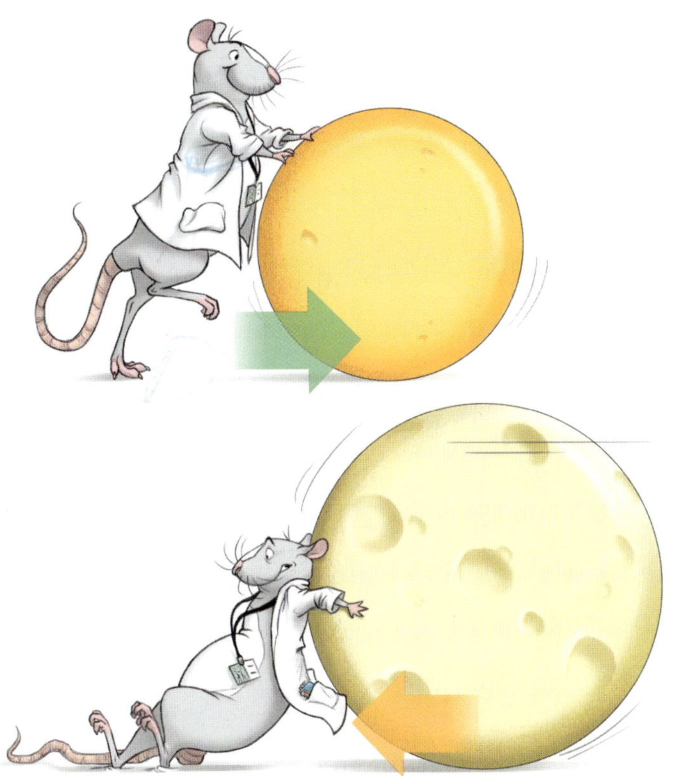

Sometimes, a push or pull can even change an object's speed. If an object is pushed *with* the direction of its movement, it will speed up. If an object is pushed *against* the direction of its movement, it will slow down.

All movement is the result of some kind of force. People, animals, and machines are able to move because of the forces they can exert. You are able to run, lift, push, and pull because of God's design for your body's mechanical energy. Energy in nature, such as wind or waves, can exert forces on objects to move them. The force of wind or water can even generate, or make, electricity. *Friction, gravity, weight, and magnetism are other kinds of forces we experience every day.*

Friction

Some forces can slow things down. **Friction** *is a force that works against motion along a surface.* Rubbing two objects together creates friction. Scientists believe that atoms stick together when this happens. Because of this, friction can wear things down and cause heat.

Things with rough textures create more friction than things with smooth textures. A hockey rink is smooth and slippery; and skates have smooth blades. Hockey players depend on their playing surface and equipment to help them overcome friction so that they can glide quickly on the ice. When they need to stop or change direction, hockey players make more friction by pushing against the ice with their skates. If you look closely, you can see a spray of ice where friction is happening. Skate blades wear down the ice. After a hockey game, the surface of the ice rink shows where skates have caused friction.

Gravity and Weight

No matter how high a basketball player jumps, he or she will always come back down because of the force of gravity. Usually, when we talk about *gravity*, we mean the force that draws objects to the earth. The *weight* of an object is measured by how strongly gravity pulls on it.

Magnetism

The word *magnetism* may have come from the Greek legend of Magnes the shepherd. While he was walking one day, an iron part of his sandal stuck to a mysterious stone. Stones that can attract iron are called magnetite. A piece of magnetite is called a lodestone. Its name comes from the words *leading stone* because it can work like a compass—always pointing north.

Magnetic metals, such as iron, can be pulled by a magnet. *A magnet can also pull, or attract, the opposite pole of another magnet. It can push away, or repel, like poles of another magnet.*

Balanced and Unbalanced Forces

When nothing is moving, it might seem that there are no forces at work. But forces like weight, gravity, and friction are always acting. Usually, more than one force is acting on an object at the same time. For example, the earth's gravity keeps the ocean on its surface. But the gravity of the sun and moon are also acting on the ocean, causing the tides that move the water.

When you come home from hiking, you probably bring your backpack inside. Perhaps you set it down on your kitchen table while you eat a snack. Your backpack remains in one place because of balanced forces. **Balanced forces** *are opposing forces which prevent work from being done.* The weight of your backpack pushes down on the table, but the table pushes upward with equal force. That is why your backpack will still be on the table when you are finished eating. Balanced forces cancel each other out; they cannot change the motion of an object. Balanced forces prevent work from being done.

Opposite poles attract.

Like poles repel.

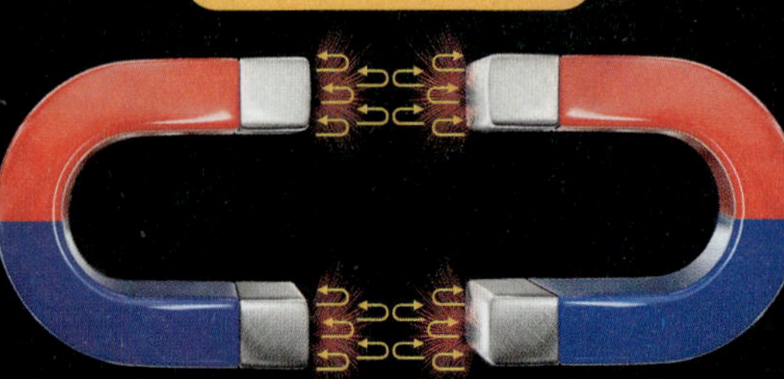

What if you had set your backpack on the edge of the table so that it was hanging off? If there was not enough table to balance the weight of the backpack, what happened to it? Your backpack tipped over and fell down. It moved from the place you set it because of unbalanced forces. An **unbalanced force** means there is *a greater force that overcomes opposing forces, resulting in work.* To move an object by lifting it, you must exert more force than gravity does on that object. If a box weighs fifteen pounds, you must exert more than fifteen pounds of force to lift it. *Unbalanced forces do work.*

Without force and motion, our world would be lifeless. And yet, God has not only designed each kind of force, but He also governs them by the way they can balance and interact with each other. This gives us the order we enjoy in the world we live in. What does this design tell you about the Designer? We can trust that our God is a wise and powerful Creator Who rules His creation.

Ah Lord God! Behold, Thou hast made the Heaven and the earth by Thy great power and stretched out arm, and there is nothing too hard for Thee. *Jeremiah 32:17*

Comprehension Check 3.1

Give the correct answer.

1. How can you define motion? ____________________
2. What is force? ____________________
3. Name a kind of force. ____________________
4. What force works against motion along a surface? ____________________
5. What kind of force can either push away (repel) or pull together (attract)? ____________________

Think and Conclude.

6. What is the difference between force and work? ____________________
7. Could balanced forces cause work to happen? ____________________
8. Why or why not? ____________________
9. Circle the pictures that show balanced forces. Box the pictures that show unbalanced forces.

Think and Predict.

10. What results would you expect after rubbing sandpaper on a block of wood for one minute? ____________________

Force and Motion

Throughout your study of this chapter, observe force and motion. Think about what forces are being exerted on the things you see. Ask yourself: What force made that move? What forces could slow it down or stop it? What force could change its speed?

laws of motion: Newton's discoveries about how motion always works

inertia: the property of matter that makes it resist a change in motion

3.2 Motion Needs Force

Galileo was an Italian scientist who was the first to study the heavens with the telescope. Besides studying heavenly bodies in space, he also studied how forces affect objects. He noticed that once an object is moving, a force does not need to keep it moving. He also noticed that an object stops if a force, such as friction or gravity, acts upon it.

Sir Isaac Newton studied motion in more detail. His idea, or theory, was that motion always works in certain ways. *Scientists call Newton's discoveries the* **laws of motion**.

Though it was Newton who discovered the laws of motion, God was their designer. He created the world to work in an orderly way. Things like gravity and movement always work in the way God designed them to work.

One of these special designs deals with inertia. **Inertia** *is a property of matter that makes it resist a change in motion. God gave all matter this special property.* Some objects have more inertia than others. *The greater the* **mass**, *or amount of matter an object has, the more inertia it has.*

Brain Stretcher

Mass and Inertia: *Study the pictures and decide which has the greater mass. Then determine which has more inertia.*

How Inertia Always Works

1. An object at rest will stay at rest unless acted upon by an unbalanced force.
2. An object in motion will move in a straight line at a constant speed unless acted upon by an unbalanced force.

Have you ever been bowling? To get the bowling ball down the lane, you use force to overcome its inertia. Remember, an object needs a force to change its movement.

Once the ball is in motion, its inertia will keep it rolling in a straight line and at a constant speed unless another force acts upon it. The bowling pins at the end of the lane also have inertia. They will remain at rest until the force of the bowling ball moves them. Inertia is a property of an object that makes it resist a change in motion.

What if you were to push a marble down the bowling lane instead? Would it take you as much force to push it? It takes much less force to push a marble because a marble has much less inertia than a bowling ball. Why does it have less inertia? It has less inertia because it has less mass.

Can you think of any kind of ball that can move without a force to start its motion? A ball cannot begin to move unless it is pushed, rolled, thrown, dropped, kicked, struck, or bounced. Each kind of ball has inertia because each ball is made of matter.

We can see the effects of inertia at a soccer game. After the ball is kicked, it will keep moving in the same direction and at the same speed, unless another force acts upon it. Friction from the grass may slow it down, or it could be blocked by the goalie. If there were no other forces in its path, it would keep moving in the same direction and at the same speed.

You might wonder, "Why then, does a soccer ball seem to slow down on its own?" After all, any ball out of play will eventually slow down and stop. Think very carefully—what force does the playing surface give to the ball as it rolls? If you guessed friction, then you are right! Friction and gravity keep objects from traveling forever and from flying into outer space.

Even *you* have inertia! Imagine you are in the car on the way home from soccer practice. Your father suddenly steps on his brakes to prevent an accident. What does your body do? It thrusts forward, doesn't it? Your inertia is resisting your change in forward motion. It makes your body keep going even though the car has stopped. Aren't you thankful you are wearing your seat belt?

Without inertia and forces that can overcome inertia, we would live in a very chaotic world. God not only planned inertia, but He also designed the forces that change and control motion. Objects move in the way we can predict they will move because of God's orderly plan.

Great is our Lord, and of great power: His understanding is infinite.
Psalm 147:5

Comprehension Check 3.2

Think and Predict: *Use what you know about inertia to make predictions about the motion in these situations.*

The Situation	Predict its motion without any other force acting upon it.	Predict the types of forces that could act upon it to change its motion.
A hockey puck is sliding across the ice.		
A hockey puck is resting on the ice.		
A football is resting on the fifty yard line.		
A baseball is moving through the air after it has been hit.		

Multiple Choice: *Circle the correct answer.*

1. Who discovered and developed the laws of motion?
   a. Franklin b. Galileo c. Newton

2. What property of matter makes it resist change in motion?
   a. density b. inertia c. weight

3. What can we measure to predict how much inertia an object has?
   a. mass b. motion c. speed

TERMS

collision: two objects bumping into each other, causing energy transfers

3.3 Motion Has Energy

Most team sports have defensive and offensive players. An offensive player is trying to score for his or her team, while a defensive player is trying to stop the other team from scoring. Often, defensive players prevent the other team from scoring by stopping the movement of their opponent. Some players are harder to stop because of their total energy.

A moving object has kinetic energy. Factors such as mass, speed, and direction give a moving object the *amount* of energy it has. If an object has more mass or is moving faster, it has greater energy.

Think about a game of tackle football. If a player catches the football, he will run toward the end zone. He will gain more energy the faster he runs. More mass gives a moving object more energy. That is why bigger football players are harder to block and tackle. It takes a bigger force to stop them.

But what if that same football player is sitting on the sidelines and you run past him? Who has more moving energy now? You have more moving energy because your speed is greater than his. *Mass, speed, and direction are all part of an object's moving energy.*

Collisions

What happens if two football players with moving energy bump into each other? When two objects bump into each other, their total energy is usually added together and shared equally between them. That means *energy is transferred from one object to the other.* We call this kind of energy transfer a **collision**. If both players had an equal amount of moving energy before the collision, they would both be thrown back a little in the opposite direction after the collision. Their speeds and directions would also change.

If two football players collide and one player has more moving energy, it causes the other player to be pushed back even farther in the opposite direction. This is one of the reasons football players wear protective gear and a helmet.

In a collision, all of this combined energy has to go somewhere. *Whatever energy does not make the objects move is converted into other types of energy,* such as the sound of both objects' impact or the heat made from friction. This is why tires leave skid marks on the street after an accident. The friction from a collision causes the tires to heat up and wear down.

Some collisions are dangerous and can cause damage. What if two racecars crash head-on into each other? If one of them has more energy, some of its energy is transferred to the other car. This combined energy can cause damage.

What if the collision was not head-on? What if one racecar crashed into the racecar in front of it? A push coming from the same direction of an object's movement will make that object speed up. That is why the car in front may even speed up enough to collide with the next car in front of it.

What Happens in Collisions

1. When two objects collide, *some energy is usually transferred* from one to the other, causing a change in speed.
2. Usually, *some energy is converted* from moving energy to other forms of energy, such as sound and heat.

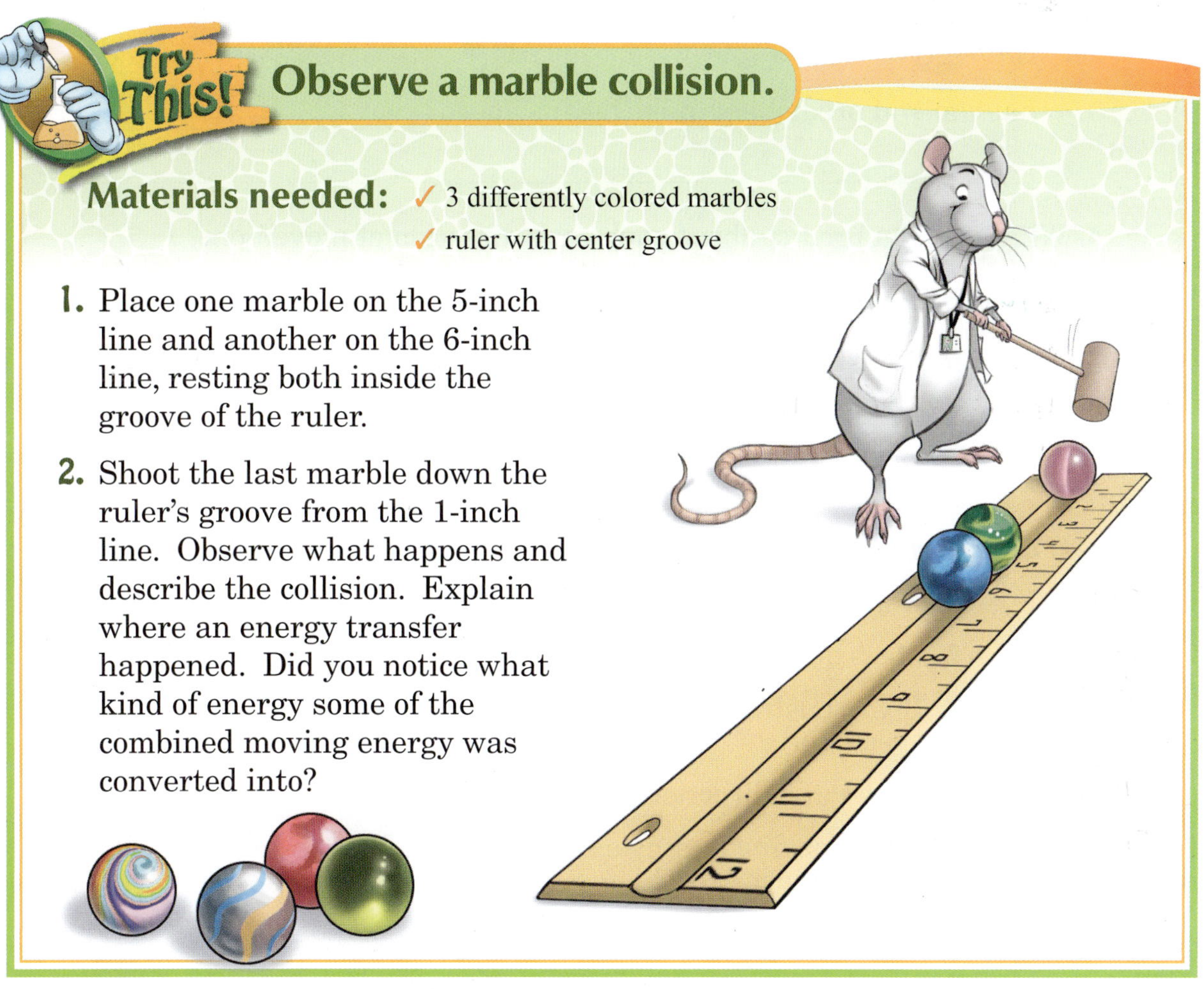

Try This! Observe a marble collision.

Materials needed:
- ✓ 3 differently colored marbles
- ✓ ruler with center groove

1. Place one marble on the 5-inch line and another on the 6-inch line, resting both inside the groove of the ruler.
2. Shoot the last marble down the ruler's groove from the 1-inch line. Observe what happens and describe the collision. Explain where an energy transfer happened. Did you notice what kind of energy some of the combined moving energy was converted into?

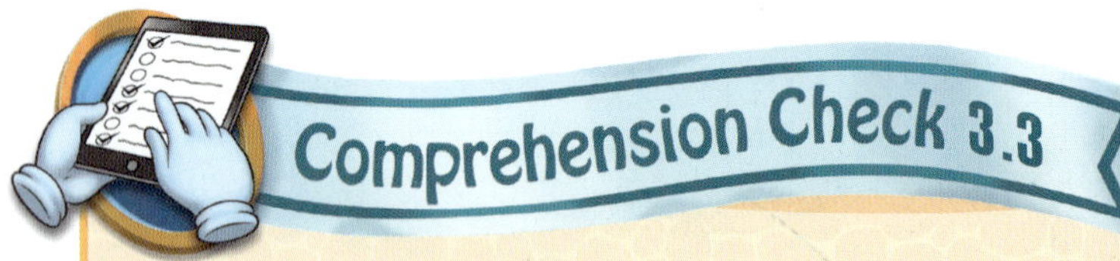

Comprehension Check 3.3

Think and Predict: *Make predictions about what would happen in each collision.*

Collision 1

A B

1. Which car will speed up? B

2. Which car will slow down? A

Collision 2

3. What can you predict about each car's speed? slow down or speed up

4. What can you predict about each car's direction of movement? they go back

5. What kinds of energy could their moving energy be converted into? propoct to Kictorce

True/False: *Write the word* true *or* false *in each blank.*

True 6. A moving object has kinetic energy.

True 7. Mass does not change an object's total moving energy.

False 8. If you push against a moving object's direction, it will speed up.

False 9. Speed and direction change after a collision.

True 10. Some moving energy can be converted into other types of energy during a collision.

Discuss: *Explain why the false answers are incorrect statements.*

magnetism: the attracting and repelling force created by a magnetic field

polarity: having two opposite sides that work differently

electromagnet: a device that generates electromagnetic force

3.4 Electromagnetic Forces

Magnetism *is a force that can both attract* (pull) *and repel* (push). A magnet works because of its magnetic field. You cannot see a magnetic field just by looking at a magnet, but there are ways to observe its effects.

A magnet has polarity. **Polarity** means *the magnet has two opposite sides, or poles, that work differently.* Every magnet has a north pole and a south pole. The force of a magnet flows from the north pole to the south pole, creating a magnetic field. That is why *the magnet's force is strongest at its poles and weakest in its center.* No one has ever been able to separate the poles of a magnet. Even if you cut a magnet in two, each piece will always have a north pole and a south pole.

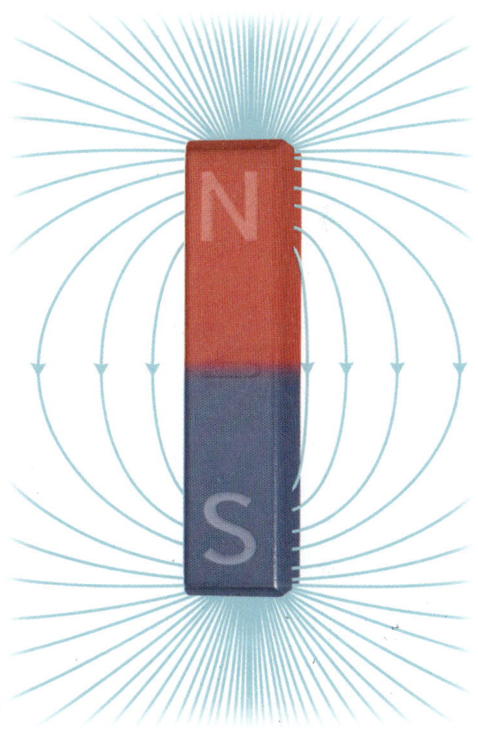

Electromagnetism

If electrical energy and magnetic force are combined, a stronger force can be made in an electrical current. We remember that atoms have electrical charges. An atom is balanced because it has a positive nucleus and negative electrons. But what happens if a negative electron is pulled away from its positive nucleus? The atom is no longer balanced. These unbalanced charges can attract and repel like a magnet, causing an electromagnetic force.

Friction is a type of electromagnetic force. Remember, friction causes atoms to gain or lose electrons. Rubbing your hair on a balloon is one example; rubbing your feet on carpet is another. You recognize this as static electricity, but because static electricity can both attract and repel, it is also called an electromagnetic force.

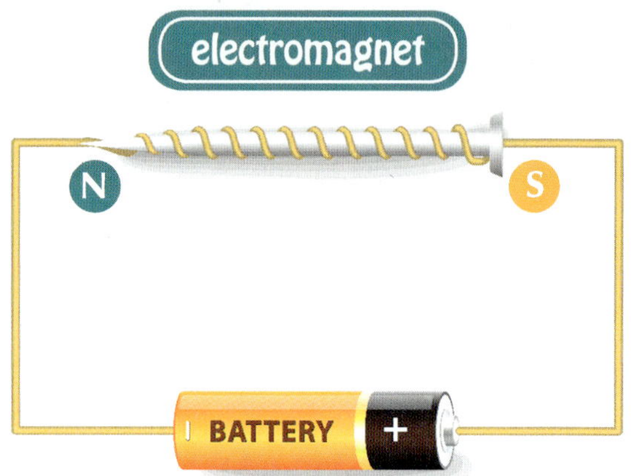

The Use of an Electromagnet

An **electromagnet** *is a device that can generate electromagnetic force* because of its coiled wire, metal core, and source of electricity. A coiled wire generates a stronger magnetic field than a straight wire does. An electromagnet can attract and repel just like a natural magnet, but it is often stronger and can be turned on and off. This makes it very useful.

Try This! Make an electromagnet.

Materials needed:

- ✓ 1 yard 18-gauge bell wire
- ✓ 3″ iron bolt with 1 nut and 2 washers
- ✓ 6-volt battery
- ✓ metal paper clips

1. Slide one washer onto the bolt, head side down. Tightly wrap the wire around the bolt, being sure to leave at least six inches of wire free on both ends. Place the remaining washer and nut on the other end to secure the wires.
2. Wrap one end of the wire to one end of the battery. Touch the free end of the wire to the other battery pole with one hand while touching the bolt tip to the paper clips.
3. Lift the nail while observing the paper clips. (If the bolt starts to feel warm, take the free wire away from the end of the battery to prevent the wires from becoming too hot.)

When electricity flows through a metal wire, its current makes the wire act like a magnet. The paper clips were attracted to the nail because coiled wires have a stronger magnetic field than straight wires.

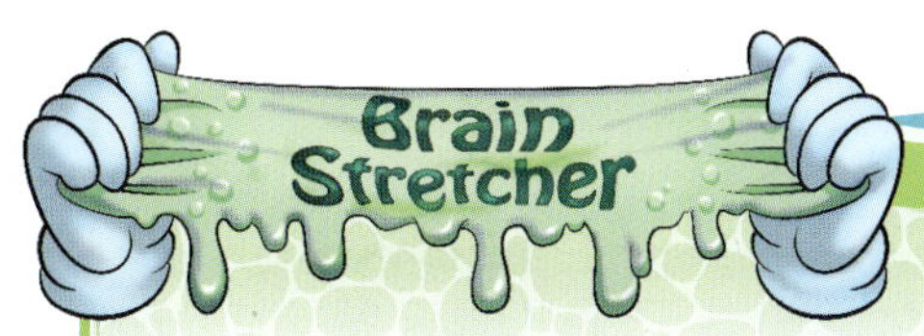

The Telegraph: The telegraph was designed with an electromagnet. The coiled wire became magnetized when electricity flowed through it. This gave the telegraph its ability to transmit messages in Morse Code. A pen or pencil attached to the telegraph would write dots and dashes onto a long strip of paper. The telegraph operator could interpret the dots and dashes as letters. Some operators became so good at listening to the telegraph's noises that they did not even need to look at the paper strip. They could listen to the sounds of dots and dashes and write down the telegram on paper.

Because there are much better ways to send and receive messages today, most people who use Morse Code use it just for fun. They are sometimes called ham radio operators because they use special ham radio equipment. During natural disasters, such as hurricanes and earthquakes, ham radio operators often help with communication when power lines or cellphone towers are down.

If you have the right equipment and are able to pass a special test, you can earn an amateur radio license through the American Radio Relay League.

A · —	I · ·	R · — ·
Ae · — · —	J · — — —	S · · ·
B — · · ·	K — · —	T —
C — · — ·	L · — · ·	U · · —
D — · ·	M — —	Ue · · — —
E ·	N — ·	V · · ·
E′ · · — · ·	O — — —	W · — —
F · · — ·	Oe — — — ·	X — · · —
G — — ·		Y — · — —
H · · · ·		Z — — · ·

A Very Big Electromagnet

Scientists believe the earth works like an electromagnet because it has a metallic core. As the earth rotates, the spinning metal creates an electromagnetic field. The earth's north and south magnetic poles are near the North and South geographic poles. If you hold a bar magnet suspended horizontally from a string, its south pole (marked "N" for north-seeking) will point toward the magnetic north pole. Opposite poles of two magnets attract each other. A compass, which has a magnetized needle, helps us find North because the needle is drawn toward the magnetic north pole.

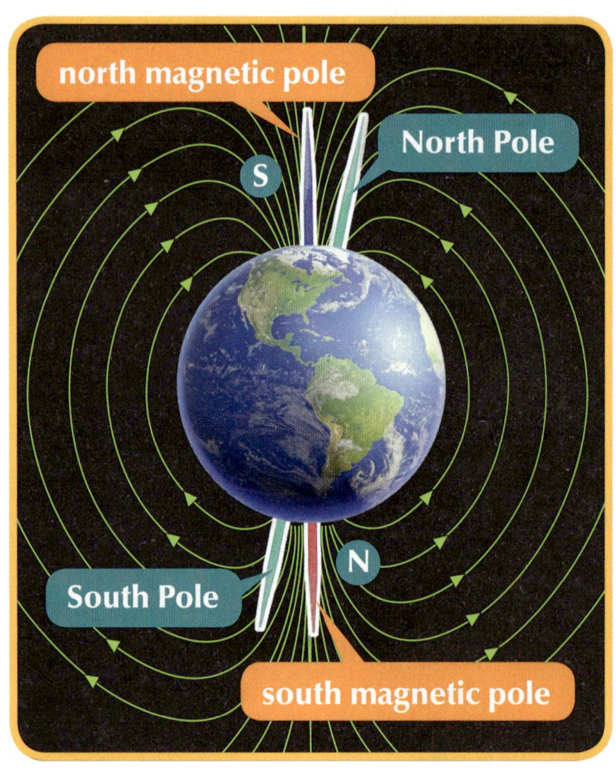

Comprehension Check 3.4

Matching: *Write the letter of the correct answer in the blank.*

C 1. two opposite sides, or poles, working differently from each other

D 2. places on a magnet where magnetic force is strongest

B 3. uses a coiled wire around a metal core to generate magnetism

A 4. a giant electromagnet

A. Earth
B. electromagnet
C. polarity
D. poles

machine: a device that makes work easier

lubricant: a substance that minimizes friction

3.5 Overcoming Friction

Friction is a force that works against motion, making it harder for work to happen. Rough surfaces can increase friction. That is how sandpaper can wear down a piece of wood. Think about sports with rough playing surfaces, such as a grassy field. Adding a rough playing surface to the game of football or soccer makes it safer. Other sports surfaces are smooth and flat, such as a bowling alley. If a bowling alley were made of grass, it would be harder to score a strike, wouldn't it? A basketball court is a very smooth surface. It would be difficult to dribble a basketball if the court were made of grass instead of wood. *A rough surface increases rubbing, which increases friction.*

It is much easier for work to happen when we find ways to overcome some of the friction. Try pushing a brick with your finger. The friction of the brick rubbing against the surface makes the brick very hard to move. Now place a few pencils under the brick and push it with the same finger. What happens? The brick moves easily because there is less friction happening between the brick and the surface. Do you see why it is easier to roll something than it is to drag something?

Machines That Overcome Friction

A **machine** *is a device that makes work easier.* The effort you use, or put in, can be greatly added to by the force a machine can produce, or put out. Machines can have moving parts and non-moving parts. Some machines can overcome friction because of their designs.

The **wheel and axle** *can help to overcome friction because of its rolling action.* As the wheel revolves around the axle, it rolls. Adding wheels to a heavy object reduces friction, making it easier to move. A racecar is designed with a powerful engine, but it would still be hard to move without its wheels. Your bicycle glides along smoothly and quickly because of its wheels. What about your roller skates? You can zip in and out of the cones at an obstacle course because of the wheels that are helping you reduce friction.

An **inclined plane** *is another simple machine*. Inclined planes can help to overcome some gravity and friction, depending on how they are used. We have many uses for this machine, such as ramps and slides.

If you go to a skate park, you will find a halfpipe. You can skate down one side of the halfpipe and skate up the other side. As you skate down this inclined plane, you can take advantage of gravitational energy. This helps you overcome friction and pick up speed. Enough moving energy can help you skate up the halfpipe—maybe even to flip in the air.

A downhill skier uses the mountainside as an inclined plane. Skiers can reduce friction even further by waxing their skis. Wax is a lubricant. *A* **lubricant** *is a substance that minimizes friction*. Oil, grease, and wax are all lubricants that can minimize friction by making a surface even more smooth and slippery. Car engines need oil as a lubricant to reduce friction and to make them run more efficiently.

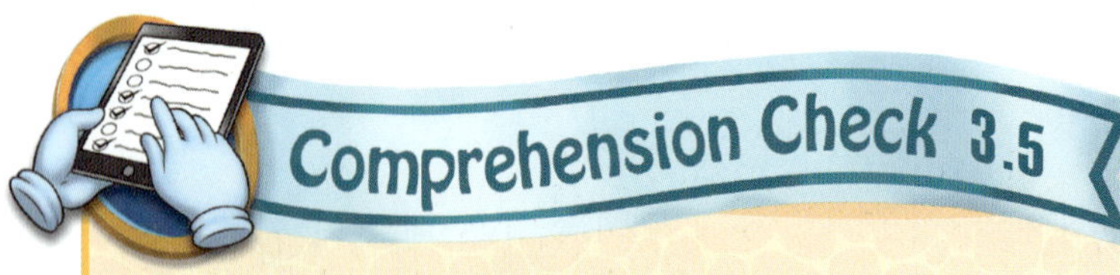

Comprehension Check 3.5

Think and Conclude.

1. Give an example of a surface that minimizes friction and describe it.

   lubricant

2. Give an example of a surface that increases friction and describe it.

3. Describe how a wheel and axle works. The axle comes to the wheel and spin it

4. Give an example of an inclined plane. it help roll something

5. Explain why a lubricant minimizes friction. it roll and help to do stuff

Try This! Use machines to overcome friction.

Materials needed:

- ✓ small rock
- ✓ toy car
- ✓ stack of books
- ✓ tape
- ✓ plank

1. Push and release the rock on a flat surface. What force was resisting its motion? What machines could overcome this force?

Friction resisted the rock's movement. Adding a wheel or using an inclined plane will help the rock overcome friction.

2. Tape the rock to the top of a toy car and push it again. Does using a wheel and axle help you overcome friction?

The rolling action of the wheels overcame friction, helping the rock to move more easily.

3. Make an inclined plane with the plank and stack of books. Experiment on your own with various inclines. Find out how to best overcome friction and add more moving energy to your rock.

TERMS

gravity: the attracting force between two objects (such as the pull of the earth to itself)

weight: the force of gravity on an object

center of gravity: the point of balance in an object that represents the effects of gravity on all its particles

3.6 Gravitational Forces

Sir Isaac Newton was the first scientist to suggest the existence of gravity. According to a very famous story, an apple fell from a tree and landed on his head. He stopped to observe and thought for a moment. His observations led to a question: "What makes an apple fall from a tree?"

He believed that the apple was pulled by a force. After hypothesizing and experimenting, he called this pull gravity. He could observe gravity everywhere. Using what he knew about gravity, Newton could predict and conclude that the sun had enough gravity to hold the solar system together.

We know that *gravity is a pulling force*. Gravity pulls a skateboarder back down on the halfpipe. It pulls a baseball back down to the ground after you hit a home run. It pulls a basketball back down to the court while you play.

We usually think of gravity as the pull of the earth to itself. This is true, but it is important to know that *all* objects exert a pulling gravitational force. The earth has gravity, but because the sun has more mass, it has more gravity than the earth.

Because the moon has less mass than the earth has, it has less gravity than the earth. The sun's gravity keeps the planets of the solar system in orbit. The earth's gravity holds the ocean to itself while the moon's gravity can pull the oceans just enough to make the tides work in the way that they do.

We can think of **gravity** as *the attracting force between two objects. Objects with greater mass pull with greater gravitational force than smaller objects.* That is why Earth's gravitational pull is what we notice. The attracting force of gravity between you and the earth keeps you from floating into outer space.

Remember, an object's mass determines its weight. We say that **weight** *is the force of gravity on an object.* There is less gravity on the moon because the moon has less mass than the earth. Free from the gravity of Earth, a person on the moon would seem to float and bounce. An astronaut on the moon still has the same mass that he had on Earth, but he weighs one-sixth of his Earth weight.

Every object has a point *that represents the effects of gravity on all its particles*. We call this part of an object, *where the pull of gravity is strongest,* its **center of gravity**. It all depends on where mass is located in an object. You have long legs but a large torso. That means your center of gravity isn't exactly in the middle of your body. Most people have a center of gravity somewhere between their chest and hips. You are finding and using your own center of gravity as you try to walk a balance beam. God designed your body exactly right to give you the best center of gravity as you work and play.

Different shapes have different centers of gravity, making them more or less stable than others. A cube is stable because its center of gravity is in its very center. *An upright pyramid shape is the most stable.* Its wide base gives it a very low center of gravity.

But what if the pyramid were turned over—would it be as stable? If a pyramid is supported only at its point, then it becomes very unstable. There is too much mass on the top of the shape and not enough at the bottom.

A bus has been designed with a low center of gravity. This keeps it from tipping over as it makes turns. A tall, narrow object is easier to tip over than a large, wide object.

Football players crouch low to the ground to give them better centers of gravity. This gives them more stability to hold their ground or prevent being tackled.

Try This! Determine the center of gravity.

Materials needed:
- ✓ clay
- ✓ 2 forks
- ✓ wide-mouthed jar
- ✓ toothpick

1. Roll a small amount of clay into a ball. Partially insert one fork into the clay; insert the second fork at a 45-degree angle to the other fork.
2. Insert the toothpick into the clay between the forks.
3. Try to balance the end of the toothpick on the side of the jar. Keep moving it inward until the forks are balanced.

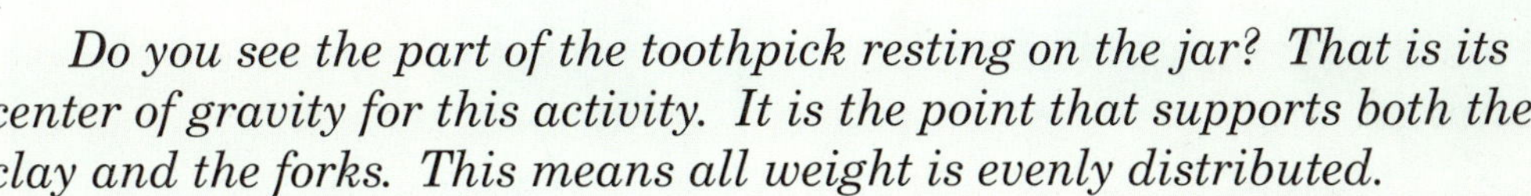

Do you see the part of the toothpick resting on the jar? That is its center of gravity for this activity. It is the point that supports both the clay and the forks. This means all weight is evenly distributed.

Fill in the Blank: *Choose the correct answer from the box below and write it in the blank.*

1. Gravity is a _____________ force.
2. Objects with more ___________ have more gravity.
3. Gravity gives things their _____________.
4. The center of gravity is the point in an object where the pull of gravity is _______________.
5. An upright pyramid is ___________ stable than other shapes.

less	**more**	**pushing**	**weakest**
mass	**pulling**	**strongest**	**weight**

Think and Predict: *Circle the shape that is more stable. Explain your answer.*

TERMS

foundation: the base of a building

frame: the skeleton of a building, giving it its shape and support

fulcrum: the part in a lever that holds and balances weight

3.7 Overcoming Weight and Gravity

Architects and engineers work together to design stable bridges and buildings. They use what they know about weight and gravity to give a structure its strength. Their goal is to design a structure that can hold weight and remain stable.

Engineering a Strong Structure

The foundation is the first part builders make when they build a structure. *A* **foundation** *is the base of a building*, acting like "roots" to keep the building strong. Most of the time, the taller a building is, the deeper into the ground the foundation needs to be. This gives it a better center of gravity.

After a foundation is laid, builders add a frame. *A* **frame** *acts like a skeleton, giving a building its shape.* It supports the rest of the building's structure. A frame can be wood, steel, or even concrete.

Sometimes, an architect will add certain support structures to give a building or a bridge more strength. Arches are often used in the design of bridges to make them stronger. The curved design is able to spread out weight. A dome is similar to an arch, though it covers a large space. It can also spread out weight. A column by itself can tip over easily, but its shape holds weight well. Columns are often used to support a structure because they are not easily bent under the force of weight.

Triangular structures, such as *pyramids*, are the strongest. Because they are wide at the base, their center of gravity is lower. Adding triangular design to many things makes them stronger.

Try This! Make a square frame more stable by adding triangles.

Materials needed:
- ✓ toothpicks
- ✓ miniature marshmallows
- ✓ small reading book

1. Make a cube frame using toothpicks and marshmallows. Carefully place a book on top of the frame and observe the structure.
2. Make a cube again, this time adding diagonal toothpicks, corner-to-corner in each square face. Carefully place the same book on your new structure and observe. Did your second structure seem more stable?

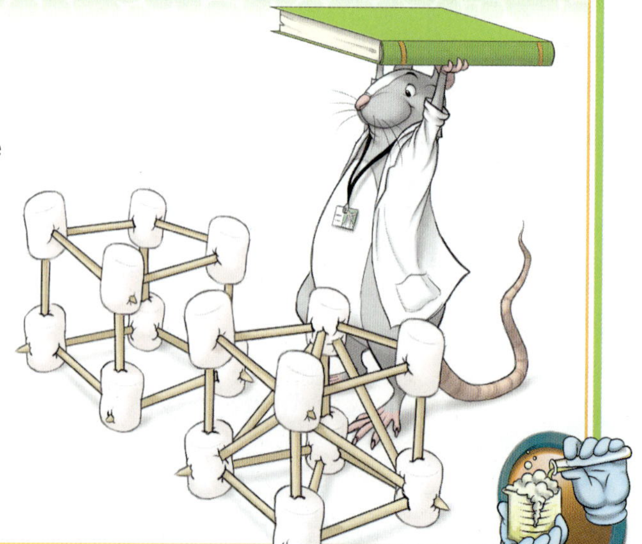

Machines That Overcome Weight and Gravity

Lifting heavy things takes a lot of effort. To lift something very heavy, you need a force that is greater than the object's weight and gravity. Some machines help make work easier.

An **inclined plane** *gives you a slope on which to push upward.* It takes less force to push something up a slope than it takes to lift something straight up. Before heavy machinery was invented, people depended on inclined planes to help them build large things. Often, we see inclined-plane designs in ancient architecture. We still use ramps today to help us push very heavy things upward. Even an axe is made of two back-to-back inclined planes. We call this kind of inclined plane a **wedge**. It can more easily split apart a piece of wood. *The* **screw** *is another type of inclined plane.*

inclined plane

You can move a screw easily from a high point to a low point because it is an inclined plane wrapped around a cylinder.

The pulley is a machine that *can be used to lift heavy things. A* **pulley** *has a wheel that moves because of the cord around it. The cord rests on the grooves of the wheel, while the wheel turns around the cord.* The weight of the load is shared between both sides of the cord.

wedge

pulley

People who rock rappel use a pulley action and their weight to pull against gravity. If they did not have a rope that could lower them gradually, they would do what all objects do when they are above the surface of the earth—fall. What force pulls them down? Gravity is the pull of the earth to itself.

Try This! Construct a pulley to overcome weight and gravity.

Materials needed:

- ✓ empty thread spool
- ✓ masking tape
- ✓ pencil
- ✓ 10–12 in. piece of ribbon
- ✓ wide-mouthed jar
- ✓ small rock

1. Insert a pencil through the center of the spool, being sure it is secure. There should be an equal amount of pencil ends for "handles" on either side of the spool. Rest your "grooved wheel" on top of the jar.
2. Attach the rock to one end of the ribbon using masking tape. Be sure it is taped securely.
3. Rest your "cord" (ribbon) over your grooved wheel, balancing the rock at the open end of the cord. Raise and lower your rock using your pulley.

Your pulley is able to help overcome the weight and gravity of the rock, lifting it upward.

A **lever** can also make lifting heavy loads much easier. Between each side is a non-moving part called a fulcrum. *The* ***fulcrum*** *holds and balances weight so that when you push one side of the lever, the other side can lift or pry the load more easily.* Think about a seesaw. You could never lift your friend over your head, but if he or she were on the opposite side of a seesaw, you could lift him or her high off the ground.

The Greek scientist Archimedes once said, "Give me a lever long enough and a fulcrum on which to place it, and I shall move the world."

Archimedes was making the point that the right machine or tool can make work much easier. Using a lever can turn an impossibility into a possibility.

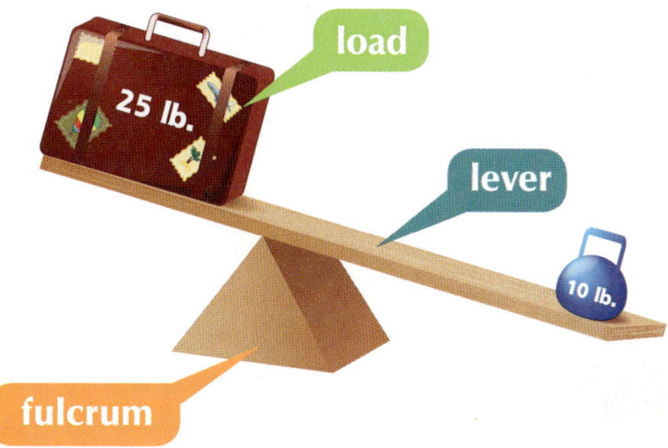

Brain Stretcher

Levers: Archimedes invented and improved many kinds of machines. The catapult was one of the inventions that he improved. *Can you find the fulcrum on this catapult?*

Try This! Construct a catapult.

Materials needed:

- ✓ 7 craft sticks
- ✓ plastic spoon
- ✓ pompom
- ✓ rubber bands
- ✓ marker

1. Stack 5 craft sticks together, securing both ends with rubber bands. Label this "fulcrum" with a marker.
2. Attach the plastic spoon to a craft stick with a rubber band. Place another craft stick underneath it and secure them together, using a rubber band. Label the spoon "bucket."
3. Slide your fulcrum between the two craft sticks under the spoon so that both main pieces intersect in the middle. Secure them with a rubber band.
4. Now get ready to launch your "projectile." Place a pompom in the spoon and push down the other side of the lever. Did you see how the fulcrum made this lever work?

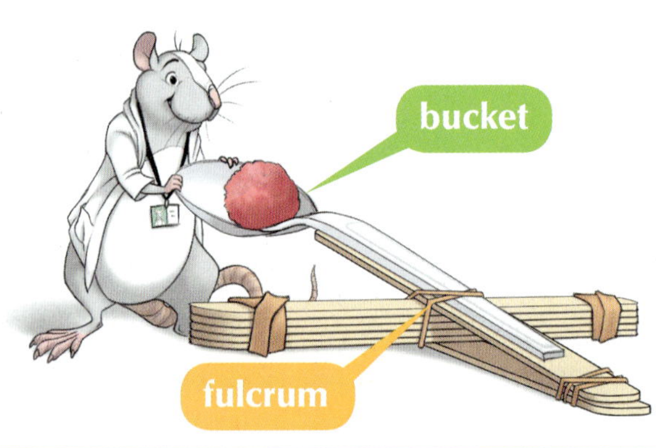

Now that you understand force and motion, you can see that God planned a specific, orderly way for force and motion to work. This design affects not only everything that moves on Earth, but also the entire universe. This perfect design shows us that God is in authority over His creation. How good it is to trust and obey our wise God!

For by Him were all things created, that are in Heaven, and that are in Earth, visible and invisible, whether they be thrones, or dominions, or principalities, or powers: all things were created by Him, and for Him: And He is before all things, and by Him all things consist.

Colossians 1:16–17

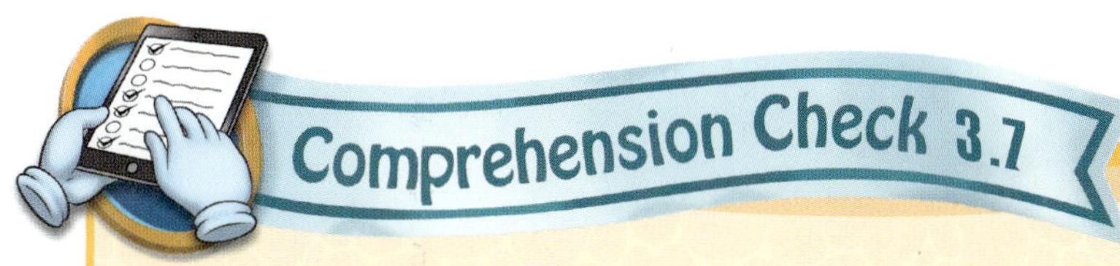

Comprehension Check 3.7

***Multiple Choice:** Circle the correct answer.*

1. A(n) __?__ is the base of a building.
   a. arch b. frame c. foundation

2. An upright pyramid has a __?__ center of gravity.
   a. low b. high c. tall

3. A seesaw is a type of __?__.
   a. fulcrum b. lever c. inclined plane

4. A __?__ is a machine which makes lifting heavy things easier.
   a. dome b. pulley c. structure

Label these support structures.

5. ____________ 6. ________________ 7. ____________

Chapter 3 Concepts Review

Force and Motion Concepts 3.1–3.3

A. Remember

True/False: *Look at the underlined word to determine if the statement is true or false. If the statement is true, write* true *in the blank. If the statement is false, write* false *in the blank. Be prepared to explain why the false answers are incorrect.*

__________ 1. Like poles of a magnet <u>attract</u>.

__________ 2. Sir Isaac Newton <u>created</u> the laws of motion.

__________ 3. An object at rest will <u>stay at rest</u> unless acted upon by an unbalanced force.

__________ 4. Inertia is a <u>property</u> of matter that makes it resist a change in motion.

__________ 5. Sound is <u>converted</u> energy during a collision.

__________ 6. Energy <u>cannot</u> be transferred in a collision.

Short Answer: *Write the correct answer in the blank.*

7. What is the movement resulting from force? ____________________

8. What is a push or pull on an object? ____________________

9. Give an example of a kind of force we experience every day. ____________

____

10. What do we call two objects bumping into each other? ____________

B. Think Like a Scientist

Identify: *Circle the correct answer and explain your answer.*

1. Which of these objects has greater inertia?

Why? ____________________

2. Which of these objects has less inertia?

Why? ____________________

Label: *Write* B *for balanced forces and* U *for unbalanced forces. Be prepared to explain your answers.*

______ 3. allow work to be done

______ 4. prevent work from being done

______ 5. a magnet pulling a nail to itself

______ 6. a book resting on a desk

______ 7. a ball falling back to Earth

______ 8. a tied game of tug-of-war

Short Answer: *List one more example of a balanced force and an unbalanced force.*

9. ____

10. ____

C. Fun with Terms

Crossword: *Study the definitions for the terms below. Fill in the answers to complete the crossword puzzle.*

Across

1. the force of gravity on an object
3. a push or pull on an object
5. a change in position
6. movement resulting from force

Down

2. the property that makes matter resist a change in motion
4. two objects bumping into each other
5. the amount of matter an object has

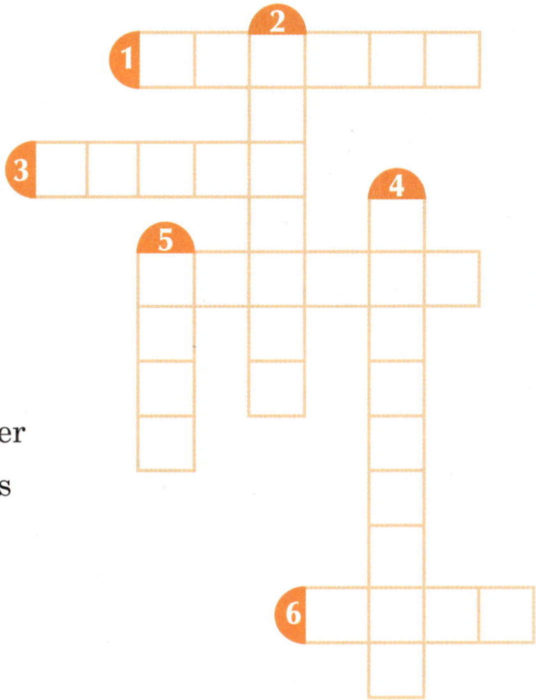

collision	**force**	**mass**	**weight**
energy	**inertia**	**motion**	**work**

2 Machine and Structure Concepts 3.4–3.7

A. Remember

Identify: *Label these machines. Be prepared to explain how each machine can help overcome weight and gravity.*

1. ______________ 2. ______________ 3. ______________ 4. ______________

Identify: *Label these support structures.*

5. ______________ 6. ______________ 7. ______________

B. Think Like a Scientist

Identify: *Circle the shape that is the most stable. Be prepared to explain your answer.*

C. Fun with Terms

Matching: *Write the letter of the correct answer in the blank.*

______ 1. having two opposite sides that work differently

______ 2. the attracting and repelling force created by a magnetic field

______ 3. the attraction between two objects (such as the pull of the earth to itself)

______ 4. a substance that minimizes friction

______ 5. a device that generates an electromagnetic force

______ 6. the base of a building

______ 7. the skeleton of a building giving it its shape and support

______ 8. the force of gravity upon an object

______ 9. the part of a lever that holds and balances weight

______10. a device that makes work easier

A. electromagnet
B. foundation
C. frame
D. fulcrum
E. gravity
F. lubricant
G. machine
H. magnetism
I. polarity
J. weight

Unit 2 Life Science

Chapter 4
Understanding How Plants Grow and Reproduce

TERMS

organism: a living thing

chlorophyll: the green matter in plants which traps sunlight energy; gives a plant the ability to produce its own food

photosynthesis: the food-making process in green plants

glucose: the simple sugar made by green plants

producer: an organism that is able to produce its own food by using sunlight energy

consumer: an organism that depends on producers for food energy

leaf litter: layer of dead and decaying leaves on the forest floor

humus: the soft organic material in soil made from the decayed remains of living things

4.1 God's Purposes for Plants

Almost anywhere you go on Earth, you will find plants. Some are able to grow in hot deserts; others live in the icy Arctic regions of the earth. Some even live inside the bodies of animals.

In the plant world, we find the largest and some of the smallest living things of the world. Have you ever seen a giant sequoia tree? Sequoias are the largest living plants on Earth. Watermeal is one of the smallest flowering plants on Earth. Hundreds of watermeal plants would fit on your fingertip.

To study life science is to study the living parts of God's creation, such as plants. Life on Earth depends on plants. If there were no green plants on Earth, we could not survive. Plants produce food and provide oxygen, shelter, fuel, medicine, and many other things. Because all life on Earth depends on plants, we will begin our study of life science with plants.

Plants for Beauty

On the third day of Creation, God spoke, and the earth was immediately filled with grass, plants, and trees. He designed these plants to produce seeds that would grow more plants.

And God said, Let the earth bring forth grass, the herb yielding seed, and the fruit tree yielding fruit after his kind, whose seed is in itself, upon the earth: and it was so. And the earth brought forth grass, and herb yielding seed after his kind, and the tree yielding fruit, whose seed was in itself, after his kind: and God saw that it was good. *Genesis 1:11–12*

God created a beautiful garden for Adam and Eve to live in and to care for. The Garden of Eden became their home. It must have been filled with marvelous variety and color. When Adam and Eve disobeyed God, sin entered the world. The loveliness of nature became scattered with the ugly results of sin. Thorns, thistles, and unwanted weeds began growing among the trees and flowers. Adam's work of tending the soil became difficult.

God graciously left the beauty of His creation for Adam and Eve to enjoy. God did not need to make plants beautiful to do their jobs, but He did. Like all living things, the plant world gives us a picture of the limitless creativity of the Creator.

Delicate petals of the lotus plant, deep pink blooms of the bleeding heart, and wing-like blossoms of the bird of paradise speak of God's amazing variety in creation. Most flowering plants bloom often, but some take years to bloom even once! A plant commonly called the Queen of the Andes can take up to one hundred years to bloom, and once it blooms, it dies.

Think of the marvelous array of colors we see in plants—golden daffodils, crimson roses, waxy white lilies, gentle blue forget-me-nots, and cool green fir trees. Some flowers have even given their names to shades of color, such as rose, violet, and lilac. Think of the many fragrances we smell from flowers. An apple blossom does not smell like a pine tree, for each has its own refreshing scent. We can enjoy thousands of designs in the plant world. In the same forest, growing side by side, we find massive oaks and tiny, velvety mosses. Lacy ferns grow at the feet of towering pines.

God has given each plant a job to do in its habitat, or natural home. Weeds that may seem useless provide food and nesting materials for birds and small animals. Even in this fallen world, God uses His creation for His glory and for our benefit.

Plants Produce Food and Give off Oxygen

All *living things,* or **organisms**, need energy to grow and reproduce. Where does this energy come from? It starts with sunlight energy.

Green plants contain **chlorophyll**, *a green matter.* Chlorophyll *traps sunlight energy, which gives the plant the ability to produce its own food. The food-making process in green plants is called* **photosynthesis**. The plant uses carbon dioxide from the air and combines it with water from the soil to make a *simple sugar* called **glucose**. Plants use glucose to store energy to grow and reproduce. *Sunlight energy, carbon dioxide, and water are the input needed for photosynthesis to occur.* As the plant uses carbon dioxide and makes glucose, it *gives off, or outputs, a very important byproduct—oxygen.*

All organisms that can produce their own food by using sunlight energy are called **producers**. *Organisms that depend on producers for food energy are called* **consumers**. Now that you understand photosynthesis, how would you classify a green plant? *All green plants are producers.* People, animals, and other organisms which cannot make their own food using sunlight energy are consumers.

Think of all the ways we use plants for food. We eat the seeds of wheat, rice, and corn plants; we also grind those seeds to make bread. Other seeds can be made into cooking oil. Have you eaten any roots lately? If you had any radishes, carrots, or sweet potatoes for supper last night, you have! Think of all the leaves we eat, such as lettuce, spinach, and cabbage. We also snack on crunchy celery stems. Your french fries came from the underground stem of a potato plant. We even eat some flowers, such as broccoli and cauliflower. Your favorite plant products might be the fruits of some plants, such as crisp apples, bananas, and juicy berries.

Most of the food we eat comes directly from plants; the remaining food comes from animals that also depend on plants for food.

Photosynthesis
sunlight
carbon dioxide molecules
oxygen molecules
glucose (sugars)
water molecules
input
output

People and animals also depend on oxygen to live. When a green plant makes its food using the process of photosynthesis, it takes carbon dioxide from the air and releases oxygen back into the air for you to breathe. When you breathe, you return carbon dioxide to the air for the plants to use.

Plants Improve the Soil

A forest floor is covered in a layer of dead and decaying leaves called **leaf litter**. As leaf litter decays, it gradually breaks down to become a *soft organic material called* **humus** [hyo͞o′məs]. Humus helps the soil because it *adds nutrients and is able to soak up water.*

leaf litter

Other Important Uses

Many homes and shelters are made of plant material. Cotton and linen fibers in your clothing come from plants. The rubber in an eraser and in the tires of your parents' car comes from a rubber tree. The paper that this book is printed on was once a plant.

Many medicines that we use today are made from plant ingredients. Native Americans used to brew tea from willow bark and leaves for people with minor aches and pains. Years later, when scientists examined willow bark, they discovered a chemical that became the main ingredient for aspirin. The Madagascar periwinkle plant is used to make important cancer-fighting medicine. We find that many life-saving medicines come from plants.

There would be no life on Earth if there were no plants. Who designed plants so that they could have life? Life can come only from life. Our life-giving God designed plants to use the sun's energy to provide what we need for life—food and oxygen. Give God thanks for His continual provision!

Comprehension Check 4.1

Matching: *Write the letter of the correct answer in the blank.*

A 1. the green matter in green plants

D 2. the food-making process in green plants

F 3. input needed for photosynthesis to occur

C 4. output products of photosynthesis

E 5. organism that makes its own food

B 6. organism that depends on producers for food energy

A. chlorophyll
B. consumer
C. oxygen and glucose
D. photosynthesis
E. producer
F. water, carbon dioxide, sunlight energy

crown: the leaves and branches of a tree

trunk: the stem of a tree

bark: the protective layer on the outside of the trunk of a tree

roots: the part that anchors the tree and draws water and minerals from the soil

annual rings: layers of wood in the trunk; each ring represents one year of the tree's life

needleleaf tree: a tree with narrow leaves, or needles

evergreen tree: a tree that keeps its leaves throughout the year

conifer tree: a tree that produces seeds in cones

Identifying Trees

The tallest plants on Earth are trees. Trees are also the oldest living things on Earth. There are bristlecone pines in California that are over 4,000 years old!

What makes a plant a tree? Most scientists classify trees as plants with one woody stem that is thirteen to sixteen feet tall when mature, or fully grown. The stem of a tree, or its trunk, is able to stand upright on its own and is at least three to four inches thick.

Bristlecone pine

Parts of a Tree

As you observe trees around you, notice that each tree has three main parts. The **crown** of the tree is its *leaves and branches.* The leaves are very important because they make food for the entire plant. At certain times of the year, you will find flowers or fruit in the crowns of most trees. The *stem of the tree* is its **trunk**, which is made of woody fiber. It holds the tree upright. A *protective layer* called the **bark** covers the outside of the trunk. Like all plant stems, a tree's trunk carries water and minerals to the leaves so that they can make food and grow. The trunk carries food down to the roots for storage.

A tree's **roots** *anchor the tree* in the ground. How large do you think a tree's root system is? Look at the crown of a tree. Most trees have a root system at least as large as the spread of its crown. You could even compare the roots of a tree to the foundation of a house. *Roots give a tree the structure and center of gravity* it needs so that it will not be uprooted when the wind blows.

Roots draw water and minerals from the soil as they help protect the soil. Just like all plant roots, tree *roots can help the soil by keeping it in place.* A river's moving water is powerful enough to wash away the soil. Planting trees along a river's bank helps absorb some of the river's force. The moist, rich soil near the water then helps the trees grow.

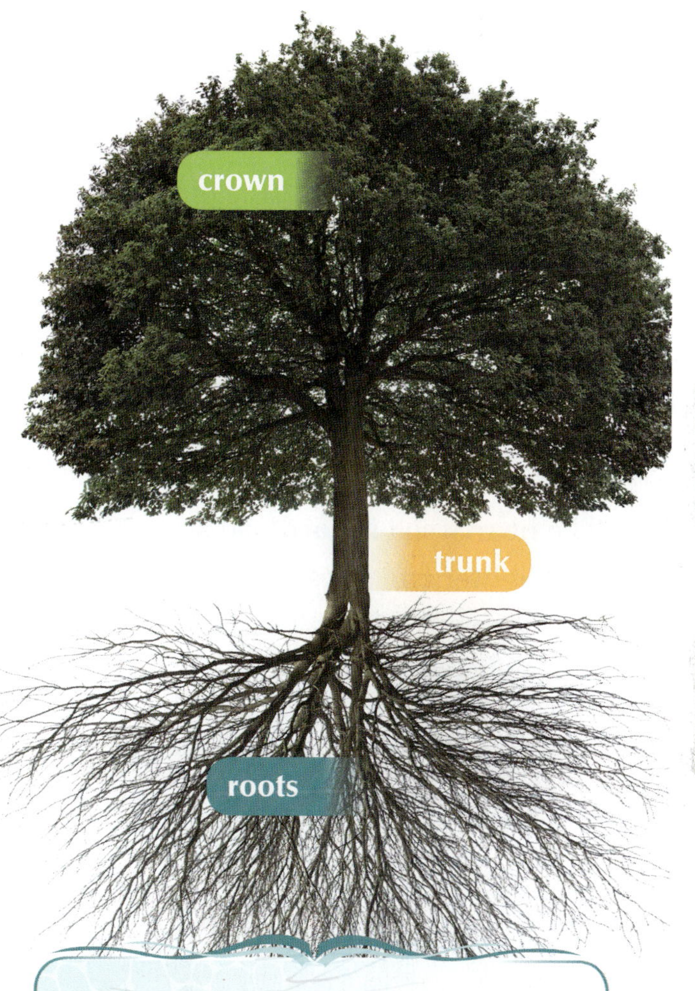

But his delight is in the law of the L*ORD; and in His law doth he meditate day and night. And he shall be like a tree planted by the rivers of water, that bringeth forth his fruit in his season; his leaf also shall not wither; and whatsoever he doeth shall prosper.*

Psalm 1:2–3

Many trees grow a new layer of wood each year. If a tree is cut down, you can see the layers of wood called **annual rings**. *Each ring represents one year of the tree's life*. You could count them and determine the age of the tree. The width of each annual ring also tells about the conditions under which the tree grew each year. If during one year the tree did not get enough water or sunlight, the annual ring for that year would be thin. In a year when there was plenty of water and sunlight, the annual ring would be thicker. The dark rings in the center of the tree are called the heartwood. The heartwood is hard and strong. The lighter rings are called sapwood. The sapwood carries nutrients and minerals from the roots to the leaves.

Although most trees are alike in these important ways, there are also many interesting differences. Learn to identify several of the most common trees in your area by observing them carefully and looking them up in field guides. You may want to begin collecting tree leaves. Do you think you can find a large variety of leaves in your area? Most of the trees that grow in North America can be divided into two main groups: needleleaf trees and broadleaf trees. Palms belong to a third group of trees.

White pine

Eastern hemlock

Balsam fir

Blue spruce

Needleleaf Trees

Trees with narrow leaves, or needles, are called **needleleaf trees**. Needles can be long or short. Pines, firs, spruces, hemlocks, larches, redwoods, cedars, and junipers are all needleleaf trees. Most needleleaf trees living in regions where seasons change are evergreen. *An* **evergreen tree** *keeps its leaves throughout the year.* Old needles turn brown and fall off a few at a time, but they are constantly being replaced with fresh, green needles. Most needleleaf trees are also conifers. **Conifers** are trees that *produce seeds in cones instead of flowers.* Conifer forests are often found in colder climates because they are well suited to extreme temperatures. This also means conifers can be found where it is hot. If you live in the southern United States, you know pine trees can thrive in high temperatures.

By examining the needles and cones, you will be able to tell which family a needleleaf tree belongs to. *Pines have long, thin needles that usually grow in bundles of two to five*. The number of needles in a bundle will tell you what kind of pine tree it is.

Hemlocks have very short, flat needles growing on little stalks on each branch. Each needle is dark green on the top side and lighter green underneath. *Fir trees have single needles that grow directly on the twig, not on a stalk*. Notice how the cones grow on most fir trees. They grow only on the top side of the branch, pointing upward. The short, four-sided needles of spruce trees grow in spirals on stalks. Spruce cones grow on the bottom side of the branch, pointing down.

Comprehension Check 4.2A

Multiple Choice: *Circle the correct answer.*

1. The _?_ is made up of the tree's leaves and branches.
   a. crown b. trunk c. roots

2. The _?_ is the stem of the tree.
   a. bark b. crown c. trunk

3. The _?_ is the protective layer that covers the outside of a tree.
   a. bark b. crown c. trunk

4. The _?_ anchor the tree in the ground.
   a. annual rings b. roots c. trunk

5. A conifer is a _?_ tree.
   a. broadleaf b. needleleaf c. palm

Think and Conclude.

6. Why do a tree's roots need to be as large as its crown? To keep Stable & balanced

7. How can a river change the soil near a riverbank? A river can wash soil away

8. Why are a tree's annual rings wider during years with plenty of sunlight and water? It would be properly nurtured to grow & become big

Trees

Materials needed:

- ✓ notebook or journal
- ✓ field guides and research sources
- ✓ marking flag (or other marker)
- ✓ sketching materials
- ✓ tape measure
- ✓ *camera
- ✓ *crayons
- ✓ *loose paper
- ✓ *paper towels
- ✓ *heavy books
- ✓ *glue

1. Find a tree in your backyard, neighborhood, or playground to observe throughout the school year; then begin a tree journal. Start by taking measurements and recording this data at the beginning of your journal.

   - *How to measure the girth of a tree:*

     Use a tape measure to measure around the circumference of your tree's trunk. This is called its girth. You may need a friend to hold one end of your tape measure if your tree is very large. Record your data in centimeters.

   - *How to measure the approximate height of a tree:*

     You can take an estimated measurement by stepping several paces away from the tree and looking through your legs behind you. Keep stepping until you can see the top of your tree between your legs. Mark the spot where you are standing with a flag or other marker. Then use a tape measure to measure from the tree to your marker. Depending on the length of your tape measure and the height of your tree, you may need to place a marker where your tape measure ends, record that number, and then measure the rest of the distance. Your measurement should be about the height of the tree. Record this data in meters.

2. Draw or photograph your tree during every season. You may also choose to record weekly journal entries. For each entry, include details, such as which birds and insects use the tree for food or shelter, when leaves bud or change color and fall to the ground, and when or if the tree blossoms. Be sure to date each journal entry.

3. You may also glue fun things into your journal, such as bark or leaf rubbings, pressed leaves or flowers, and seeds.

*optional

broadleaf tree: a tree having flat, broad leaves and producing seeds

deciduous tree: a tree that loses its leaves and is bare for a part of the year

fruit: a plant structure that holds and protects seeds

oak: a tree that produces acorns

acorn: the fruit of an oak tree

elm: a tree with toothed leaves that produces small, flat seeds encased in paper-like wings

maple: a tree with hand-shaped leaves that produces winged fruits

palm: a branchless tree or shrub; does not have annual rings or bark

Broadleaf Trees

Have you ever jumped into a pile of leaves? The leaves were probably not needle-like; they were broad, rippled, and crunchy. At one time, they were soft and green, but now that they have fallen from the tree they are likely brown. Perhaps they even changed color from green to a brilliant orange, red, or yellow before they fell from the tree.

We can classify types of trees by looking at their leaves. A tree can have needle-like leaves or broad leaves.

A **broadleaf tree** has *flat, broad leaves.* Some common broadleaf trees are oaks, maples, ashes, elms, walnuts, and willows. *Most broadleaf trees are deciduous,* which means "to fall off." **Deciduous trees** *lose their leaves and are bare for a part of the year.* Do you live in a region where leaves change color in autumn? Perhaps it is your job to rake the leaves that have fallen in your yard.

Instead of cones, most broadleaf trees form flowers and a type of fruit. Tree flowers are usually tiny and hard to see, but some are very beautiful. Magnolia, dogwood, cherry, and apple trees have very attractive flowers for us to enjoy. Part of each flower becomes **fruit** that *holds and protects seeds.* Some fruit structures are the kind you eat, but most are not. The seeds produced by the apple blossom are protected inside the apple.

Oak *trees are most easily distinguished from other trees by their fruits, which are called* **acorns**. Acorns are hard-shelled fruits topped with scaly caps. Most oaks have leaves that are indented along the edges, but the shape of oak leaves varies greatly, even among leaves growing on the same tree. Some kinds of oak trees are evergreen, keeping their leaves through the year. When new leaves develop, the old leaves are pushed off the tree. Oak leaves do not grow opposite each other on the branch. They grow in an *alternate pattern*, with a leaf on one side of the branch, a space, and then a leaf on the other side of the branch.

Elm trees can also be easily recognized by their fruits, which are *small, flat seeds encased in paper-like wings.* Oval-shaped elm leaves are toothed, or notched, along the edges.

Many varieties of the popular and useful **maple** tree grow in North America. *The shape of the maple leaf is unique; it is a broad leaf which resembles a hand.* The leaves grow in pairs that are always opposite each other on the branch. The maple leaf is one of the national emblems of Canada and appears on the Canadian flag. The fruit of the maple is also easy to recognize. *Maple trees produce winged fruits* which *grow in pairs*, each bearing a seed.

Palms

One kind of tree that is neither a broadleaf nor a needleleaf is the **palm**. Not all palms are trees; some are shrubs or vines. Palms do not grow in the same way that other trees do. *They have no branches, annual rings, or bark.* A bud in the center of the crown controls the palm's growth. If the bud is removed, the palm dies. Palms are evergreens with fan-shaped or feather-like leaves. The raffia palm has the biggest leaf in the world—over eighty feet long! Palms produce fruits that may be smaller than peas or as large as coconuts.

The coconut palm is probably the most useful kind of palm. There are many uses for its products. People eat the sweet meat of the coconut. Dried coconut meat is also processed into an oil which is used in cooking and making products like soap. The date palm is also important for the delicious fruit it produces. The sabal palm and the Washington palm both grow in the United States.

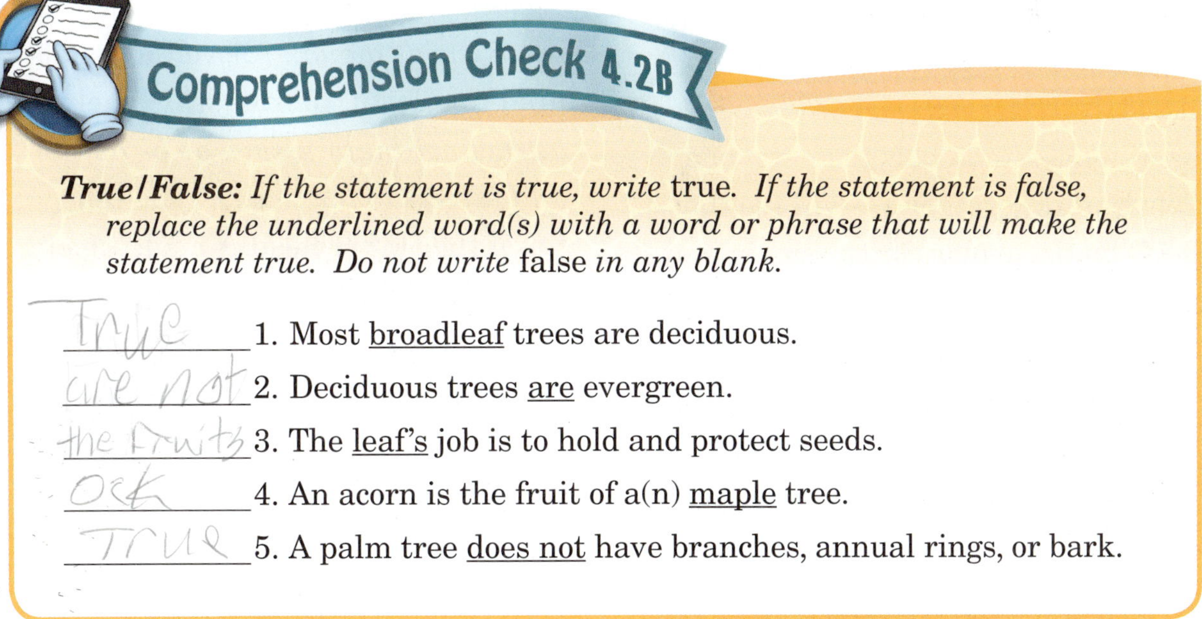

Comprehension Check 4.2B

True/False: *If the statement is true, write* true. *If the statement is false, replace the underlined word(s) with a word or phrase that will make the statement true. Do not write* false *in any blank.*

_______ 1. Most broadleaf trees are deciduous.

_______ 2. Deciduous trees are evergreen.

_______ 3. The leaf's job is to hold and protect seeds.

_______ 4. An acorn is the fruit of a(n) maple tree.

_______ 5. A palm tree does not have branches, annual rings, or bark.

TERMS

sepal: leaf-like structure that surrounds the base of a flower's petals; protects the flower's bud

petal: the colorful part of many flowers that attracts pollinators

stamen: the center part of the flower that makes pollen

pollen: yellow dust on flowers made by the stamens

pistil: the vase-shaped center part of the flower that makes undeveloped seeds

pollination: the moving of pollen from a stamen to the pistil of the flower

sperm cells: cells in the pollen produced by the stamen

4.3 Observing Flowers

No matter what time of the year it is, there are flowers blooming somewhere. Do you know the names of the flowers that are blooming where you live? Do you know which flowers are most common in your area? God has created a wide variety of beautiful flowering plants, and yet most of them are variations of the same basic design. This is because *all flowers have one purpose for the plant—to produce seeds*. As we understand how the parts of a flower work together, we will see that God's intricate plan for flowers is well engineered.

The Parts of a Flower

The stem that holds up the head of the flower and supplies it with water and food may be short or tall, thick or thin. *Surrounding the base of the flower's petals are leaf-like structures* called **sepals**. The sepals *enclosed and protected the flower when it was a bud*. Above the sepals are the colorful **petals**, which help *attract pollinators, such as birds and insects to the flower.* The center of the flower head is made up of stamens and pistils. The **stamens** *make* **pollen**, the fine, *yellow dust* you often see on flowers. The **pistil** *makes undeveloped seeds in the vase-shaped center part of the flower.*

Try This! Dissect a lily.

Materials needed:
- ✓ freshly cut lily
- ✓ sharp knife
- ✓ cutting board
- ✓ paper towels
- ✓ *goggles
- ✓ *gloves
- ✓ *mask

1. Use laboratory safety equipment, such as gloves, a mask, or goggles to prevent any irritation from pollen.
2. If the bloom is closed, carefully use a knife to cut open the petals, as well as each part of the flower.
3. Use the picture above to identify the sepals, petals, stamens, and pistil. As you identify each part, lay it on a paper towel.
4. Be sure to wash your hands after dissecting, or cutting apart, your flower.

*optional

Pollination

Who gave the rose its beautiful fragrance? God gave some flowers a sweet scent to attract insects to them. The bright petals also attract pollinators.

When honeybees, bumblebees, wasps, butterflies, and other insects visit flowers in search of nectar, pollen brushes from the tips of the stamens onto their bodies. As the insect travels from flower to flower, *it moves pollen from the stamens to the pistils.* Scientists call this process **pollination**.

Farmers are so dependent upon honeybees to pollinate their crops that they rent hives of honeybees during the time their crops are in bloom. Pollination is such an important job that God designed other animals to help, too. Insects, such as gnats, beetles, and ants, as well as certain birds and bats, are also designed to be pollinators.

Each pollen grain contains two tiny **sperm cells** that have been *produced by the stamen.* After pollination, sperm cells unite with the undeveloped seeds at the base of the pistil. Now the seeds will be able to grow into new plants, each after its own kind.

God provided flowers with their beautiful colors and variety of scents. He knew what each plant would need. As you observe them, think of the ways God provides for you. God knows your needs and takes care of you.

Consider the lilies how they grow: they toil not, they spin not; and yet I say unto you, that Solomon in all his glory was not arrayed like one of these. If then God so clothe the grass, which is today in the field, and tomorrow is cast into the oven; how much more will He clothe you, O ye of little faith? Luke 12:27–28

Identify, research, and observe your state flower.

Materials needed:
- ✓ field guides and research sources
- ✓ sketching materials
- ✓ *camera

1. Every U.S. state has chosen a flower as one of its state emblems. Identify your state flower by using the field guide at the back of this book. (See pages 513–529.)
2. Research why your state chose that flower as an emblem and if any other state has the same state flower. Discover what birds, insects, and other animals benefit from, help, or harm the flower.
3. Try to find your state flower in your own community for observation. As you observe, sketch or write down a description. Notice the size of the plant, the leaves, the flowers, when the plant blooms, and where it grows. You may even choose to photograph your state flower in its natural habitat. Write a brief report about what you observed and read, and make a colored sketch or painting of the flower.

You might also enjoy learning more about other flowers that grow in your area. Look at the flower field guide on pages 513–529 in the back of this book if you would like to know what flowers you can learn to recognize.

*optional

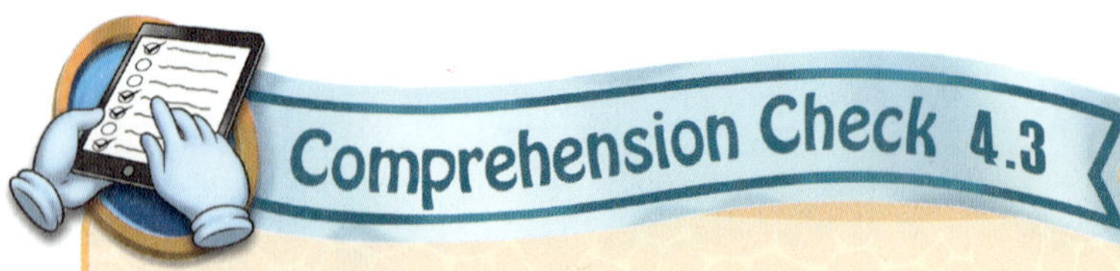

Label the parts of a flower.

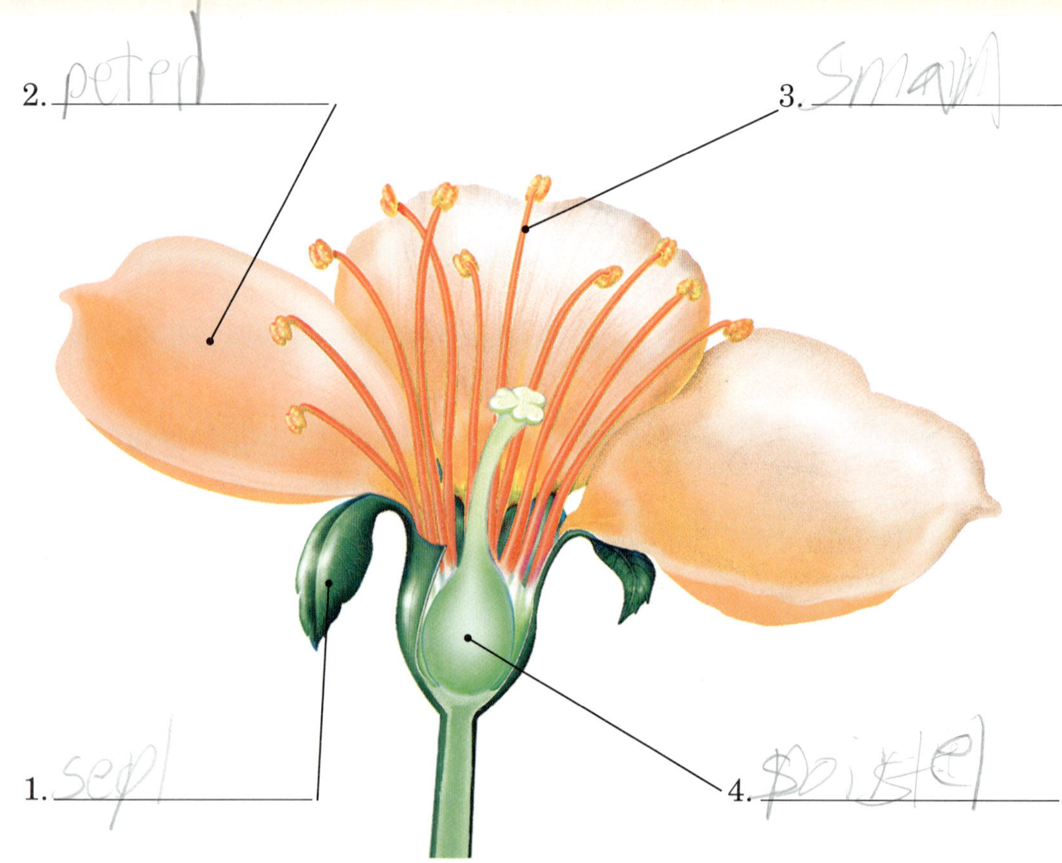

Give the correct answers.

5. What is the flower's job? ____________________

6. What part of the flower makes pollen? ____________________

7. Which part of the flower contains undeveloped seeds? ____________________

8. Pollination happens when pollen is moved from the ____________________ to the ____________________.

Practice the scientific method with photosynthesis. Record your data on Worksheet 22.

Materials needed:
- ✓ a potted plant
- ✓ a sunny place
- ✓ water
- ✓ dark construction paper
- ✓ tape

What to do:

1. Observe and ask questions.
   - What form of energy controls the process of photosynthesis?
   - What could happen to a plant if it cannot make food for itself?

   Question to answer:
   What kind of energy does a leaf need to keep making food?

2. Form a **hypothesis**. Complete the statement below.
   - Leaves need ________________ for photosynthesis.

3. Experiment and gather data.
   - Decide on your **independent variable** (what will change).
   - Decide on your **controlled variables** (what stays the same).
   - Determine your **dependent variable** (what is observed/measured).
   - Complete the experiment.
     - Gently fold construction paper over three of the leaves and tape securely in place. Be sure all the leaves are on the same side of the plant.
     - Water the plant and place it near a window. For consistency, be sure the side of the plant with covered leaves is facing the sun.
     - After one week, remove the construction paper for evaluation. Record your data.

4. Study data and reach conclusions.
   - Compare your results and data.
   - Discuss the results.
   - Did you support your hypothesis?

ray flower: flowers that radiate from the disk of a composite flower

disk flower: flowers that form a disk at the center of a composite flower

composite flower: flowers made of a combination of many smaller flowers

weed: a plant that grows where it is not wanted

Identifying Flowers

Composite Family

Have you ever picked a bouquet for someone special? The gift of a hand-picked bouquet is sure to make someone you love smile. Perhaps the first bouquet you picked for your mother was a plastic cup full of dandelions. Even this bouquet took a place of honor in the center of the family dinner table.

Perhaps you have created a garland or crown of daisies from a field of wildflowers. Did you know that when you pick a daisy, you are actually picking more than one flower? The part of the daisy that we usually call the flower is made up of hundreds of tiny flowers. The white parts that we often call petals are not true petals at all. Each one is an individual flower! Since these flowers *radiate from the center*, they are called **ray flowers**. The yellow disk, or center, of the daisy contains many tiny tube-shaped flowers. Because these flowers *form a disk*, they are called **disk flowers**.

Many of our favorite flowers are in the **composite** family. Some composite flowers have only ray flowers, such as the dandelion; others have only disk flowers, such as the burdock. Like the daisy, the sunflower has both ray and disk flowers.

Weeds

Have you ever wondered why some flowering plants are called weeds? *A* **weed** *is any plant that grows where it is not wanted.* Weeds are often unwanted because they compete with garden or field crops. Most weeds are very hardy and grow easily. Nutrients that could help crops help the weeds instead.

Some weeds arrived in North America accidentally in shipments of other seeds. Other weeds were brought here on purpose. For example, the early colonists brought burdock with them because they ate the root. Burdock root is not a popular food today, but the burdock plant continues to thrive.

Though gardeners and farmers have to work hard to remove them, God uses even pesky weeds to help natural environments. The root systems of weeds help prevent soil erosion. Nutrient-rich topsoil can remain safely intact where many plants are growing. Roots also bring minerals that are deep in the earth up to the surface where other plants can use them. Some weeds are useful as medicines, dyes, or foods. Many weeds produce great quantities of seeds. This is very important to the birds that depend on weed seeds for food.

4.5 Seeds Designed for Travel

After a plant is pollinated, a part of its flower begins to grow into a piece of fruit. Remember, *fruit is the structure in a flowering plant that holds and protects seeds.* God also designed many fruits so they can help the seed travel to a place where it can sprout and grow. If seeds could not travel away from their parent plants, they would not be able to grow into strong, healthy plants, because the older, taller parent plants would take most of the sunlight, water, and minerals available in each spot. God designed several ways for fruits to help the seeds move away from the parent plant.

Airborne and Windblown Seeds

Seeds that are designed to be spread by the force of wind are usually lightweight. Some plants produce seeds that are as fine as dust. *Because the seeds are so light, the wind can easily carry them.* The beautiful poppy flower produces a capsule fruit. After the petals fall off, the capsule splits open in dry weather. The many tiny poppy seeds are released when the capsule shakes in the wind.

poppy capsule

Even before people invented the parachute or helicopter, God had engineered some seeds to travel in these special ways. Have you ever seen a dandelion that has gone to seed? Each tiny seed has its own little parachute. Dandelions, cattails, milkweed, and thistles are a few of the plants that are equipped with downy tufts, or plumes, that help the seeds float away from the plant by a gentle breeze.

Some trees, such as maples, have seeds designed to "fly" very far away from their parent plant because of their propeller-like wings. If you think of a helicopter's design, you can see how these seeds travel.

What about very tall pine trees? Each seed in a pine cone has one delicate wing that is caught by a gentle breeze. God designed them this way so they can fly away and find an open area with plenty of sunlight.

In the central and western parts of the United States, there are several different kinds of plants which are called tumbleweeds. When their seeds are ripe, tumbleweeds become so dry that they break off from their roots. As the wind rolls the tumbleweeds across the prairie, seeds are scattered wherever they go.

Water-Traveling Seeds

Water is another distributor of seeds. Rain washes seeds out of open seed pods and carries them to new locations.

Plants that grow in or near a body of water often depend on it to scatter their seeds. Some waterborne seeds, or seeds that are carried by water, are covered with a waxy coating that keeps them from being soaked. Some coconuts fall into the ocean and float to islands or other shores where they grow into coconut palm trees. Seeds of the red mangrove tree sprout roots while they are still on the tree. When the seeds drop into the water where the trees grow, the heavy roots keep them upright in the water. When the rootlet reaches a suitable spot, it anchors there, and the tree grows.

coconut sprout

mangrove seeds

mangrove

mangrove sprouts

Hitchhiker Seeds

Animals move many kinds of seeds. Some seeds are designed with little hooks or barbs. These structures help the seeds to latch onto an animal's fur, a bird's feathers, or even your sweater or socks. Cockleburs, burdocks, and beggar's ticks are a few *hitchhiking seeds*. Other seeds, such as flax, are covered with a sticky substance when they get wet that helps them to hitch a ride on a passing animal.

Birds and small animals move seeds by collecting them and storing them for food. Squirrels and blue jays often store seeds for the winter, but sometimes they forget about them. When fruits with large seeds are eaten by animals, the seed is dropped to the ground after the animal eats the juicy fruit. When fruits with small seeds, such as blackberries, are eaten, the animal often eats the seeds right along with the fruit. The seeds then pass unharmed through the animal's digestive system and grow where they are dropped.

Have you ever given a seed a ride on purpose? If you have planted a garden, you have! People often move seeds deliberately when they travel from one place to another and carry plants or seeds with them.

squirting cucumber

Other Ways Seeds Travel

Some plants are designed to scatter their own seeds. When the fruit of the jewelweed ripens, it explodes, shooting its seeds in all directions. Wisteria and other members of the pea family produce seeds in pods. When the pods dry out, they split, and the seeds are scattered. When the squirting cucumber is ripe, it drops from its stalk, leaving a hole in the end of the fruit. The seeds are shot out of this hole. Some seeds are shot as far as twenty feet away from the plant.

Give the correct answer.

1. What structure of the plant holds and protects the seeds? ____________
2. Name a seed that can travel because of the wind. ____________
3. Name a seed that can travel by water. ____________
4. How do some seeds travel with the help of an animal? ____________

Think and Predict.

5. Why can a maple seed travel far from the maple tree? ____________

embryo: the living part of the seed that becomes the new plant

stored food: nourishment for the young plant until it can make food on its own

seed coat: the covering that protects the embryo until it is time for the seed to sprout

dormant: alive but resting

germinate: to sprout

primary root: the first root

shoot: a new sprouting plant

4.6 Seed Design and Germination

God had many plans for flowers when He designed and created them. Like any part of creation, flowers glorify their Creator.

Not only are flowers beautiful, but they also have a very important job. They were designed to make seeds so that the plant can reproduce after its kind.

Parts of a Seed

Every seed has three main parts: *the embryo, the stored food, and the seed coat. The* ***embryo*** *is the living part of the seed that becomes the new plant.* It contains information the seed needs to grow. An apple seed contains a new apple tree; a corn plant is inside a seed of corn. You will never get corn from apple seeds, nor apples from corn seeds. You can plant apple seeds and corn seeds side by side in the same garden where they will share the same soil, sunlight, and water. The apple seed will always grow into an apple tree that produces apple seeds. It will never grow corn, squash, tomatoes, or any other fruit.

And God said, Let the earth bring forth grass, the herb yielding seed, and the fruit tree yielding fruit after his kind, whose seed is in itself.
Genesis 1:11

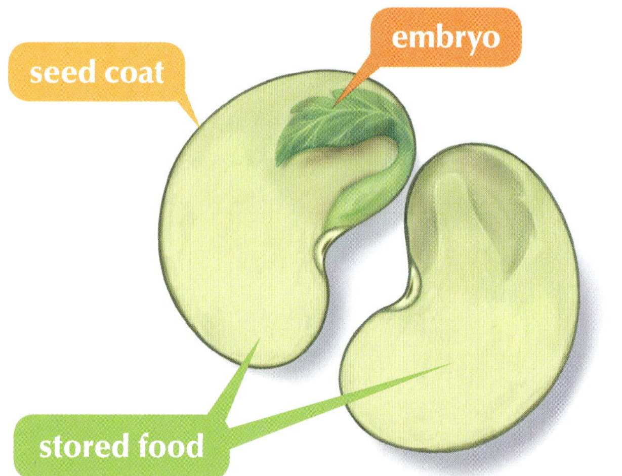

Seeds contain **stored food** *to nourish the young plant until it can make food on its own.* This food is stored in a special structure called the cotyledon. Lima beans have two cotyledons. The third part of the seed is the seed coat. *The* **seed coat** *is a covering that protects the embryo until it is time for the seed to sprout.*

Try This! Observe the three parts of a seed.

Materials needed:
- ✓ lima bean seeds
- ✓ glass of water
- ✓ paper and pencil

1. Soak some lima bean seeds overnight in a glass of water. The seed coat will absorb enough water to make it easy to remove.
2. Remove the seed coat by carefully splitting it along the outside edge of the seed.
3. Observe the three parts of a seed.

Food is stored in the cotyledon. Do you see the baby plant, or embryo? Find the parts of the embryo that will someday become roots and leaves. You may choose to sketch the cotyledons and the embryo of your lima bean seed.

Germination

Most seeds are carried away from their parent plants in autumn. In areas that have cold winters, most seeds remain **dormant** throughout the winter. This means that *though the seed is alive, it is resting.* Why do you think God designed seeds to be dormant for a time? What would happen if the seed sprouted and began to grow into a new plant at the end of autumn? The freezing temperatures would kill the new plant. When spring comes and conditions are right, the seed is ready to **germinate**, *or sprout.* The seed, which looked no more alive than a pebble, begins to change. *There are three things that all seeds need to germinate: water, oxygen, and the right temperature.*

Water

Seeds absorb water when they are ready to germinate. Some seeds need to be soaked in water; others need only a small amount of water. Water causes changes to take place inside the seed. Water softens the seed coat so that the embryo can break through and continue to grow. The seed coat then dies and returns valuable materials to the soil. The embryo contains all the seed's information and is ready to grow into a new plant that can produce fruit. The growing plant depends on water to carry nutrients from the soil into the roots, stems, and leaves.

Oxygen

All living things require oxygen. Even when a seed is buried beneath the soil, it contains oxygen that helps it to use its stored food. As the seed sprouts and grows, it begins the process of photosynthesis and continues to use oxygen as it begins making its own food.

The Right Temperature

Without the proper temperature, a seed cannot grow and mature as it should. Different kinds of plants need different amounts of heat to germinate. Some seeds must also have just the right amount of light to germinate. The sun provides the light and heat plants need in their growth process.

The Growing Seed

As the seed germinates, it sends its *first little root,* the **primary root**, downward into the soil. Soon more roots push their way into the soil, and a network of roots begins to form. God has designed seeds that sprout to send their roots down into the soil, not up into the air. Then a *new sprouting plant*, or **shoot**, begins to grow upward in search of sunlight. The seed is perfectly designed to send its leaves up toward the sun, not down into the soil.

The stem and first leaves break through the soil as they continue to grow. When the stored food is gone, the roots and the leaves are ready to do the jobs for which they were created. The roots begin drawing water and minerals from the soil, and the leaves begin using sunlight energy to make food.

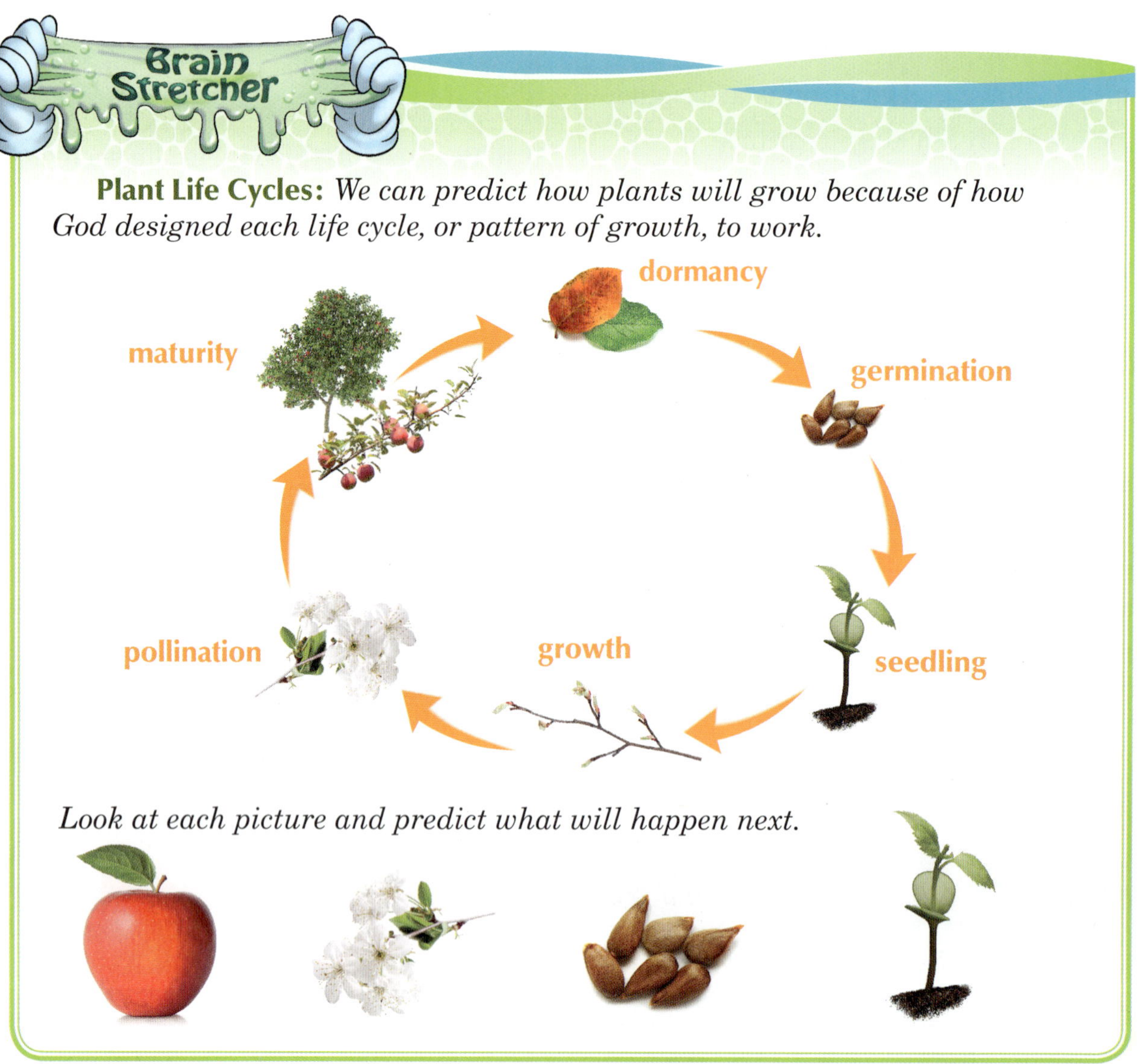

Try This! Observe germination.

Materials needed:

- ✓ lima bean seeds
- ✓ gallon-size resealable plastic bag
- ✓ 2 paper towels
- ✓ water
- ✓ transparent plastic cup

1. Dampen the two paper towels, fold them into fourths, and arrange them inside the plastic cup so that the paper towel presses flat against the side.
2. Tuck the seeds neatly between the towels and the side of the cup, about one to two inches apart. Place the cup inside the resealable bag and seal it.
3. Set your cup in a warm place and observe the seeds as they sprout. If the towels appear dry, open the bag, moisten the towels, and carefully reseal the bag. What happens in a few days?

Comprehension Check 4.6

Label: *Label the parts of a seed.*

1. empbo ____________
2. Food ____________
3. seed coat ____________

Fill in the Blank: *Choose the correct answer from the box below and write it in the blank.*

4. alive but resting dormant ____________
5. to sprout germinate ____________
6. the first root; grows downward primary root ____________
7. tiny stem and leaves; grows upward shoot ____________

dormant	**germinate**	**pollinate**	**primary root**	**shoot**

List: *Give the three things a plant needs to germinate.*

8. water ____________
9. oxenge ____________
10. tem. light ____________

spore: a single cell that is able to reproduce some organisms

algae: non-flowering green plants without roots, stems, or leaves; important producers in water habitats (alga, *sing.*)

ferns and mosses: non-flowering green plants which grow from spores instead of seeds

fungi: organisms that produce spores instead of seeds, such as mushrooms, lichens, molds, or yeasts (fungus, *sing.*)

decomposer: an organism that breaks down dead things to help the soil

bacteria: single-celled organisms (bacterium, *sing.*)

4.7 Plants without Seeds

Could a plant grow without a seed? At first thought, this might seem like an impossibility. But, because the plant world was planned by an imaginative Designer, we see variety in His creations all over the earth. God displays His limitless creativity in the plant world.

Not all plants have flowers. Can you think of some plants that we have already learned about that do not have fruit or flowers? A conifer tree does not make flowers; its seeds are held in cones instead of fruit. Other plants do not have seeds at all.

We can divide the plant world into two main groups: *plants with seeds* and *plants without seeds*. Study the chart below to help you understand basic plant classification. You probably already know a little bit about some plants without seeds, such as algae, ferns, and mosses. **Algae** [ăl′jē] *are important non-flowering producers in water habitats.* Algae can use the process of photosynthesis without roots, stems, or leaves. **Ferns** and **mosses** are non-flowering green plants.

Plants	without seeds		algae ferns mosses
	with seeds	*has flowers*	fruit trees
		no flowers	conifers

Green Plants Grown from Spores Instead of Seeds

Ferns and mosses do not grow from seeds, but begin as a tiny spore. *A* **spore** *is a single cell that is able to reproduce some organisms*, such as green plants and plant-like living things. Unlike a seed, a spore does not have a seed coat, embryo, or stored food inside of it.

Your eyes could not see one little spore all by itself; it is only one tiny cell. However, you can see a group of spores. Often, spores can be found in spore cases on the leaves of some ferns and mosses. Look for spore cases on the back of a fern leaf. Each spore case contains dozens of spores. After the case breaks open, spores float through the air to a new home. If a spore falls in a moist, shady place where it can get food and air, it will grow.

Ferns

Ferns are often found in shady areas, such as a forest floor. Tropical ferns can grow eighty feet high and can live a long time. The main part of a fern plant can live up to 100 years. Ferns do not have flowers, or cones, but they do have chlorophyll and can make food through photosynthesis.

Mosses

Moss can grow in any habitat except salt water. Mosses are tiny green plants that grow closely together. They have chlorophyll and can make food through photosynthesis. They do not have flowers or cones, but they do have structures that are leaf-like and stem-like. If you examine them, you will see tiny "leaves" and "stems," though they work differently from other green plants. Mosses cling to rocks and other surfaces with tiny root-like structures.

Plant-Like Organisms Grown from Spores

Some living things remind us of plants because of the way they grow, *but we cannot classify them as green plants because they do not have chlorophyll and cannot make their own food.* They have their own classification: **fungi** [fŭnj′ī] and **bacteria**. *Fungi and bacteria are mostly microscopic organisms and can grow from spores.*

Fungi

Fungi often grow where plants do. If you have seen a mushroom or a toadstool, then you have seen a fungus. Where do fungi get energy to grow and reproduce? Because a fungus does not have chlorophyll, it cannot make its own food. It must depend on other sources. Some depend on other living plants for food while others depend on dead things for food. Fungi that live off dead things, such as decaying leaves, are decomposers. *A* **decomposer** *is an organism that helps the soil by breaking down dead and decaying things.* As fungi feed on decaying things, they break it down into nutrient-rich humus.

Besides mushrooms and toadstools, some common fungi are lichen, yeast, and mold. Lichens grow on the bark of trees. Lichens are a mystery to scientists because they are part fungi, part algae. The algae feed the fungi with the sugar they make, while the fungi give the algae a place to live.

lichen

puffballs

toadstool

There are many types of interesting fungi, and they come in all shapes and sizes when their cells grow together. A puffball can grow to be the size of a watermelon. Another kind of fungi living underground can be over two miles across. Yeast is a very tiny, microscopic type of fungi. It is the ingredient in bread which helps it rise, making it soft. As it grows, yeast gives off a gas that makes bubbles. The tiny carbon dioxide bubbles in yeast cause the dough to puff up and become spongy. Mold is another microscopic fungi. Some molds are harmful; others are helpful.

In 1928, Alexander Fleming, a British scientist, discovered penicillin, a life-saving antibiotic that is produced from a green mold. Although the mold from which penicillin is made is called a green mold, it does not contain chlorophyll and is not a true green plant. It is a helpful type of fungus. Other molds, such as black mold, are toxic and harmful to your lungs.

Bacteria

Bacteria are *single-celled organisms*. You could fit millions of bacteria on the space the period takes at the end of this sentence. These organisms can be found everywhere—in cold places and hot places, in dry places and wet places. Some bacteria are harmful and can cause sickness, but *most are very helpful*. Good bacteria in your body help to protect you from harmful bacteria. Living things depend on bacteria as a decomposer, breaking down dead and decaying things to bring nutrients back to the soil. Some bacteria can even decompose harmful soil pollution or oil spills in the ocean.

God created bacteria to be complex and diverse. Because of the mysterious varieties within bacteria, they are sometimes challenging to classify. Some bacteria can make their own food, using the sun's energy or their environment; others cannot and depend on other living things for food. Bacteria is different from a green plant because of its cell structure, and it does not have roots, stems, or leaves.

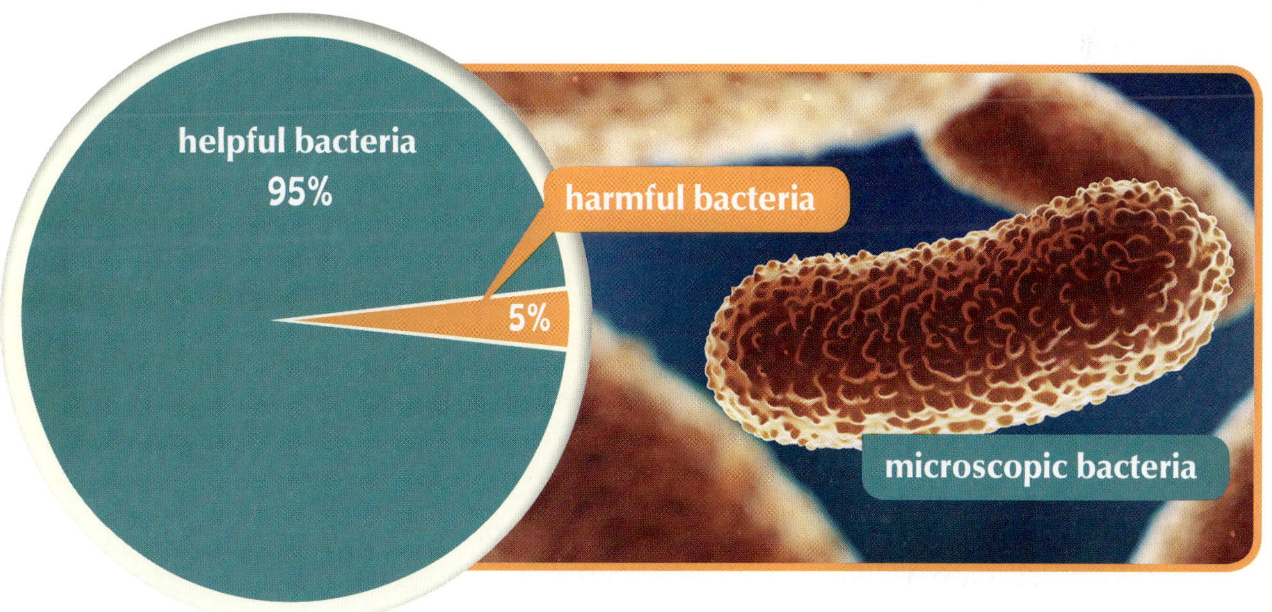

What does the diversity and complexity of living organisms show you about the Creator? In His wisdom, God constructed the intricate workings of each living cell. Organisms on Earth work together for our benefit and God's glory. Each plant, or producer, has its own way of getting sunlight energy to grow and reproduce. In turn, animals and people, or consumers, use plants for food energy to live and grow. Without the plant world, there would be no life at all.

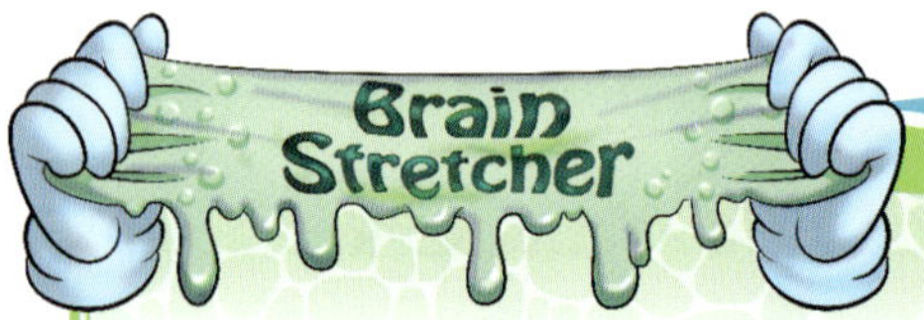

Bacteria: Although harmful bacteria are the cause of some sicknesses, most bacteria are helpful. God, in His wisdom, created bacteria to work inside our bodies as a defense against disease. Scientists have learned that *microbes, microscopic cells that include bacteria, fungi, and viruses*, greatly outnumber the cells that make up the human body! Trillions of bacterial cells form their own communities that work together to keep our bodies functioning properly.

Bacteria are an important protection for our skin, our airways, and our digestive system. They help break down the food we eat. They help the body to use and store energy from food. Good bacteria assist the body's immune system by constantly fighting against harmful bacteria both on our skin and inside every part of our bodies. The more good bacteria we have living inside us, the better health we will enjoy!

Try This! Observe bacteria decompose lettuce.

Materials needed: ✓ lettuce leaves ✓ container ✓ *gloves

1. Place lettuce leaves into a container and leave it in the refrigerator for a week or two.
2. Using gloves, examine the lettuce leaves. Are the leaves slimy?

A type of decomposing bacteria can live in your refrigerator. It causes plant matter to break down. As the lettuce leaves start to decompose, they become slimy.

*optional

Give the correct answer.

1. What single cell can reproduce some living things, such as ferns, mosses, and fungi? ______________________

2. What do we call fungi, bacteria, and other organisms that break down dead and decaying things? ______________________

Think and Classify: *Sort the living things by writing them in the correct columns. Circle the answers that are fungi.*

algae	**grass**	**mold**	**pine tree**
bacteria	**lichen**	**moss**	**yeast**
fern	**maple tree**	**mushroom**	

Green Plant with Seeds	Green Plant without Seeds	Not a Green Plant

Dr. George Washington Carver

The Plant Doctor

From the time he was a very young boy, George Washington Carver loved plants. He wanted to know all he could about them. What made some plants remain healthy while others withered and died? Why were some flowers pink and others blue? George was filled with curiosity and a thirst for knowledge. Whenever he had free time, he roamed the woods. He learned to use his eyes and ears. Because he was a careful observer, George became familiar with all living things; however, the study of plants was his special love. He recorded his observations accurately in the beautiful pictures he drew. George had a talent for helping sick plants to become healthy again. He used this God-given ability to help any neighbor with an ailing plant. Even though he was only a boy, he earned the nickname the Plant Doctor.

It sounds as though George had a wonderful childhood, doesn't it? But, George was an orphan who had been born a slave. When he was young, he was small and not very strong. George overcame these obstacles by learning. He worked hard and did without things so that he could go to grade school, high school, and then college. George never stopped learning. Many years later, to honor him for his outstanding work, his college presented him with a Doctor of Science degree. One of his professors said that George was a brilliant student, the best scientific observer he had ever seen. How do you think George developed those qualities? He looked carefully, thought about what he saw, and asked God to help him see and understand even more.

Dr. Carver was a superb artist and an excellent musician; he could have pursued a career in either of those fields. Because he was determined to use his talents to help people, he became a botanist. A botanist is a scientist who studies plants. For many years, most southern farmers had grown only cotton or tobacco. These crops had removed important minerals from the soil. The soil was poor—and so were the farmers! The soil would not produce enough for them to make a living. Dr. Carver knew he could help the farmers by teaching them better methods of agriculture, the science of farming. He saw his chance when, in 1896, he was asked to teach at Tuskegee Institute, a college in Alabama.

Dr. Carver used a model farm at Tuskegee so that his students could learn to investigate and study plants as he did. Even people who were not his students were welcome to visit, observe, and ask questions. To help those who could not visit, Dr. Carver began a "movable school." He traveled and taught people better ways to farm. He taught them to rotate crops. Instead of always growing cotton, farmers learned to grow peanuts, vegetables, or soybeans. These crops enriched the soil. But what would farmers do with bushels and bushels of peanuts?

So Dr. Carver went to work again. He set up a laboratory with whatever odds and ends he could find for equipment. His most valuable piece of equipment, however, was his Bible. He kept a copy in his laboratory at all times. Dr. Carver often said that he relied on Genesis 1:29:

> *"And God said, Behold, I have given you every herb bearing seed, which is upon the face of all the earth, and every tree, in the which is the fruit of a tree yielding seed; to you it shall be for meat."*

Dr. Carver's laboratory became known as God's Little Workshop.

When Dr. Carver began his investigation of plants, he formed hypotheses, experimented, and then recorded his observations. Over the years he found 75 uses for pecans, 115 uses for tomatoes, 118 uses for sweet potatoes, and over 300 uses for peanuts! He also made a marble-like product from wood shavings, concrete from cotton stalks, dyes from clay, and rubber from goldenrod milk. Many times people offered him large sums of money to leave Tuskegee and work for them instead. One of these men was his good friend Henry Ford. Dr. Carver always refused, however, choosing to use his marvelous talents to help others. By his life and teaching, Dr. George Washington Carver showed that science is a wonderful tool to unlock and use the wonders of God's creation.

plant cells

Chapter 4 Concepts Review

Photosynthesis and Tree Concepts 4.1–4.3

A. Remember

Identify: *Label the* input *and* output *of photosynthesis.*

Input:

1. ______________________
2. ______________________
3. ______________________

Output:

4. ______________________
5. ______________________

Fill in the Blank: *Choose the correct word from the box below and write it in the blank.*

6. A ______________ is an organism that can produce its own food using sunlight energy.

7. Green plants contain ______________, a green matter.

8. A green plant can make a simple sugar called ______________.

9. An organism that depends on producers for food energy is called a ______________.

10. Leaf litter becomes ______________, which helps the soil by adding nutrients and soaking up water.

chlorophyll	**glucose**	**predator**
consumer	**humus**	**producer**

B. Think Like a Scientist

Label: *Write* B *for broadleaf tree,* N *for needleleaf tree, and* P *for palm. Be prepared to explain your answers.*

______ 1. date	______ 3. spruce	______ 5. elm
______ 2. oak	______ 4. fir	______ 6. sabal

Compare and Contrast: *Write the similarities and differences of* needleleaf, broadleaf, *and* palm trees.

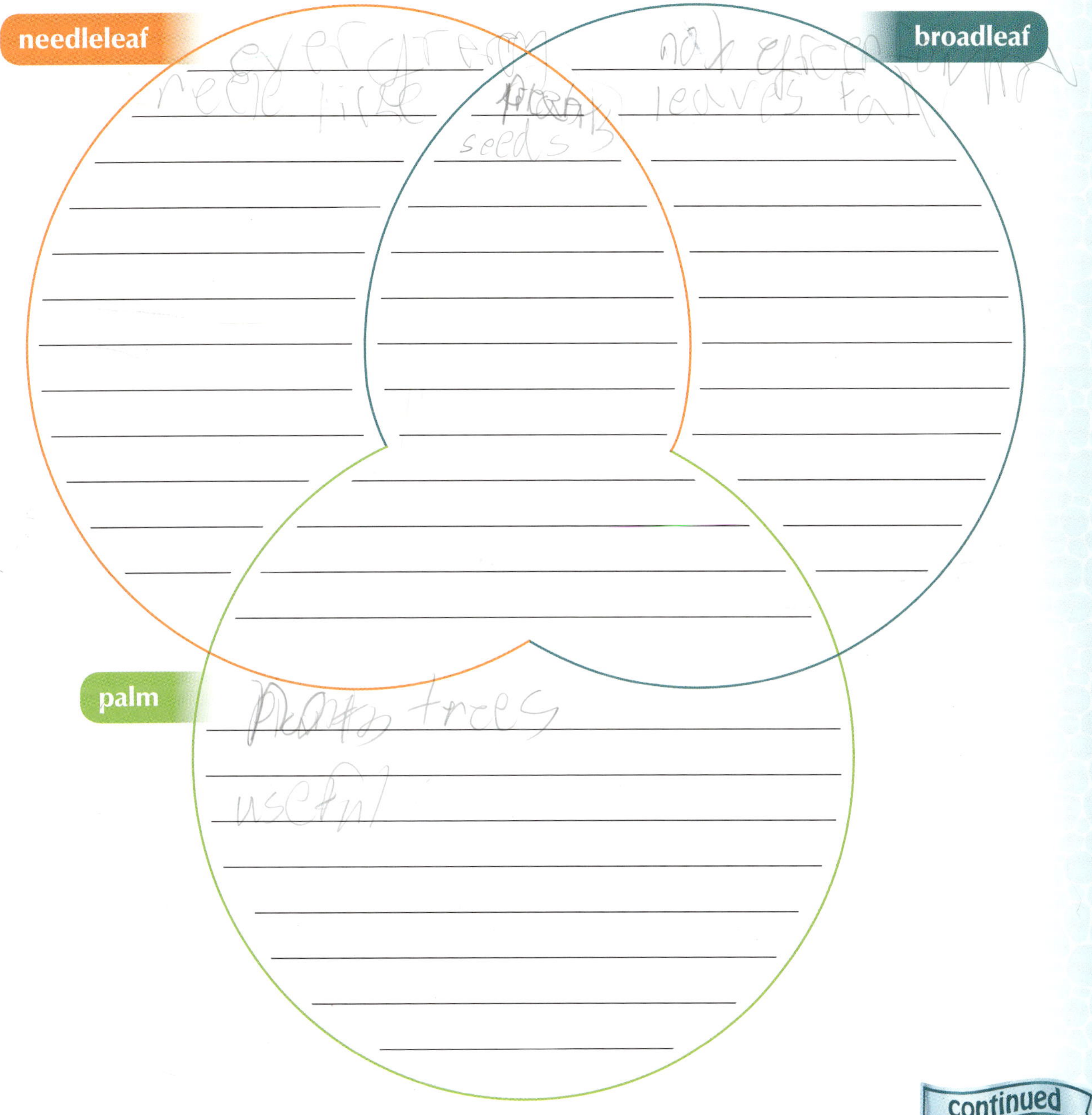

continued

C. Fun with Terms

Puzzle: *Fill in the answers to find the word in the starred column.*

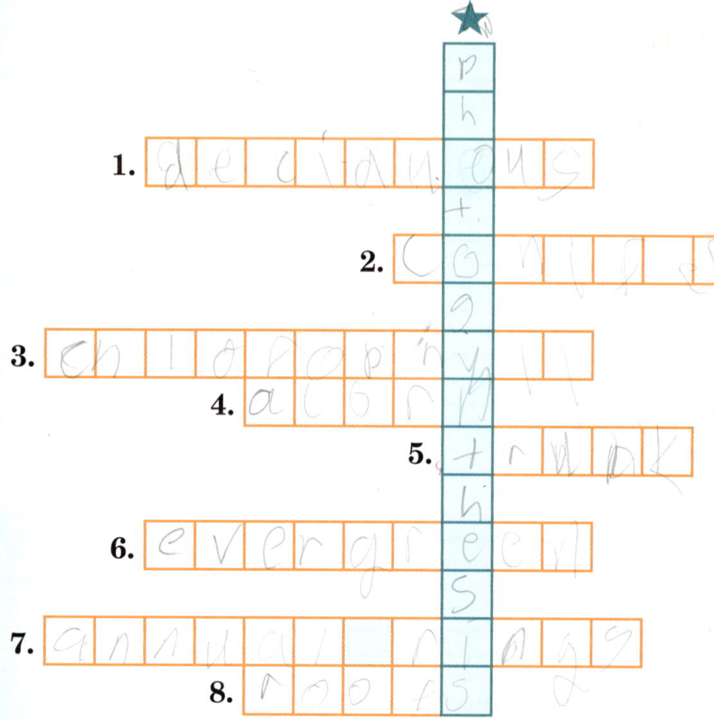

1. a tree that loses its leaves and stays bare for part of the year
2. a tree that produces seeds in cones instead of flowers
3. a green matter found in plants
4. the fruit of an oak tree
5. the stem of the tree
6. a tree that keeps its leaves throughout the year
7. layers of wood in the trunk of a tree
8. anchor the tree and draw water and minerals from the soil

★ the food-making process in green plants

acorn	**chlorophyll**	**deciduous**	**roots**
annual rings	**conifer**	**evergreen**	**trunk**

2 Plant Life Cycles Concepts 4.3–4.8

A. Remember

Identify: *Label the flower using the clues provided. Write the name for each part on the line.*

1. protects flower's bud
2. attracts pollinators
3. makes pollen
4. contains undeveloped seeds

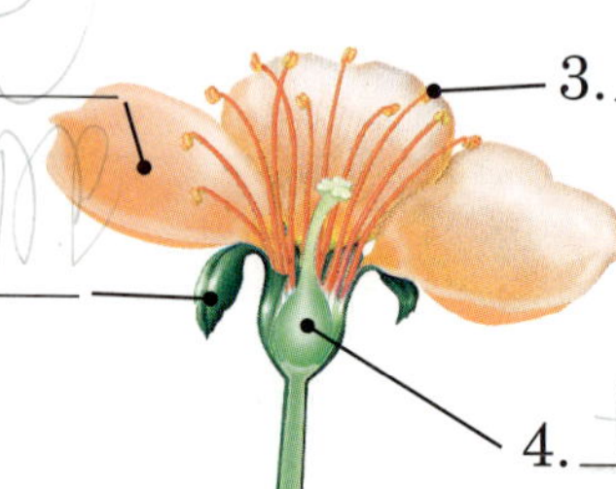

B. Think Like a Scientist

Identify: *Label the parts of a seed and the composite flower.*

Short Answer: *Write the correct answer in the blank.*

6. Give an example of a composite flower. ______________________

7. Which of the following is *not* needed for a seed to germinate: water, soil, oxygen, and the right temperature? ______________________

C. Fun with Terms

Crack the Code: *Find the letter of the correct term and write it on the numbered line below.*

____	1. a single cell that is able to reproduce some organisms	C. primary root
____	2. produces spores instead of seeds	D. pollination
____	3. nonflowering green plants of water habitats	E. shoot
____	4. the first root	F. spore
____	5. single-celled organisms	L. algae
____	6. alive but resting	O. weed
____	7. a plant that grows where it is not wanted	P. dormant
____	8. breaks down dead things to help the soil	R. fungi
____	9. a new, sprouting plant	S. decomposer
____	10. moving pollen from a stamen to the pistil	T. bacteria

Chapter 5 Understanding Animal Design

TERMS

vertebrate: an animal with a backbone

invertebrate: an animal without a backbone

warm-blooded: having a body temperature that stays the same temperature no matter the environmental temperature

lungs: organs that use oxygen from air for breathing

cold-blooded: having a body temperature that changes with the environmental temperature

gills: organs that filter dissolved oxygen from water for breathing

streamlined: designed to move easily through water or air

instinct: a God-given ability or behavior that is inherited rather than learned

5.1 A Variety of Vertebrates

Think of the ways we observe God's creativity in nature. Out of the countless snowflakes that fall, no two are exactly alike. Each person is unique from the inside out. The plant world alone is filled with millions of designs that glorify their Creator. The variety found in the animal world puts God's creativity on display. Each creature was precisely engineered for the life and purpose God designed it to have.

Scientists can organize, or classify, animals into groups. They do this by examining each animal's design—what kinds of parts each animal has and how its body works.

Animal classification begins with dividing the animal world into two groups. *Animals with backbones, or vertebrae, are called* **vertebrates**. Each vertebrate has an *inside skeleton*. *Animals without backbones are called* **invertebrates**. Some invertebrates have an *outside skeleton*.

Vertebrates are divided into five different groups, or classes: *mammals, birds, fish, amphibians, and reptiles.* An animal's vertebrate class is determined by its body design, skin covering, life cycle, and body temperature.

Warm-Blooded Vertebrates

Some animals can make their own heat energy even if it is very cold outside. *The two classes of vertebrates that have warm-blooded bodies are mammals and birds.* **Warm-blooded** animals *stay at the same temperature no matter the environmental temperature.* God designed animal bodies to use potential energy from food. Some of the food energy animals eat gives them mechanical, or moving, energy, and some gives them thermal energy. Warm-blooded bodies can make and maintain thermal energy.

Mammals

Mammals are easy to classify. They usually have four limbs including legs, arms, or flippers. They also have hair or fur. Many mammals depend on hair and fur for warmth and protection, as well as for feeling or sensing. Most mammals are born alive, not hatched from eggs. Warm-blooded animals take care of their young. A mammal mother gives her babies milk and teaches them how to survive.

Mammals *breathe oxygen from the air* with **lungs**. Land mammals breathe air the way you do, steadily inhaling oxygen and exhaling carbon dioxide. Marine, or water, mammals, such as whales, dolphins, and porpoises, also have lungs. They breathe through blowholes at the top of their heads. When a marine mammal swims to the surface, it exhales through its blowhole; then it inhales oxygen before diving back into the ocean. Marine mammals can stay underwater from less than ten minutes to over two hours, depending on their type.

Birds

Birds are also easy to classify because they are the *only animals that have feathers*. They also have wings, though not all birds can fly. All birds have two legs and a beak or a bill.

Like mammals, birds can maintain their own thermal energy because they are warm-blooded. Their temperature stays the same no matter the environmental temperature. Birds also have lungs for breathing oxygen. Even swimming and diving birds get all the oxygen they need from the air.

Unlike most mammals, all birds hatch from eggs. When a baby bird hatches, it is fed and cared for until it can take care of itself.

Cold-Blooded Vertebrates

Other animals cannot maintain their own thermal energy but depend on the sun's heat energy to warm them. *A* **cold-blooded** *animal has a body temperature that changes with the environmental temperature. The three classes of vertebrates that have cold-blooded bodies are fish, amphibians, and reptiles.*

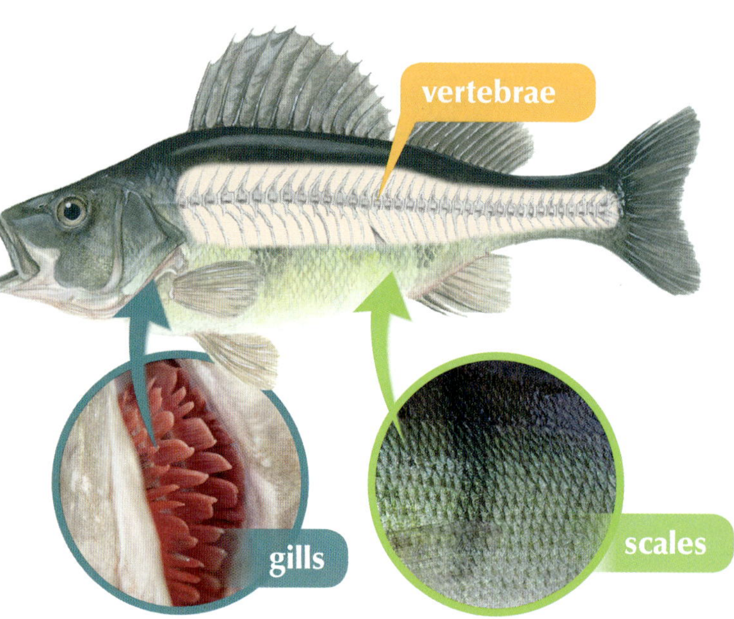

Fish

Like all other vertebrates, fish have a skeleton on the inside of their bodies, where their backbone is found. Some fish are born alive, but most hatch from eggs. Most fish do not need to swim to the surface to breathe because they get all the oxygen they need from the water using vent-like gills. **Gills** *filter oxygen from the water for the fish to use as it breathes.*

Fish swim using tail fins. God designed their bodies to be **streamlined**, or *to move easily through water.* Most fish are covered with protective, armor-like scales.

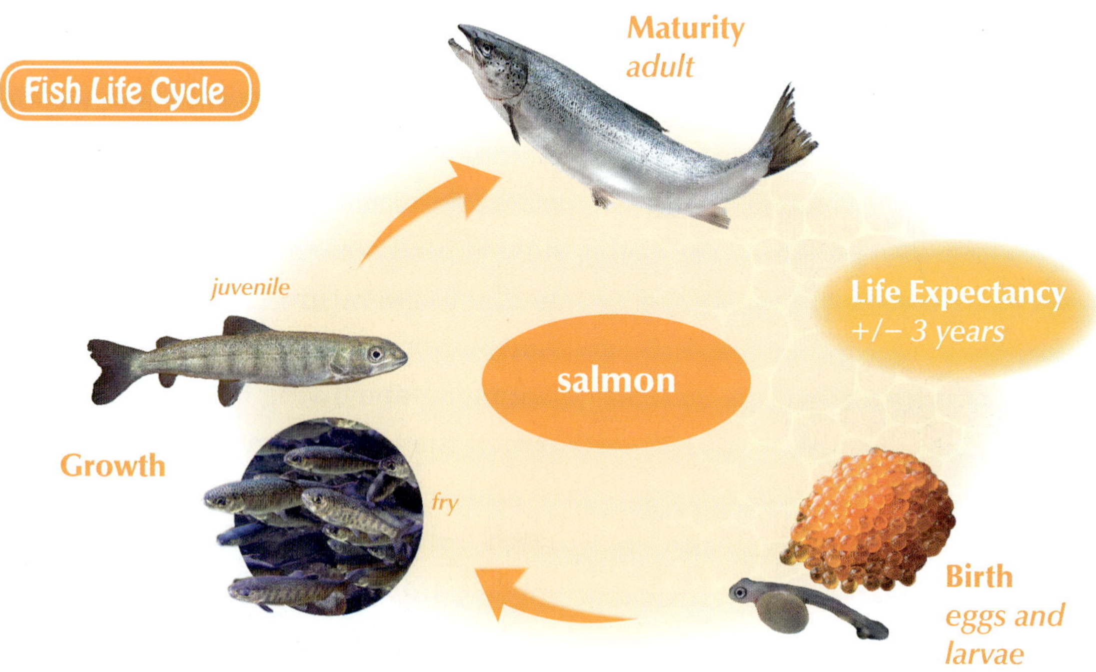

Amphibians

Amphibians begin their lives in water, hatching from slime-like eggs. Baby amphibians know how to swim and find food without a parent to teach them. How is this possible? Like all animals, amphibians have been given instincts. *An* **instinct** *is a God-given ability or behavior that is inherited, or passed down from parent to baby, rather than learned. Cold-blooded animals depend completely on instincts* instead of their parents when they are born or hatched. Their mothers usually leave as soon as they lay their eggs.

When amphibian eggs hatch, *larvae, or baby amphibians, do not look at all like their parents* because they are not fully developed. A tadpole looks very different from a frog. It does not have limbs, and swims using a tail. God designed its tail with fins on the top and bottom. Unlike a fish, tadpole fins are not part of its skeleton because a tadpole does not use fins its whole life.

Tadpoles breathe oxygen by using gills. As they grow, they will not need gills anymore. Frogs develop lungs when they are ready to live on land.

Amphibians have *smooth, moist skin*. They do not have scales like a fish does. Their skin is very important; amphibians get most of their water through their skin. Some amphibians also breathe through their skin.

Do mammals and birds have instincts?
Yes, warm-blooded animals instinctively know when to hibernate or migrate. Who taught the first mammal and bird mothers how to teach their young? God gave all animals instincts to help them know how to survive and do the job He planned for them.

Reptiles

Most reptiles hatch from leather-like eggs. When they hatch, they look like their parents right away. All reptiles breathe with lungs their entire lives because they spend most of their time on or near land.

Some people confuse reptiles with amphibians because most are small animals that have similar body designs. Looking closely at the skin of each animal is the easiest way to classify them. *While amphibians have smooth, moist skin, reptiles have dry, scaly skin.*

Each animal has a unique design because each animal has a unique life and purpose. Who gave each animal its design and purpose? As God created each animal, He also equipped it for what He created it to do. As animals fulfill their purposes, they are glorifying their Creator.

Warm-Blooded Vertebrate Characteristics

Mammals

- ✓ breathe with lungs
- ✓ have hair or fur
- ✓ produce milk
- ✓ most have four limbs
- ✓ most give birth to live young

Birds

- ✓ breathe with lungs
- ✓ have feathers
- ✓ lay eggs
- ✓ have two legs
- ✓ have bills or beaks
- ✓ most can fly

Cold-Blooded Vertebrate Characteristics

Fish

- ✓ breathe through gills
- ✓ most have scales
- ✓ have fins
- ✓ live in water
- ✓ most lay eggs

Amphibians

- ✓ young breathe through gills in water
- ✓ most adults breathe with lungs on land
- ✓ have smooth, moist skin
- ✓ lay eggs

Reptiles

- ✓ breathe with lungs
- ✓ most live on land
- ✓ have dry scales
- ✓ most lay eggs

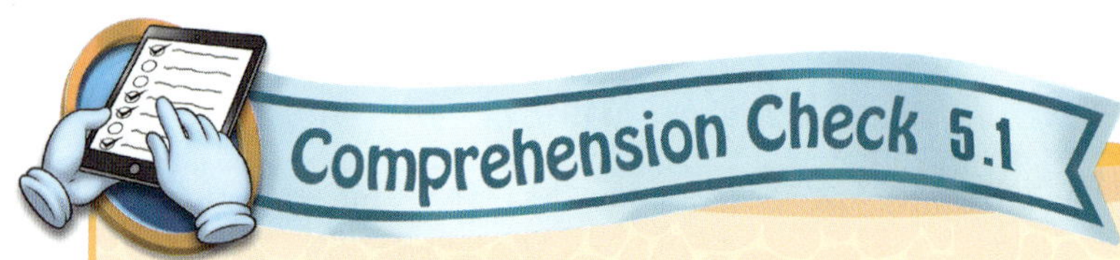

Comprehension Check 5.1

Fill in the Blank: *Choose the correct word from the box below and write it in the blank.*

1. An animal with a backbone is classified as a(n) vertebrate.
2. An animal without a backbone is classified as a(n) invertebrate.
3. An animal that can stay at the same temperature no matter the environmental temperature is warmblooded.
4. lungs are organs that help some animals breathe oxygen from air.
5. An animal whose body temperature changes with the environment is cold-blooded.
6. gills are organs that filter dissolved oxygen from water for some animals to breathe.

cold-blooded	**invertebrate**	**vertebrate**
gills	**lungs**	**warm-blooded**

Think and Conclude.

7. Give an example of an instinct or inherited behavior. birds nest
8. Give an example of a learned behavior. bird flying
9. Look at the pictures below. Which is an amphibian? Which is a reptile? Explain the difference. salamander alligator back

alligator

salamander

evolution: the unproven theory that animals slowly turned into other animals over millions of years instead of being created by God

species: different types of animals

predator: an animal that hunts other animals for food

prey: the animal a predator hunts for food

habitat: the natural home of an animal or plant

range: continental area

5.2 Observing Bird Design

The belief that animals slowly turned into other animals over millions of years is called **evolution**. Evolutionists believe that smaller organisms turned into larger organisms, which turned into invertebrates, which turned into vertebrates, which eventually turned into humans. Does this theory sound logical? You have already learned a little bit about the design and diversity of organisms. Every design must come from a Designer. The Bible tells us this miraculous Designer is God.

Evolutionary Theory

> *And God said, Let the waters bring forth abundantly the moving creature that hath life, and fowl that may fly above the earth in the open firmament of heaven. And God created great whales, and every living creature that moveth, which the waters brought forth abundantly, after their kind, and every winged fowl after his kind: and God saw that it was good.* *Genesis 1:20–21*

One common evolutionary theory is that birds evolved from reptiles. But as we understand bird design, we can see that the abilities and structures of each type, or **species**, were planned by a wise Creator. Each bird has been engineered by God for the life it has.

Designer Beaks and Bills

God wisely gave each bird the mouth structure it would need to eat its particular food. You can often tell what kind of food a bird eats just by looking at its beak or bill.

Birds with short, hard, cone-shaped beaks usually eat seeds. Seedeaters use their sharp beaks and powerful muscles to crush their food. Birds belonging to the finch family are seedeaters. Goldfinches, grosbeaks, cardinals, and sparrows are a few common finches.

Insect-eating birds usually have long, slender beaks. Some insect eaters, such as flycatchers and warblers, capture insects in midair. Creepers use their delicate, pointed beaks to pluck insects from cracks in the bark of trees. Swifts, swallows, and whippoorwills are insect eaters that have wide, gaping mouths. These rapid-flying birds sweep through the air with their mouths wide open, trapping small insects.

goldfinch

vermilion flycatcher

pileated woodpecker

Woodpeckers are specifically designed to live in trees or cacti. All woodpeckers have sharp, pointed bills to bore holes in trees, strong claws to hang on to the bark, and stiff tails to prop themselves up as they climb.

The woodpecker's head is designed for its job as a living hammer. Its skull is thick, and its beak is designed to withstand the pounding that would injure another bird. Extremely strong neck muscles also help support the woodpecker's head.

swallow

Some birds eat nectar. Honeycreepers, sunbirds, and hummingbirds *have very slender, needle-like bills* that allow them to reach deep into flowers for nectar.

Eagles, owls, and hawks hunt smaller animals for food. *An animal that hunts other animals for food* is called a **predator**. *The animal a predator hunts* is called its **prey**; that is why predator birds are called birds of prey. You can recognize other *birds of prey* by their *sharp, hooked beaks designed to hold their prey and to tear flesh.*

Some birds that live near water, such as herons and bitterns, can *spear fish with long, dagger-like beaks.* Other water birds use their mouth structures for net fishing. The pelican uses its *oversized bill and huge pouch* to scoop up its catch. Ducks and flamingos that eat small plants and animals have *special strainers* in their bills to filter out water and mud.

flamingo

See how birds use their God-given mouth structures.

Materials needed:

- ✓ tray of birdseed
- ✓ miniature marshmallows
- ✓ pan of water with gummy-fish candy and large marshmallows
- ✓ pistachios
- ✓ tweezers
- ✓ toothpicks
- ✓ skewer
- ✓ small net

1. Try to mimic how birds use their tools with the food on the tray.
   - ✓ Use the tweezers to crush and pry open a pistachio the way seedeaters use their cone-shaped beaks. Try to pick up birdseeds and nut meat.
   - ✓ Use two toothpicks to pluck a marshmallow the way insect eaters use their delicate, pointed beaks.
2. Try to mimic how water birds use their tools with the pan of water.
   - ✓ Use a skewer to spear a marshmallow like herons and bitterns.
   - ✓ Use the net to scoop up gummy-fish candy like a flamingo or other water bird.

God wisely equips each living thing with what it needs to live.

Designer Feet

Can you imagine a duck trying to perch like a chickadee or a woodpecker trying to swim like a swan? Each kind of bird has the type of feet it needs for the life God planned for it.

Chickens and pheasants have powerful claws with strong, blunt nails to scratch the ground and find food. *Perching birds,* such as robins, sparrows, and orioles, *have toes designed to hang on to branches.* Did you ever wonder why birds do not fall off their perches when they fall asleep? Their legs and feet are designed to lock in place to keep them from falling off.

Birds designed to live in flat, open country are often better runners than fliers. God gave the roadrunner strong feet with long, flat toes to help it move quickly. The flightless ostrich is another swift runner. Its powerful, two-toed feet are also designed for kicking its enemies. But swallows and hummingbirds, which are designed to spend most of their time in flight, have small, weak feet. How do you think large feet would affect the flight of these birds?

Woodpeckers' feet have two toes facing forward and two toes facing backward. This unique design enables the woodpecker to hold on to the bark of trees.

Swimming birds always have their paddles with them. *The webbed feet of ducks and geese are perfectly designed to propel them swiftly through the water.*

Wading birds have long legs; some seem to walk as though on stilts. Herons, egrets, and flamingos belong to this family of birds. Their long, thin toes distribute their weight over a large surface area, keeping them from sinking in mud and sand as they feed.

Eagles, hawks, and other *birds of prey have talons, or sharp claws, on their feet that help them catch, kill, and carry away smaller animals.*

Do you think these amazing structures could have evolved by themselves? The complexity and diversity of the design of each type of bird points to the Master Designer. Each bird is equipped with the right structures for its life.

Observe to Understand: Birds

Throughout your study of this chapter, use your senses to observe birds in your local park or backyard. The best time of day for birdwatching, or birding, is early in the morning or at dusk. Birds are very active beginning at sunrise and continuing for the next two hours.

Remember to move quietly as you look for birds. Try to keep the sun to your back; you will be able to see the birds much better. When a bird appears, watch it until it leaves. Concentrate on the whole appearance of the bird and notice what makes it special. Observe its feet and mouth structures for clues. Use a notebook to write down characteristics or sketch the bird you are observing. A camera or binoculars are also good observation tools.

Here are some other things to look for:

Size

Is the bird large like a crow (almost 20 inches), small like a sparrow (5–7 inches), or medium-sized like a robin (10 inches)? Becoming familiar with the sizes of these three common birds will help you to recognize others.

Shape

Would you describe the bird as plump or slender? Also, notice the size and shape of the bird's tail and wings. Is the tail pointed, rounded, forked, or squared?

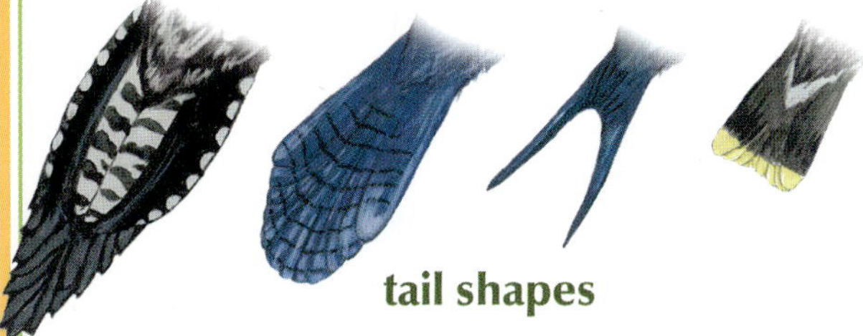

tail shapes

Color and Markings

Some birds, like the cardinal, can be quickly recognized by their color. To recognize other birds, you may have to look very closely for special patterns in their feathers, called field marks. Does the bird have a crest on its head? Is the bird's breast spotted, streaked, or plain? Are there patches of white on its wings, back, or tail? Does the bird have an eyebrow or a ring around its eyes? Is there a band on the bird's tail or bars on its wings?

crest

breast patterns

patches of white

eye patterns

continued

Behavior

Many birds have unusual ways of flying, walking, or perching. For example, nuthatches climb down the trunk of a tree headfirst. Some birds, such as goldfinches and sparrows, travel mostly in flocks. Kingfishers and flycatchers are more solitary, or alone. Some birds move in peculiar ways. Mockingbirds flash their wings as they walk; doves bob their heads. Notice also if the bird strolls, hops, or runs as it walks. How does it fly? Some birds fly in straight lines; others dip like a roller coaster.

Habitat

An animal's **habitat** is its *natural home*. The forest, prairie, seashore, and wetlands are all different kinds of habitats. Two different kinds of birds may live in the *same continental area*, or **range**, but are never seen together because each lives in a different habitat. For example, hairy woodpeckers and mallard ducks share the same range, but you would not find them living side by side because each needs a different habitat. Hairy woodpeckers live in the woods; mallard ducks live near marshes or ponds. Notice the habitat of your new bird. Is it in an open field? Is it perched on a tree in the woods? Or is it hiding in the reeds by a pond? Birds can and do travel, but each species has a favorite place to live—the place God designed for it. Noticing the kind of surroundings where the bird lives will help you to identify it.

When you return from birding, compare your notes and pictures with field guides. Try to identify and learn about the birds you observed.

Once you have trained yourself to be quiet and observant, you may enjoy birding in other places.

Think and Predict: *Based on body design, can you predict what kind of habitat and food each bird prefers?*

Habitat Choices: water, trees, open land, most time in flight

Food Choices: insects, seeds, fish, small animals, nectar

cardinal

Habitat: Ken.

Diet: seeds

pheasant

Habitat: I don't know

Diet: I don't know

heron

Habitat: Tell.

Diet: fish

mockingbird

Habitat: Ken.

Diet: seeds

owl

Habitat: forest

Diet: Meat

hummingbird

Habitat: trees

Diet: water

streamlined: designed to move easily through air or water

lift: the force that keeps a bird or airplane in the air

down: small, fluffy feathers

gizzard: the part of a bird's stomach used for grinding food

5.3 Engineered for Flight

A bird's body is designed so that all parts work together for flight. The bird's body is **streamlined** so that the bird can *move easily through the air.* The shape of the bird's wing is designed for air to move quickly over the curved upper part. The rapid movement of the air creates low air pressure above the wing. Air rushes upward to fill this low-pressure area. The air rushing upward against the lower side of the wing causes the wing to lift. **Lift** *is the force that keeps a bird or an airplane in the air.* Engineers studied the design of birds to understand how the force of lift works. After many, many tries, the airplane was invented in 1903—less than 150 years ago. Engineers still study birds to help them design more efficient ways for airplanes to fly. Do you think a bird could invent flight itself by evolution? *God, the Master Engineer, designed the force of lift and gave most birds the gift of flight.*

Bird Bones

The bones of a bird, especially the spinal column and the ribs, are joined together firmly. This provides a solid foundation for the bird's wings and legs. The bones in the bird's neck, however, are very flexible. This allows the bird to move its neck freely. *Most birds have hollow bones* strengthened inside by tiny cross ribs. This design makes the bird's body much lighter than it would be if the bones were solid. This design also makes a bird's bones strong. Some of the hollow bones are connected to the bird's lungs by air sacs. This connection allows the bird to carry an extra-large air supply. The air which is stored in the bones and air sacs also helps to control the bird's body temperature. God designed the arrangement of the bones and the air sacs to give the bird's body the perfect balance it needs for flight.

Feathers

Birds are the only animals with feathers. Feathers are very important, even for birds that do not fly. The *feathers trap a layer of air close to the bird's skin.* This layer of air *insulates* the bird to *prevent a loss of heat energy*—even in freezing temperatures. For extra warmth, some birds have a layer of *small, fluffy feathers* called **down** underneath their outer body feathers.

As the bird flies, it often needs to change the position of its feathers to take advantage of air currents. It can adjust its feathers automatically in flight because there are nerve endings in the bird's skin where flight feathers attach. These nerves carry messages to the bird's brain to let the bird know which feather needs to be moved. In response, the brain sends a message to one of the thousands of tiny muscles. Then the bird is able to move the feather into just the right position.

You have probably heard the expression "as light as a feather." Individual feathers are very light, but if you weighed all of a bird's feathers, you might be surprised by how heavy they are. A 9 pound eagle has over 1¼ pounds of feathers. All of the eagle's bones put together weigh less than half as much as its feathers.

Why Don't Birds Get Tired?

A bird can fly long distances without getting tired because of its efficient muscular system. The power for flight comes from the bird's two large breast muscles. The same muscles that help birds fly also work to make birds breathe. As a bird beats its wings, it is also filling and emptying its lungs. The bird's body is uniquely designed so that it takes in and releases air at the same time. It has two breaths of air moving through its system at once.

Bones, feathers, and muscles are not the only parts of a bird's body that help it fly. God designed parts of a bird's body that you cannot see to help the bird to fly, too. *The bird's digestive system is designed to make flight possible.* A bird is able to turn its food into energy very quickly. It takes only thirty minutes for a thrush to digest its dinner of berries. How can a bird do this so quickly? A bird does not have teeth, powerful jaws, or a large intestine. It depends on its stomach to do the work. The first part of the stomach softens the food with digestive juices. The second part, or **gizzard**, has hard, ridged walls that *grind up the softened food*. Sometimes, a bird will eat gritty material like sand to make its gizzard work more quickly. How does this help the bird to fly? Since the bird has no teeth, jaws, or large intestine, the bird's body is light. Since food does not remain in the body long, the bird does not have the extra weight of undigested food to slow it down.

Seeing and Hearing

Animals depend on their sense organs to help them survive in their environment. Most birds do not have sharp senses of taste and smell, but their *senses of sight and hearing are very sensitive.* Have you ever seen a bird's ears? You can't; *they are inside the bird's head.* Why do you think God put them there instead of on the outside? Having ears on the inside of its head makes a bird more streamlined for flight, swimming, or running.

Some birds have a sharp sense of sight. A bird is able to focus its sight on two things at once. Birds are also able to clearly see small objects from far away. Some birds are even able to see markings on flowers that humans cannot see. These markings guide them to where the nectar is. Birds have a transparent third eyelid. As the bird is flying, it is able to close this eyelid to protect its eyes from dust and wind.

Try This! Demonstrate lift.

Materials needed:
- ✓ notebook paper
- ✓ scissors

1. Cut a sheet of notebook paper in half vertically. With one hand, hold one end of the paper just below your lower lip. Let the other end of the paper hang down over your hand.
2. Blow as if you were whistling. What happens to the paper?

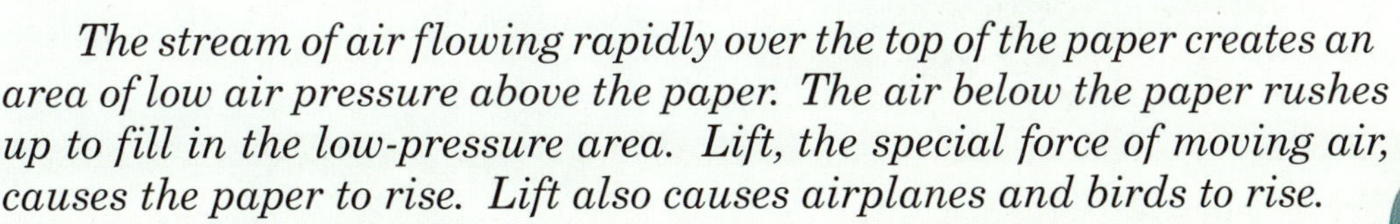

The stream of air flowing rapidly over the top of the paper creates an area of low air pressure above the paper. The air below the paper rushes up to fill in the low-pressure area. Lift, the special force of moving air, causes the paper to rise. Lift also causes airplanes and birds to rise.

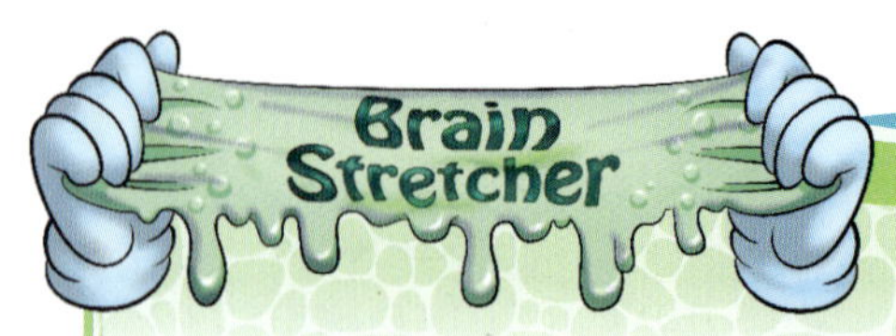

How Birds Stay Warm: When it is cold outside, you reach for your favorite coat or sweater to help you keep warm. Birds do not wear coats, yet we see them cheerfully hopping about on very cold days. Birds are warm-blooded, just as mammals and people are. How can they endure the cold weather?

God has provided each bird with a coat of feathers. Feathers help the bird to fly, and they also work in a marvelous way to keep the bird warm. Some materials are good conductors of thermal energy while others are not. *Remember, heat energy always moves from the warmer space to the cooler space.* If much of the heat produced by the bird's body traveled into the cold, outside air, the bird would become very cold. Instead, the bird's feathers act like a blanket. The feathers trap a layer of air next to the bird's body. *Since air is a poor conductor of heat, the trapped layer of air insulates the bird's body.* Body heat is trapped by the air under the feathers, and the bird stays warm. Most birds have an extra layer of small, fluffy down feathers. Down provides extra insulation. For more warmth, birds fluff up their down feathers so they can trap more air.

Now that you know how a bird's feathers keep it warm, can you explain why a coat or sweater keeps you warm on a cold day? How does the fur of your cat or dog insulate its body?

black-capped chickadees

Comprehension Check 5.3

True/False: *If the statement is true, write* true. *If the statement is false, replace the underlined word(s) with a word or phrase that will make the statement true. Do not write* false *in any blank.*

__________ 1. A bird's <u>body design</u> helps it to be more streamlined for flying.

__________ 2. <u>Lift</u> is the force that keeps a bird in the air.

__________ 3. Most birds have <u>solid</u> bones to make them lighter for flight.

__________ 4. <u>Feathers</u> work together to insulate the bird, keeping it warm.

__________ 5. A bird's digestive system, including its gizzard, makes it <u>heavier</u> for flight.

__________ 6. A bird's senses of <u>sight and hearing</u> are very sensitive.

Think and Predict.

7. What would happen to a bird's flight abilities if its bones were more solid? ______________________________

8. Why is a bird able to fly long distances without getting tired? ______________________________

5.4 Birds in Your Backyard

A sharp whistle and a flash of red in a nearby pine tree alert you to the presence of a cardinal. Across the damp lawn runs a robin searching for its earthworm breakfast. The screeching of a blue jay may be a warning that danger is at hand—or it may only mean that it is claiming the birdbath for itself! In the flower bed, you think you see a large moth, but looking more closely, you discover a tiny hummingbird.

These are a few of the birds that commonly visit our backyards. Each one is a unique part of God's creation. How many can you recognize by sight or by their song?

The Friendly Robin

Perhaps the first bird you learned to recognize was the robin, a friendly bird with an enormous appetite. It feasts on wild fruit and earthworms, its two favorite foods. Robins also eat insects and some seeds. You may see one running across your lawn in search of its next meal. Then it stops, standing absolutely still with its head cocked to one side. It is not listening for anything; it is using its eyes to hunt earthworms.

robin

Because the eyes of most birds are on the sides of their heads, they have to cock their heads and look at an object with just one eye. The other eye keeps busy looking at things in the opposite direction.

A robin will often return to the same spot each spring to build its nest. It may choose a favorite tree or the eaves of a house. The female robin is responsible for most of the nest building. She uses grass, twine, and twigs solidly held together with mud and then lines the nest with dry grass. There are three to five blue eggs in each set, or clutch, and robins often have two or three clutches each year. When the young birds hatch, both parents help feed them.

male & female cardinals

The Brilliant Cardinal

Is the cardinal singing about himself when he sings "purty, purty, purty, purty"? He is a gorgeous bird that is a little smaller than a robin, with brilliant red feathers and an impressive feathered crest, or cap, on his head. Would you know the female cardinal if you saw her? She does not share the bright plumage of the male but is a soft brown color. Look for a few touches of red among her feathers, and notice her crest. She and her mate both have bright red or orange bills. Both male and female cardinals are enthusiastic singers. In the wild, they visit thickets, woodlands, and swamps. Weed seeds, wild fruit, and insects make up the main part of their diet; but they will visit a bird feeder supplied with sunflower seeds, one of their favorite foods.

The Talented Mockingbird

As you sit in your favorite outdoor spot, you think you hear a choir of birds. There must be fifteen different kinds that are taking turns singing. You look closely and discover that the choir is made up of one bird. You have been listening to a mockingbird. This cheerful songster is about the size of a robin but more slender and with a longer tail. It is not especially colorful in its gray suit with a few white patches, but its song is amazing. It sings day and night and at almost any time of the year. Although its own song is nice, the mockingbird also enjoys mimicking the songs of other birds and the noises made by cats, dogs, cars, and people. Since it eats many harmful insects, the mockingbird is helpful as well as entertaining. You will find this bird throughout most of the United States except in the Northwest and in a few parts of southern Canada.

mockingbird

The Sassy Blue Jay

The noisy, bossy blue jay thinks it rules the backyard. It takes first place at the bird feeder and goes to the head of the line at the birdbath. Its song is loud but not very musical. Its large size, its bright blue plumage marked with black and white, and its blue crest catch your eye. Both male and female are beautifully colored.

The blue jay is found primarily in the East, and the Steller's jay lives in the West. This western jay is bluish gray with a crested black head and no white in its plumage.

Jays are members of the crow family, and many people who have studied birds believe that jays and other members of the crow family are the most intelligent of all the birds. Like mockingbirds, blue jays are able to mimic the songs of other birds. They are also very alert and warn other birds if they see a predator come near. Perhaps the smaller songbirds put up with the blue jay's bossy ways because they appreciate its alarm system. The blue jay eats just about anything—nuts, seeds, fruit, insects, small fish, and small amphibians, such as frogs and toads. Acorns are probably its favorite food. Jays have a habit of storing acorns by planting them in the ground. Sometimes, the jay fails to return for its acorns, and they grow into beautiful oak trees.

The Tiny Hummingbird

The ruby-throated hummingbird is only 3½ inches long and weighs about as much as a penny. The hum you hear comes from the vibration of its wings, which beat about fifty times per second. The hummingbird is in constant motion, and it is the only bird that can fly backward. This bird, which might be mistaken for a large moth, is the only type of hummingbird found throughout the Midwest and Eastern parts of Canada and the United States. There are many kinds of hummingbirds found in the West, among them are the broad-tailed and the rufous.

Hummingbirds are attracted to water and to flowers, especially red flowers. Their long, needle-like bills are perfectly designed for sipping nectar, and they are the chief pollinators of plants whose flowers are too long and narrow for insects to reach into. Hummingbirds also snatch tiny insects from flowers or in midair and will rob a spider's web for food.

You might walk by a hummingbird's nest and not know it because it is about the size of a golf ball, and the tiny eggs are pea-sized. A newly hatched hummingbird is no larger than a bee. The nest is made of plant down, or soft, short hairs, and spider webs. It is also camouflaged by bits of moss and lichen that the hummingbird has stuck to the outside.

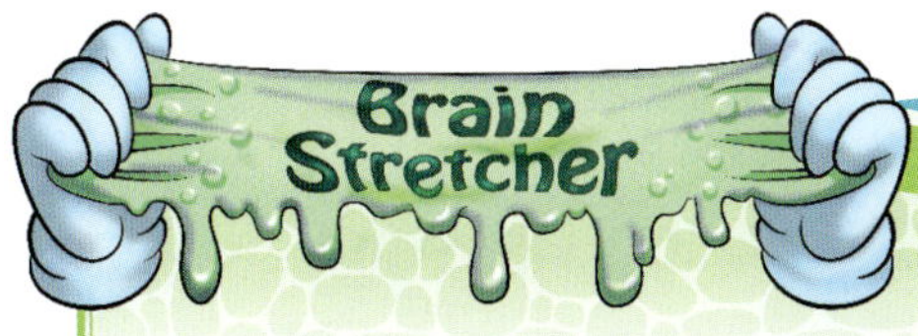

Bird Calls and Songs: Almost every kind of bird in the world has a voice that it uses to communicate with other birds. Most kinds of birds can call. A call is a single note. A baby bird calls to let its parents know when it is hungry, injured, or frightened. An adult bird calls so that its mate will recognize it. Birds also call to warn each other of danger and to let each other know that food is available. Each individual bird has its own distinctive voice, just as each person does.

Some kinds of birds sing as well as call. A bird's song is a unique pattern of notes. In most cases, only the male bird sings. Most birds are very territorial. Males use their song to establish their territory and to attract a mate.

Birds are able to call, or sing, because of a special structure in the windpipe that is called the syrinx. The syrinx is the bird's voice box. It is located at the lower end of the windpipe, just before the windpipe branches and enters the lungs. Air coming out of the lungs passes over membranes inside the syrinx, vibrating the membranes and producing sound. The pitch and volume of the sound are determined by muscles that control these membranes.

Identify, research, and observe your state bird.

Materials needed:
- ✓ field guides and research sources
- ✓ sketching materials
- ✓ *camera

1. Every U.S. state has chosen a bird as one of its state emblems. Identify your state bird by using the field guide in the back of this book. (See pages 530–540.)
2. Research why your state chose that bird as an emblem and if any other state has the same state bird. Discover what your state bird eats and if it has any predators. Determine what habitat and range it can be found in.
3. Try to find your state bird in your own community. You may choose to photograph it or draw it in its habitat.

*optional

Comprehension Check 5.4

Label: *Name these backyard birds. Can you think of something you have learned about each one?*

1. ____________________

2. ____________________

3. ____________________

4. ____________________

5. ____________________

6. ____________________

incubate: to keep an egg warm until it hatches

extinct animal: an animal that no longer exists

Birds of the World

The Peculiar Penguin

When you think of the penguin, what comes to mind? Are you thinking of the snow and ice of the Antarctic? Would it surprise you to know that some tropical penguins live close to the equator? There are approximately seventeen species of the penguin, most of them living in the Southern Hemisphere. Their size, field marks, and preferred range distinguishes them from each other.

Penguins

Penguins have two very important traits in common—they are all excellent swimmers and cannot fly. In fact, penguins spend most of their time in the ocean, using their wings as flippers. Their streamlined body design helps them to dive and swim while they look for their favorite food—fish. Because they were not designed to fly, God gave them heavier, more solid bones than other birds. This helps them dive deeply; some penguins can dive down over 1,500 feet. They are camouflaged from predators while swimming. Predators such as seals, looking for prey from above, cannot see the penguins because of the dark coloring on their backs. Sharks, predators looking for prey from below, have a hard time seeing penguins because of their white bellies. Penguins use their short legs and tail feathers like a rudder, helping them to steer.

A penguin is able to maneuver like a dolphin because it is an excellent swimmer. It builds up enough moving energy to leap out of the water over the waves and dive back into the water. This leaping is called porpoising. Penguins can also use their body design for "tobogganing," or sliding across icy surfaces.

Though penguins cannot fly, they do have feathers. Their feathers are so densely placed that they almost look like fur. This helps them to be insulated and "waterproof," keeping their bodies at the right temperature. Penguins only come on land to nest and lay eggs. *Penguins choose to nest on islands where they are safest from predators.* This is why we see penguins visiting Antarctica. In freezing temperatures, both emperor penguin parents take turns **incubating** their one egg *to keep it warm*. When the penguin eggs hatch and the babies have developed adult feathers, they follow their parents to the water and learn to fish and swim.

Because all penguins have only one or two babies each year, some penguin species are more likely to become extinct. *An* **extinct animal** *is an animal that no longer exists.* Animals can become extinct for several reasons. Loss of a habitat or food supply can make it more difficult for a species to replenish its numbers by having many babies every year. To help prevent extinction, it is illegal to hunt penguins. This protection may give some penguin species a chance to increase in population again.

The Outstanding Ostrich

The ostrich is the largest bird in the world. Some ostriches can measure up to nine feet tall and can weigh over 300 pounds. Their body design was not intended for flight but for powerful speed. At more than forty miles per hour, an ostrich can outrun many predators.

The ostrich needs speed where it lives in the open grasslands of African savannas. Grazing animals want ostriches around them because they raise the alarm when predators are near. At a distance, this alarm can sound like a lion's roar.

One of the reasons the ostrich can run so quickly is its two-toed foot design. Other birds have three or four toes for perching. If an ostrich needs to defend itself, it uses powerful legs to kick. An ostrich kick has enough force to seriously hurt a predator.

Some people think the ostrich buries its head in the sand when it is afraid, but this is not true. If it cannot outrun a predator, an ostrich may choose to lie down with its neck stretched out on the ground. This body position helps it blend into the savanna surroundings.

The Perky Parrot

The parrot family of birds includes macaws, lories, and cockatoo species. These birds are often brightly colored. *They have four toes on each foot for perching.* A parrot can use its claws for grasping while it climbs the very tall trees of a rainforest. Most types of parrots live in tropical climates, but there are some parrots that live in cold climates where there are forests or mountains.

Parrots are most known for their ability to "talk." Some species can *mimic human sounds.* Perhaps you know someone who has a parrot or a cockatoo and have enjoyed observing the bird's perky personality. Though a parrot can mimic words, it does not have the gift of language that God gave to you. You are able to think original thoughts in your own language and create your own sentences. Parrots who mimic are copying what they hear—whether it be a person, another bird, or even a car alarm. They do not understand the meaning of what they mimic.

The Famous Flamingo

The flamingo belongs to a family of birds that includes the heron and the egret. All of these birds are coastal birds. Though their beak and bill designs differ slightly, *they have one very important thing in common—their long, wading legs.* Some flamingos may be as tall as you! There are six species of flamingos; they live in different parts of the world, such as Africa, Asia, and Europe.

A flamingo uses its feet to stir up the water and mud it wades in. Then it uses its unique mouth structures to scoop up some mud while filtering out food. Did you know the flamingo gets its famous pink color from the food it eats? Plankton, some algae, and shrimp have beta carotene which can give their consumer a red-orange coloring. If a flamingo is raised in captivity and does not eat this special diet, it loses its color and is simply a dull white. In fact, when flamingo chicks hatch, they have white or gray feathers. Their mother feeds them a type of "milk" she makes in her throat. This milk is very different from the milk mammal mothers make for their babies.

Comprehension Check 5.5

Describe the habitat where each bird can be found.

1. penguin

2. ostrich

3. flamingo

4. parrot

Give the correct answer: *Beside each bird, write one of its amazing abilities or unique design features.*

5. penguin

6. ostrich

7. parrot

8. flamingo

Think and Predict.

9. Each year, if an ostrich lays twenty eggs and a penguin lays one egg, which species would be more likely to become extinct? What other factors can cause extinction?

5.6 Bird Feeders and Birdbaths

There are several things that you can do to attract birds to your backyard. If you do some or all of these things, you will be able to observe many kinds of birds at close range, and you will also be helping the birds by providing things that they need.

One of the most enjoyable and helpful activities is feeding birds, especially during the winter months when it is hard for many birds to find food on their own. If you supply a variety of foods, you will be rewarded with your own backyard bird show.

Your family may already have a bird feeder. If so, you may want to take the responsibility of checking it regularly. Make sure that there is plenty of food in the feeder.

When to Feed Birds

You should begin feeding early in the fall so that birds can learn where to find your food. Once you have begun feeding birds for the winter, it is important that you continue to do it faithfully. Late winter is a dangerous time for birds that have become dependent on you for food. There may not be enough wild food to feed the number of birds that stayed in the area because you began to feed them. Once birds discover that you are supplying food for them, they usually stop searching for other food in the area. Whenever there is a heavy snow or ice storm, it is very important that you check the food in the feeder so that the birds do not go hungry or even starve. You may continue limited feeding through the spring and summer. This may give you the chance to get very near to your visitors and observe their habits at close range.

Simple Feeders

Many birds will feed from a shallow container or tray filled with birdseed. When placed on a platform, such as a picnic table, bench, or windowsill, it will attract many kinds of birds. Be sure to make holes in the bottom of your container. The holes allow moisture to drain off, keeping the food from becoming soaked and spoiled. Sunflower and safflower seeds are favorites of cardinals, chickadees, wrens, and grosbeaks.

You may also want to try filling an empty grapefruit rind with small chunks of citrus fruit, apples, and berries. Place it on your platform, and you may be rewarded with the appearance of orioles, thrushes, thrashers, and warblers.

Some birds prefer to feed on the ground. Sparrows, mourning doves, juncos, and bobwhites will come to a cleared spot protected by a few trees. They enjoy sunflower seeds, millet, and cracked corn.

You can purchase several types of feeders. Certain types of feeders attract specific birds. Place your feeder where you will be able to enjoy seeing many birds up close. Chickadees, nuthatches, cardinals, and woodpeckers will feed quite closely to humans. Be careful not to startle them by any sudden moves or loud noises.

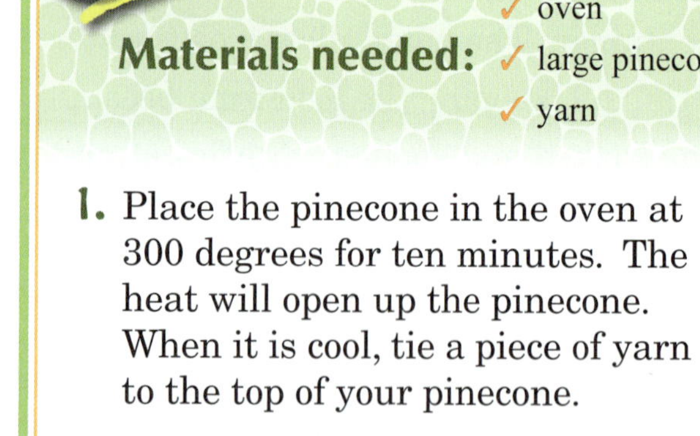

Make a suet feeder.

Materials needed:

- ✓ oven
- ✓ large pinecone
- ✓ yarn
- ✓ birdseed
- ✓ bowl
- ✓ craft stick
- ✓ suet

1. Place the pinecone in the oven at 300 degrees for ten minutes. The heat will open up the pinecone. When it is cool, tie a piece of yarn to the top of your pinecone.
2. Using a craft stick, place your suet, animal or plant fat, inside the crevices of your pinecone.
3. Roll your pinecone in birdseed and tie it to a tree branch.
4. Enjoy watching woodpeckers, chickadees, wrens, and nuthatches come for a little snack. Check your pinecone from time to time and replenish seeds as needed.

Birdbaths

Water attracts birds just as much as food does. Birds need water to drink and to keep their feathers in good condition. Their feathers will not work correctly in flight or serve as insulation if they dry out.

If you already have a birdbath, you could make it your job to keep it full of clean water. If you do not have a birdbath, you can make a simple one from the lid of a plastic garbage can. Avoid using a metal lid; it will become too hot for the birds. Choose a place that is fairly open with some trees nearby. Birds want to bathe in areas that are out in the open so that predators cannot sneak up on them. They also want some cover nearby to escape to if they feel threatened. Birds do not feel secure when their feathers are wet. After you choose the location, turn the lid upside down, surrounding it with stones to keep it in place. An even simpler birdbath can be made by filling an old glass baking dish with one to two inches of water and setting it out on a platform.

Birds love moving water. You may see some birds flying through the sprays of water shot out by your lawn sprinkler. Try hanging a dripping bucket over your birdbath. An ordinary wooden or plastic bucket with a very tiny nail hole in it can be filled with water and hung from a tree limb. The sound and sight of dripping water will attract birds to your birdbath.

As you care for your backyard birds, think of the ways your Heavenly Father cares for you. As a person, you are His most treasured creation. Give Him praise for His loving care for you!

Consider the ravens: for they neither sow nor reap; which neither have storehouse nor barn; and God feedeth them: how much more are ye better than the fowls? *Luke 12:24*

Try This! Build a birdhouse.

Check your local library or home improvement store for birdhouse crafts. You will find complete instructions and a list of materials for the birdhouse you would like to build. Here are a few guidelines to follow:

- ✓ Wood is the best material; do not use metal.
- ✓ Birds prefer rough surfaces on the inside.
- ✓ Do not add a perch to your birdhouse. Most birds do not need them, and the perch could make it easier for a predator to rob the nest.
- ✓ Each fall, clean out your birdhouse, take it down, and store it with the entrance hole plugged.
- ✓ Have your birdhouse ready by mid-February in southern regions; mid-March in northern regions.
- ✓ *Most birds choose a territory and defend it against other birds of the same species.* Avoid placing several birdhouses for the same kind of bird in one area.

Notice these things as you observe your birdhouse or a nesting bird in a backyard tree:

- ✓ Do both parents incubate the eggs, or is this the job of just the female?
- ✓ Does the male bring food to his mate?
- ✓ How often does the female leave the nest?
- ✓ How long does it take for the eggs to hatch?
- ✓ Do both parents care for the babies?
- ✓ What kinds of sounds do the baby birds make?
- ✓ What happens when the babies leave the nest?

If you are successful in attracting birds to your yard, you will learn much about a special part of God's creation.

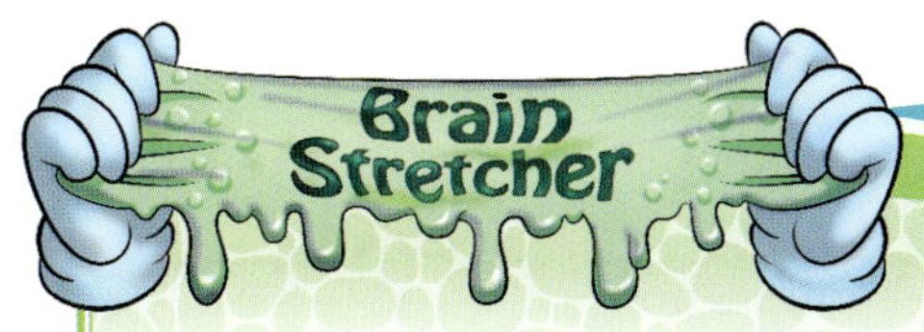

Thermal Energy and Birdhouses: When you build a birdhouse or a birdbath, use a material that is safe for the birds. Why is metal an unsafe material?

Thermal energy can transfer from one object to another. Heat energy always travels from the warmer space to the cooler space. The speed that heat travels depends upon the kind of material it is traveling through. Remember, some materials are good heat conductors while others are poor heat conductors. *Any kind of metal is a good conductor of heat.*

How does thermal energy affect the birdhouse you might build? Since metal is a good conductor, a metal birdhouse will quickly absorb heat energy from the sun. Heat from the metal transfers to the cooler air inside the birdhouse, making the air too hot for the birds. *Because wood is a poor conductor of heat energy, it is an ideal material to use for a birdhouse.*

Think and Conclude.

1. What materials should you avoid when making a birdhouse or a birdbath? Why? metal, heat ______________________________

Think and Predict.

2. What might happen if you stop feeding a bird midwinter? ______________ it dies ______________________________

exoskeleton: an outside skeleton

molting: the shedding of an exoskeleton, skin, or other body covering

antennae: "feelers," or sense organs, on the heads of some invertebrates (antenna—*plural*)

segment: a body section

setae: tiny sharp projections on an earthworm

jet propulsion: a special way some invertebrates have been engineered to maneuver in water

5.7 Interesting Invertebrates

Invertebrates are animals without backbones. If you study the circle graph, you will see that there are many more invertebrate species than vertebrate species in the world. There is such a large variety of invertebrates, such as jellyfish, starfish, squid, octopuses, clams, oysters, snails, spiders, and worms. What about insects? Because they are animals without backbones, all insects are invertebrates, too. Each invertebrate is uniquely designed and equipped for its life.

Invertebrates are usually small animals with soft bodies and weak muscles. Most invertebrates have an *outside skeleton* called an **exoskeleton**. An exoskeleton acts like a hard coat of armor that some invertebrates need for strength and protection.

Animals of the World

vertebrates
2%
invertebrates
98%

These percentages are approximate. As technology improves, scientists continue to identify new species.

As you grow, your bones grow, but an exoskeleton cannot grow with the invertebrate inside it. When the invertebrate outgrows its exoskeleton, it crawls out of the old skeleton and grows a new exoskeleton. *Shedding the exoskeleton* is called **molting**.

On a warm day, invertebrates can move very quickly; but on a cold day, they move very slowly. This is because *all invertebrates are cold-blooded* and become the same temperature as their surroundings. If the temperature becomes too hot or too cold, they may die.

Vertebrates feel with their skin, but invertebrates feel using different body parts. Some invertebrates feel with little hairs on their legs or feet; others feel with "feelers" called antennae. **Antennae** *are sense organs* that help some invertebrates taste, smell, touch, maintain balance, or find direction. *Invertebrates are able to respond to their environment using these and other sense organs.* This unique design helps them to survive.

Invertebrate Characteristics

- ✓ no backbone
- ✓ may have an exoskeleton
- ✓ usually have weak muscles
- ✓ usually are small animals with soft bodies
- ✓ cold-blooded

Scientists classify invertebrates in the same way they classify vertebrates, according to characteristics. They look for clues to answer questions, such as

- ✓ Does this animal have legs? How many?
- ✓ How many main body parts does this animal have?
- ✓ Does it have body segments?
- ✓ Does it have an exoskeleton or shell?
- ✓ What kind of habitat does it need?
- ✓ Does it have stinging cells?

longhorn beetle

Segmented Worms

Some soft-bodied invertebrates have repeated *body sections* called **segments**. If you examine an earthworm, you can see the many segments that make up its body. A leech has a similar design. Earthworms and leeches do not have legs or an exoskeleton.

God designed earthworms to help the soil. An earthworm's long tunnel, or burrow, allows air and water to get into the soil. As the worms tunnel, they make humus by eating tiny pieces of dead plants. Do you remember what we call organisms that break down dead and decaying things to help the soil? Earthworms are valuable decomposers.

If you are birding in the early morning or after a rain, you will see birds finding worms for breakfast. Though some birds depend on worms for food, God has given earthworms a special protection so that there will always be enough worms to help the soil. If you feel the sides of an earthworm, you will notice *tiny sharp projections* called **setae**. Setae help the earthworm hold tightly to the wall of its burrow. When a bird is pulling a worm from the ground, the worm will sometimes break in half. The bird will eat the half it pulled out, but the other half stays in the ground. Because of its segmented design, the worm can grow back the missing part.

Unsegmented Slugs, Snails, and Sea Creatures

Octopuses, squids, mussels, snails, and slugs also have soft bodies, but their bodies are not segmented. Most of the time, they live in very damp or aquatic habitats. Some of these invertebrates have a shell for protection.

Slugs and Snails

Perhaps you have seen a slug's silvery trail across your sidewalk. If you observe a slug, you will see that its body design is very different from an earthworm. Notice first that a *slug does not have segmented body sections.* A slug has *tentacles on its head* that it uses to see and sense things. It is able to chew with over 25,000 teeth, making it a good decomposer. However, if a garden has too many slugs, the slugs will become garden pests, eating too many living plants.

Like a slug, a snail has a similar body design, but it has a *spiral* shell for protection. Many of its vital organs are contained in its shell. A snail needs this extra protection because it is one of the slowest-moving animals on Earth. God designed some snails with lungs, while others have gills. Most snails are rather small, but the largest sea snail can be over thirty-five inches long.

Octopuses

Long ago, one of the most feared sea creatures was the mysterious octopus. If you study octopuses, you will see they are fascinating but not usually scary. *An octopus has eight long tentacles*, each covered with many round muscles that act like suction cups. The octopus uses its tentacles to swim, to move about on rocks and hard surfaces, and to catch its prey. After the octopus has made its catch, it cuts up its prey with the hard beak in its mouth. *Octopuses can secrete venomous saliva to paralyze prey;* some octopuses are more venomous than others.

In spite of its fierce reputation, the octopus is actually a shy creature that would rather flee than fight. When threatened, it changes colors, squirts an inky "smoke screen," and uses jet propulsion to quickly dart away from its predator. **Jet propulsion** *is a special way God has engineered some invertebrates to maneuver in water.* If the octopus is attacked and loses one of its tentacles, the tentacle can grow back.

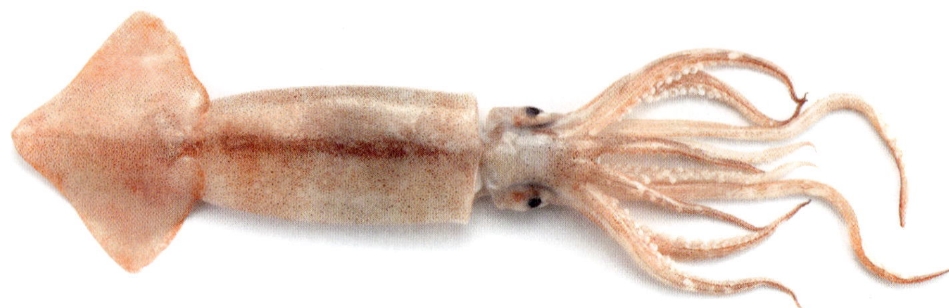

Squids

The squid is an invertebrate like the octopus, *but it has ten tentacles instead of eight.* Squids are more aggressive than octopuses, although they defend themselves in a similar way. When threatened, the squid ejects a blob of blue-black ink. Then, while the predator attacks the ink blob, the squid changes its color and escapes using jet propulsion. Squids, as well as octopuses, are able to change into a wide variety of colors to blend in with their surroundings.

Comprehension Check 5.7

Matching: *Write the letter of the correct answer in the blank.*

B 1. an armor-like protection for some invertebrates; an outside skeleton

A 2. a small, soft, segmented invertebrate equipped with setae

D 3. a small, unsegmented invertebrate equipped with tentacles on its head

G 4. what an octopus can secrete in its saliva to paralyze its prey

C 5. how some invertebrates can maneuver in water

F 6. a sea invertebrate with ten tentacles

A. earthworm
B. exoskeleton
C. jet propulsion
D. octopus
E. slug
F. squid
G. venom

Think and Predict.

7. What would happen if the earthworm population did not have setae? it world not be to proet it self

Try This! Observe jet propulsion.

Materials needed:

- ✓ balloon
- ✓ clothespin
- ✓ plastic toy truck
- ✓ tape

The octopus has no fins for swimming. Instead, God has equipped it with a special means of moving through the water—jet propulsion. You can see how jet propulsion works by making a jet-propelled toy.

1. Blow up a long balloon and clip it closed with a clothespin. Tape the balloon to a lightweight plastic truck with the clothespin end of the balloon facing the rear of the truck.
2. Remove the clothespin and notice the direction that the truck moves. As the air escapes the balloon, the truck moves forward in a direction opposite the escaping air.

Jet airplanes work on the same principle by moving forward when hot gases escape through the rear of the engines. The octopus uses water rather than air for jet propulsion. Water is forced through a tube under the octopus's head. The water escapes forward, forcing the octopus to move through the water backward.

colony: a group of animals living together

symbiotic relationship: two different organisms that live closely together and benefit from each other

arthropod: an invertebrate having jointed legs

5.8 Unusual Invertebrates

Stinging Sea Creatures

Invertebrates, such as jellyfish, corals, and sea anemones, have stinging cells for protection. The following invertebrates have tubular, stinging tentacles.

Corals

In warm, shallow seas, tiny invertebrates called corals *live as a group,* or **colony**. Each coral organism, or polyp, is tiny and soft-bodied with a limestone skeleton. As each polyp dies, its limestone skeleton hardens; then more living corals grow on top, forming a reef. Many ocean animals depend on the reef for food and protection.

fire coral

stinging hydroid

Jellyfish

There are many types of jellyfish. Jellyfish are *invertebrates whose bodies are shaped like bells or upside-down bowls.* All jellyfish *contain stinging cells in their tentacles* that hang down from their bodies. The stinging cells are like tiny harpoons. When jellyfish are touched by fish or other small marine animals, they uncoil and shoot a stinging venom into their prey. The tentacles then pass the prey up to the jellyfish's mouth.

One jellyfish found in Australia, the sea wasp, is more venomous than any snake. The sea wasp's venom is so potent, or strong, that it can kill a person in three to eight minutes.

Some jellyfish are the size of peas, but the huge Arctic jellyfish can be seven feet across. You should avoid picking up a jellyfish with your bare hands even if it is dead because the tentacles can still sting.

Sea Anemones

Some ocean animals are very mysterious. The sea anemone looks more like a plant than an animal. In fact, when discovered, it was named for the anemone flower. The petal-like projections surrounding the mouth are actually stinging tentacles. Anemones attach themselves to rocks or a coral reef. Because the anemone cannot chase its prey, it uses its tentacles to release venom that paralyzes food that passes by.

There are over one thousand species of sea anemones. They can be found all over the ocean, though the most common species live in the warm coral seas. An anemone can live up to eighty years or until a predator eats it.

Clownfish take shelter in anemones. They are not hurt by the anemone because they have a layer of protective mucus on their scales. The anemone helps the clownfish by keeping predators away; the clownfish helps the anemone by bringing it food that it cannot catch. *When two different organisms live closely together and depend on each other*, it is called a **symbiotic relationship**. Both living things *benefit from each other.*

clownfish and anemone

Spiny Sea Creatures

Some invertebrates, such as sea urchins and starfish, are covered with pointy spines.

Starfish

Starfish walk on special structures called tube feet. They are able to open clams with their tube feet and eat with their mouth, which is located on their underside.

starfish

A starfish has at least five arms, but some starfish have as many as forty. If a starfish loses one of its arms, it can grow a new one. Many years ago, before scientists learned this, fishermen used to cut starfish up and throw them into the sea. They wanted to kill the starfish to keep them from eating the clams and oysters that they wanted to catch and sell. But instead of killing the starfish, the fishermen found that many of the parts grew into new starfish.

Sea Urchins

Sea urchins are also covered with spines, but their spines are much longer than a starfish's. Instead of having a flat body with arms, sea urchins have a round body.

sea urchin

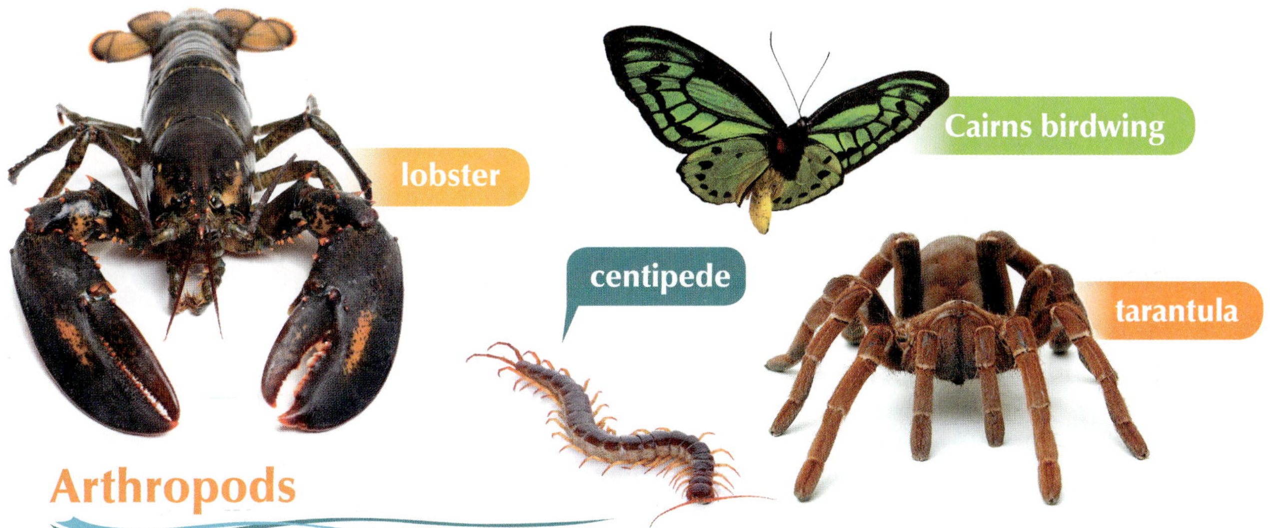

Arthropods

One very large class of invertebrates is arthropods. *An* **arthropod** *is an invertebrate with jointed legs.* Arthropods get their name from the Greek words *anthron*, meaning "joint," and *pous*, meaning "foot." Look at your elbow and wrist joints. Do you see how they bend? Arthropods have jointed legs for walking.

Arthropods also have a segmented body and an exoskeleton. An arthropod can be a crustaceous animal, like a crab or lobster. It can also be a multi-legged animal, such as a centipede. Are spiders arthropods? Yes, because spiders have an exoskeleton, segmented body parts, and jointed legs, they are all classified as arthropods. *The largest group of arthropods, however, is insects.* If you look at this circle graph, you can see that there are more species of insects than any other living thing.

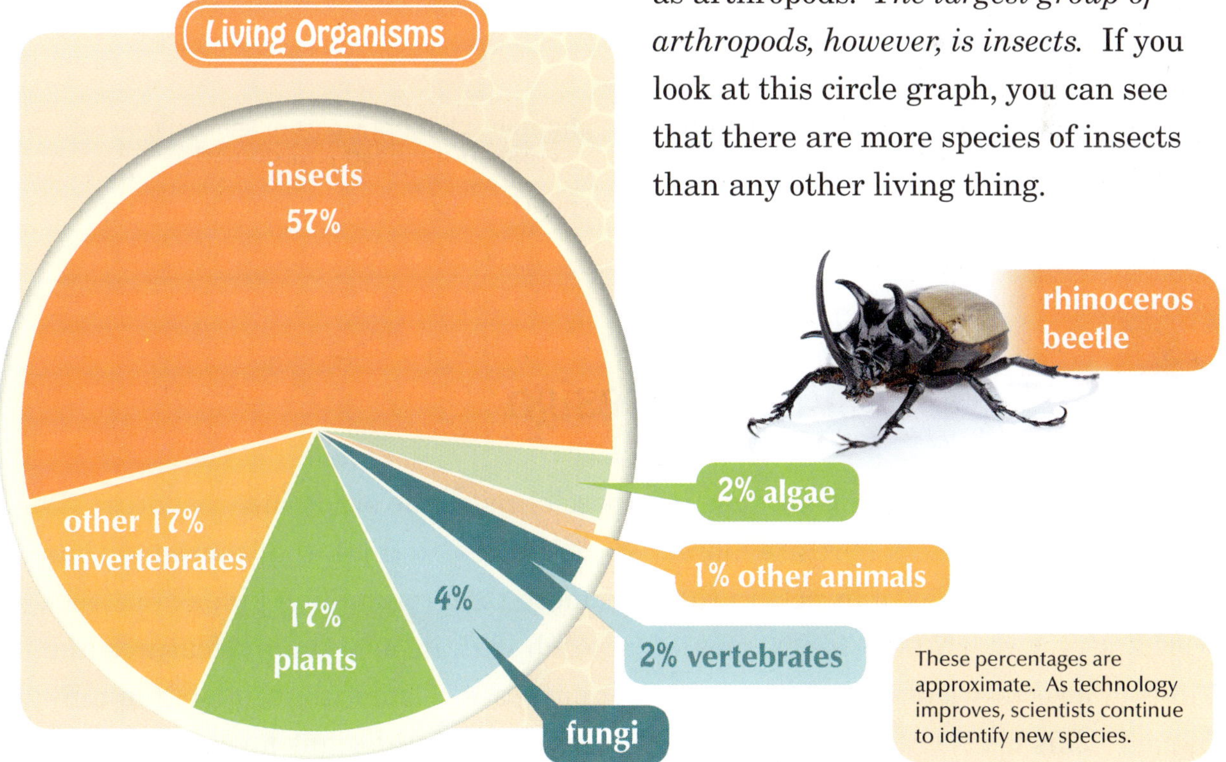

These percentages are approximate. As technology improves, scientists continue to identify new species.

Matching: *Write the letter of the correct answer in the blank.*

A. arthropod
B. colony
C. corals
D. jellyfish
E. starfish
F. symbiotic relationship

______ 1. group of animals living together

______ 2. bowl-shaped body; stinging tentacles

______ 3. two different organisms living closely together that benefit from each other

______ 4. has at least five spiny arms

______ 5. an invertebrate with jointed legs

Think and Conclude.

6. Give an example of a symbiotic relationship found in the sea.

____

____

7. Give an example of an arthropod. ______________________

TERMS

compound eyes: large eyes that are made of many small eyes

thorax: middle body part of an insect where wings and legs are attached

abdomen: last body part of an insect; contains the heart and stomach

spiracles: holes insects use for breathing

5.9 Observing Insect Design

More than half of all the animals in the world are insects. There are more than a million species of insects in the world. Every year 7,000–10,000 new kinds of insects are discovered and named. This leads entomologists, scientists who study insects and similar creatures, to think that there may be more than ten million kinds of insects, each with its own unique features and purposes. Could you think of ten million different designs for insects?

God did. What makes it even more wonderful is that each difference in design is for a special purpose.

All insects have an exoskeleton, three body parts, and six jointed legs. These three characteristics make them different from other arthropods.

Insect Characteristics

- ✓ an exoskeleton
- ✓ three body parts
- ✓ six jointed legs

An insect's three body parts are its head, thorax, and abdomen. Each part serves a different purpose.

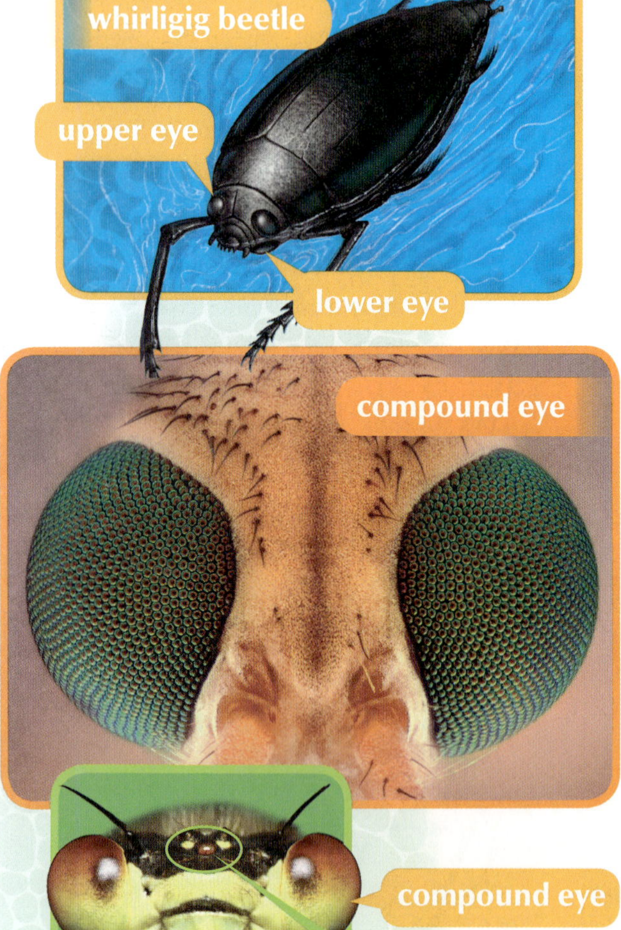

An Insect's Head

God created each kind of insect with a head to fit its needs. The head contains the insect's brain, antennae, eyes, and mouthparts.

Two Antennae

The two antennae are the insect's "feelers." If you look through a hand lens at the antennae of several types of insects, you can see how they are all designed differently. Some antennae are "feathery"; others are long and thread-like. Some insects have knobs at the end of their antennae.

Eyes

Most insects have a pair of **compound eyes**, or *large eyes that are made up of many small eyes.* These eyes enable insects to see in almost all directions at once. Insects usually have several simple eyes between their compound eyes. With these eyes they can see light and movement.

The eyes of insects vary according to their needs. An insect that spends most of its time on the ground feeding on vegetation has smaller eyes than an insect that eats other insects. The dragonfly, one of the fastest insects, has especially large eyes which enable it to catch insects while in flight. The dragonfly can see moving insects eighteen feet away. The whirligig beetle has two pairs of compound eyes. The upper eyes are designed for seeing above water and the lower for seeing under the water. The whirligig can look up and down at the same time!

Mouthparts

God designed a wide variety of mouthparts for insects, depending on the kind of food they eat. Grasshoppers and crickets need a chewing mouth to chew plants. Beetles, whose name means "biters," use their chewing mouths to chew plants and other insects. Termites and cockroaches also have chewing mouthparts.

The long-sucking mouth of the butterfly and moth is specifically designed as a "straw" for reaching the nectar in flowers. When the insect is finished drinking, it can coil up its straw for storage.

Some insects have a piercing-sucking mouth. The weevil's mouth is at the end of a long, slender snout. This design enables the weevil to make deep, narrow holes in fruits, grains, and nuts. The large milkweed bug uses its tube-like mouth to pierce milkweed pods. It injects a special saliva into the pod and sucks out the dissolved contents. Flies and mosquitoes have tiny lances, or spears, around their sucking tube. The male mosquito eats the juices of plants and flowers, but female mosquitoes would rather eat your blood!

beetle's chewing mouth

butterfly's long-sucking mouth

weevil's piercing-sucking mouth

fly's lancing mouth

An Insect's Thorax

An insect's thorax is its motor, or "engine room." The **thorax** is the *middle body part where the wings and legs are attached.*

Six Jointed Legs

Every adult insect has six legs attached to its thorax. Insect legs are marvels of design. When pursued by an enemy, the mole cricket uses its strong forelegs as shovels for burrowing quickly into the ground and making its escape. The forelegs of the mole cricket also have a pair of "scissors" for cutting through small roots in the earth. The backswimmer swims on its back under the water, using its hind legs for oars. Bees have special "baskets" on their legs for collecting pollen and carrying it back to their hive. The grasshopper's very powerful legs enable it to jump as far as thirty-nine inches. The strong legs of a flea allow this creature to jump up to 150 times its own height. The praying mantis, which holds its forelegs in what looks like a position of prayer, is actually holding them ready to grab insects to eat. God has given each insect the kind of legs it needs for the job it has to do.

Wings

In addition to three pairs of legs, most insects also have one or two pairs of wings attached to the thorax. A set of muscles in the insect's thorax moves the entire thorax up and down, moving the wings with the thorax. Other muscles directly attaching the wings to the thorax control the direction of the insect's flight. Some insects have two wings; others have four. A few insects, such as silverfish, springtails, fleas, and lice, have no wings at all.

An Insect's Abdomen

An insect's abdomen contains its vital, or important, organs. An insect could not live without these organs. The **abdomen**, the *last body part, contains the insect's heart and stomach.* Insects do not have lungs or gills. Instead, God has designed them with breathing tubes which carry air containing oxygen to their body tissues. The *air enters the tubes through tiny holes called* **spiracles** that are found along the side of the body.

Try This! Create an insect zoo.

Materials needed:

- ✓ jars
- ✓ moist sand for the bottom of each jar
- ✓ rocks
- ✓ twigs
- ✓ leaves
- ✓ plastic wrap
- ✓ rubber bands
- ✓ timer or stop-watch
- ✓ *field guides

1. Collect several insects for observation. Using the supplies above, create a jar habitat, or terrarium, for each insect. Be sure to provide the right food.
2. Place each insect in its own jar. Cover the jar with plastic wrap, securing it with a rubber band.
3. Label each jar with the name of its insect. Use a field guide if you are not sure of the name.
4. After observing for a day, release your insect zoo creatures outside.

Crickets

- ✓ Crickets are excellent insects to observe. Crickets are *nocturnal*, or active at night. Some evening in late summer or fall, sit outside with your flashlight ready. Listen. If you hear a cricket chirping, move slowly toward the sound.
- ✓ Shine your flashlight around slowly. If you find the cricket, capture it quickly by cupping the palm of your hand over it. Be very gentle. You do not want to harm it as you transfer it to its prepared jar.
- ✓ Observe your cricket eat. What kind of *mouthparts* does a cricket have? Notice the way the mouthparts move. Our mouths work up and down, but the mouthparts of a cricket work from side to side.
- ✓ Observe the way your cricket breathes. You have lungs; the cricket takes in air through the *spiracles* on the sides of its thorax or abdomen. If you have a hand lens, use it to look for the spiracles. Watch for the parts of its body that move as it breathes.
- ✓ Crickets, like all other insects, are *cold-blooded*. Notice the effect this has on your cricket. What happens when the room gets warm? Your cricket will be more lively and chirp more when the air around it is warm. You can even estimate the temperature of the air by counting the number of chirps. Use a stopwatch or timer to count the number of times your cricket chirps in fourteen seconds. Then add forty to the number of chirps. The sum of those numbers will be very close to the temperature on the Fahrenheit scale.

Grasshoppers

- ✓ You may also want to capture a grasshopper for your zoo to compare it with your cricket. When a grasshopper is not flying, its wings meet in a peak on the grasshopper's back like the roof of a house. A cricket's wings lie flat, one wing on top of the other. The cricket's *antennae* are very long and slender; often, they are longer than the cricket's body. Grasshoppers may have long or short antennae. Crickets are busy after dark; grasshoppers are usually quiet at night and active during the day.

*optional

Comprehension Check 5.9

Label: *Write the parts of the insect. Circle the names of the three main body parts.*

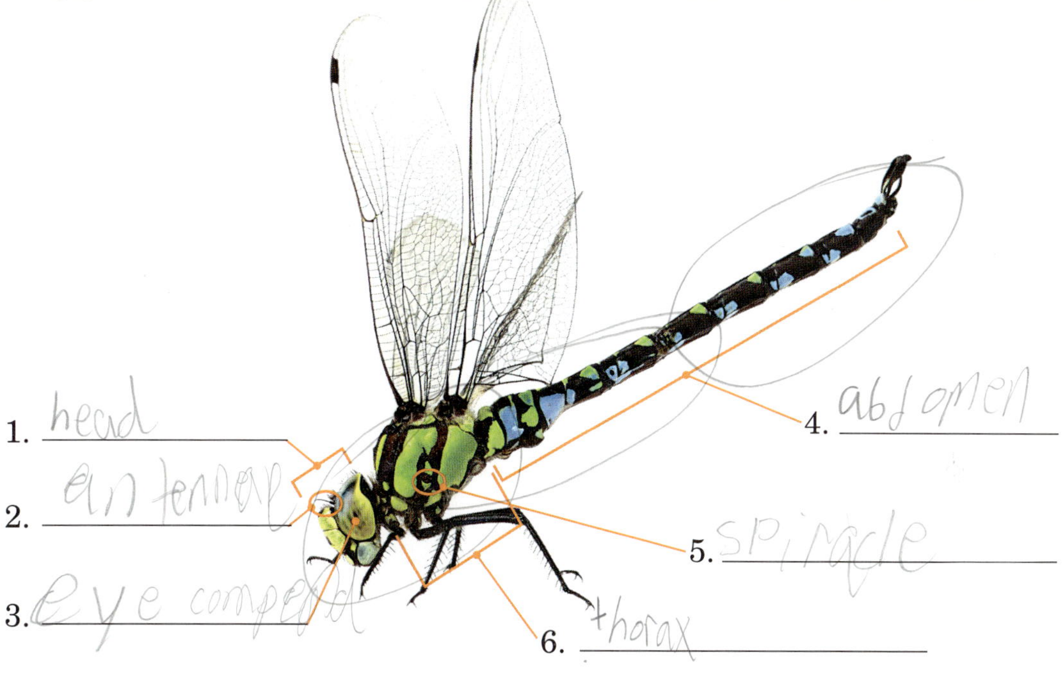

Think and Conclude.

7. Are insects warm-blooded or cold-blooded? cold-blooded

metamorphosis: a change in form

complete metamorphosis: four stages of growth in insects—egg, larva, pupa, adult

incomplete metamorphosis: three stages of growth in insects—egg, nymph, adult

parasite: an organism that attaches itself to another organism and feeds on it

host: the animal or plant on which a parasite feeds

5.10 The Miracle of Metamorphosis

Nearly all insects lay eggs. Many insects go through a number of striking changes before they become adults. We call this *change in form* **metamorphosis**. Some insects have *four stages of growing* called **complete metamorphosis**. Others have only *three stages of growing* called **incomplete metamorphosis**.

Complete Metamorphosis

The four stages of complete metamorphosis are egg, larva, pupa, adult. The **eggs** hatch into larvae. The **larvae** can be either caterpillars, grubs, or maggots. The larvae are always hungry and spend most of their time eating. As they grow, they molt because the exoskeleton cannot grow or stretch.

After molting a number of times, the larvae are ready for a rest. A moth larva spins a cozy little silk case, or cocoon, for itself. The butterfly larva

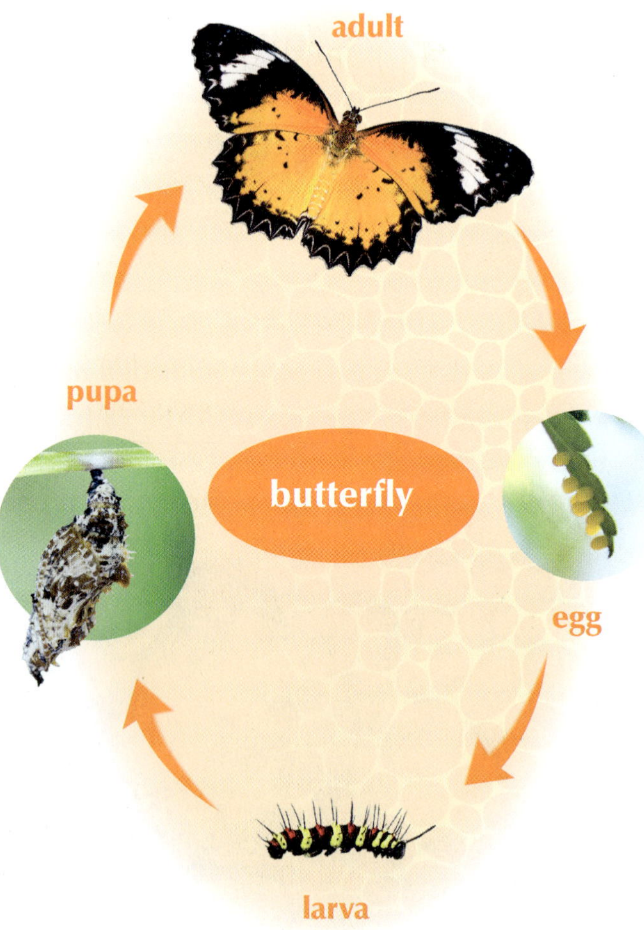

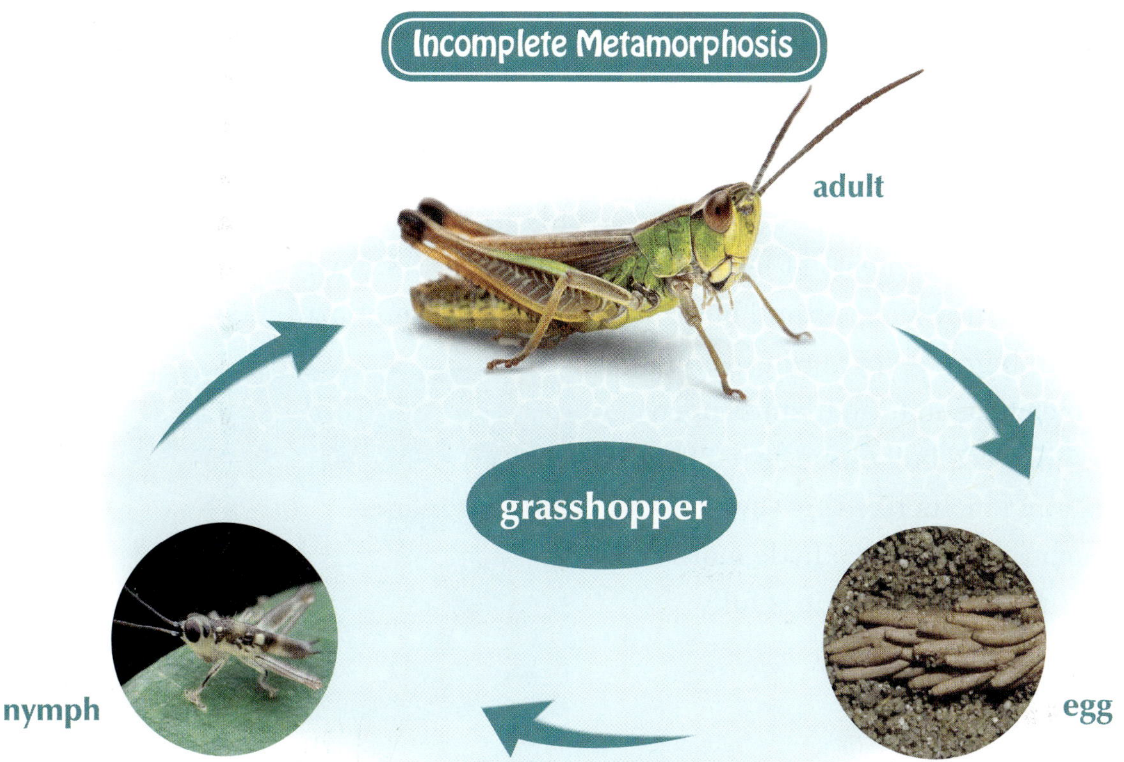

grows a hard shell covering called a chrysalis. Now the insect is a **pupa**. Hidden in its case, the pupa goes through a miraculous transformation.

When the pupa has developed into an **adult**, the case splits open. The adult insect looks very different from the larva that went into the case. The caterpillar has changed into a beautiful butterfly or moth! After resting for a few minutes to allow its wings to straighten and to dry, it flies lightly away. Butterflies and moths are not the only insects that go through complete metamorphosis; beetles, flies, mosquitoes, bees, and wasps do as well. *About 90% of insects undergo complete metamorphosis.*

Incomplete Metamorphosis

Some insects—including crickets, grasshoppers, cockroaches, mantises, dragonflies, cicadas, true bugs, and termites—go through incomplete metamorphosis. *The three stages of incomplete metamorphosis are egg, nymph, and adult.* The eggs hatch into baby insects called nymphs. Nymphs look like miniature versions of the adult insects they will become. As a nymph grows, it molts several times. Once fully grown, it is called an adult.

The large milkweed bug is one of the insects that has three stages of growth. In the spring, when the milkweed-bug egg hatches, out crawls the nymph, looking like a tiny milkweed bug without wings. How it can eat! It feeds on milkweed and grows bigger and bigger. As the nymph grows, its hard exoskeleton becomes too small, so it splits open and the milkweed bug discards it. The milkweed bug molts like this several times, each time looking a little more like an adult, until finally it is fully grown with a pair of wings.

A few insects, such as silverfish and springtails, do not go through either complete or incomplete metamorphosis. They undergo direct development changing only in size by molting several times after they hatch. A newly hatched springtail looks like a miniature adult springtail.

springtail

Brain Stretcher

Incomplete Metamorphosis: Color the graph to show approximately what percent of insects undergoes incomplete metamorphosis.

10% | 10% | 10% | 10% | 10% | 10% | 10% | 10% | 10% | 10%

praying mantis

locust

cicada

Insect Homes and Hatchings

Insects usually lay their eggs on, in, or near a food supply so the young can immediately become hearty eaters. The monarch butterfly always lays her eggs on milkweed leaves because that is the food her offspring needs. When the caterpillars hatch from the eggs, they can begin feeding at once.

Acorn weevils insert their eggs into acorns. The baby weevils feed on the acorns as soon as they hatch. Many kinds of beetles lay their eggs in the bark or wood of trees. The larvae have a safe place to grow and plenty of wood tissue to eat.

The praying mantis finds a suitable branch and secretes foam from her body onto it. She lays her eggs in the foam and then covers the eggs with more foam. The foam hardens into a snug little case, making a safe, warm place for the eggs to stay until they are ready to hatch.

A **parasite** *is an organism that attaches itself to another organism and feeds on it.* A parasitic relationship is a type of symbiotic relationship in which *only one organism benefits*—the parasite. *The animal or plant on which the parasite feeds is called its* **host**. Many insect parasites lay their eggs on other insects. When the eggs hatch, they feed on the insects upon which they were born. The host will be greatly weakened and eventually die. Parasitic wasps that lay their eggs on caterpillars are one example of insect parasites.

Lice are homebodies, traveling only a short distance during their entire lifetime. They are attached to the source of their food, which is an animal's feather or fur or human hair. Is this a symbiotic or a parasitic relationship? *Lice are parasites; some eat their host's blood.* They do not need to travel—all the food they need is right where they live!

Have you ever watched a ladybug crawling along the stem of a plant? The ladybug, or ladybird beetle, was probably hard at work. This insect is one of man's best friends because it preys on harmful insects, such as aphids. Aphids multiply very rapidly and do tremendous harm to plants with their piercing-sucking mouthparts. These garden pests suck sap from plants, taking away the nutrients they need to grow. You can tell when aphids are attacking your garden by paying close attention to your plants. Are the leaves curling or turning yellow? Has one plant stopped growing, while others in your garden are flourishing? These signs, or symptoms, match the effects aphids have on your garden. But don't worry—ladybugs are coming to the rescue. A ladybug will lay her eggs near an aphid colony so that the tiny ladybug larvae have food as soon as they hatch. A ladybug larva can eat up to 400 aphids before it becomes an adult. What a help to farmers and gardeners!

You can find ladybugs all over the world. Most species have a red shell with spots decorating the top. However, some can be yellow or orange and can have fewer or more spots on their shells. A few have stripes on their shells. These plant protectors are smaller than a paper clip, yet they are the most popular predator of aphids. Surely the little ladybug is a wonderful provision of God for man's well-being!

Identify: *What type of metamorphosis is pictured in each life cycle?*

1. *Label the life cycle of a silk moth.*

_______________ Metamorphosis

c. _______________

b. _______________

d. _______________

a. _______________

2. *Label the life cycle of a mole cricket.*

_______________ Metamorphosis

b. _______________

c. _______________

a. _______________

Multiple Choice: *Circle the correct answer.*

3. In a __?__ relationship, only one organism benefits.

   a. metamorphic b. parasitic c. symbiotic

4. A __?__ is the animal or plant on which the parasite feeds.

   a. host b. predator c. prey

camouflage: the ability to blend in with surroundings

mimicry: one animal looking like another animal for protection

5.11 Insect Instincts and Equipment

We see that even the tiniest creatures have complex designs. Insects have ways of sensing things, communicating with each other, and defending themselves. Though a compound eye has a different function from a human eye, it is just as perfectly engineered. What perfect Engineer designed both? God gives each kind of insect instincts and the right equipment for the life He planned for it.

Communication

All insects have ways of communicating with other insects of their own kind. This communication is important for finding food, for defending the nest, and especially for finding a mate.

Sight

Scientists believe that dragonflies can recognize females by their color. The males of the blue dasher, an American dragonfly, are blue and green, while the females are yellow-green with brown stripes. Female fireflies see the flashing lights of males and flash a reply. Each of the 2,000 species of fireflies has its own secret code.

Smell

Because many insects are very small, they are not able to find their mates by sight. Their powerful sense of smell comes to their rescue. When an insect wishes to attract a mate, it usually gives off a chemical from its glands that produces a certain odor. Insects sense odors using their antennae and tiny hair-like structures on other parts of their bodies.

Touch

Some insects, such as ants, produce a chemical but do not release it into the air. In such cases, the insect that wishes to receive a message does so by touching the other insect with its antennae.

Dancing

When honeybees find a good source of food, they come back to the hive and share this information with the other bees by doing a dance. A round dance means the source of food is nearby. A waggle dance, done in a figure eight, means that the food is farther away.

Sound

Although many insects are silent, others relay messages by sounds. You have heard crickets and cicadas singing, haven't you? The males use this sound to attract a mate.

Defense

Most insects are no match in size for the larger animals and birds that prey on them for food. God has given insects ingenious ways to protect themselves from predators.

bombardier beetle

Weapons

The Jerusalem cricket has strong spiked legs. When attacked, it turns over on its back and kicks its enemy. The giant water bug defends itself with its piercing mouthparts. Some beetles bite when frightened. An earwig has two pincers at the end of its abdomen for defense. Bees and wasps protect themselves by the stingers at the ends of their abdomens. Some insects, such as *grasshoppers*, escape enemies by *jumping great distances.* Other insects play dead or use their wings to fly away from danger.

Some insects, when disturbed, secrete a burning acid or spray an irritating gas or liquid on their enemies. These chemicals are made in the insect's body or come from the food the insect eats. A termite soldier shoots a disagreeable liquid out of a horn on the front of its head to discourage enemies from invading the nest. The *bombardier beetle* has an unusual defense. Inside its body, it has two storage spaces for *chemicals.* When an enemy approaches, the chemicals are combined to form a hot, bad-smelling acid which the bombardier beetle blasts from an opening in its abdomen. God has given the bombardier beetle the ability to aim its "cannon" in all directions with great accuracy.

Jerusalem cricket

The *monarch butterfly* and the milkweed bug do not need to hide from their enemies. Their *bad taste* comes from the bitter milkweed on which they feed. Most milkweed eaters are bright orange or red. This coloring serves as a warning to all around them. Because birds and other creatures that eat milkweed eaters become extremely ill, they soon learn to leave milkweed eaters alone.

Fear

Some insects look scarier than they are to intimidate, or frighten, predators. The tiger-swallowtail caterpillar's large eyes make the insect look like a snake. But, don't be fooled. Those startling "eyes" are only patches of color. Its real eyes are quite small and are hidden beneath its monstrous head. After going through metamorphosis, the tiger-swallowtail caterpillar will turn into a beautiful butterfly.

Disguises

As a dead-leaf butterfly sails through the air, a bird spots its bright wings. Swooping down, the bird is about to snatch the butterfly for dinner when it suddenly disappears. The bird is confused. Where has its dinner gone? The butterfly is still there, sitting on a branch, but the bird cannot see it. When it folds its wings and stays very still, it looks like a leaf instead of a butterfly because the undersides of its wings are the same color and shape as a dead leaf. An animal that can *blend in with its surroundings* is using **camouflage**.

The coloring of many moths blends so well with the bark of trees that when the moths flatten themselves against a tree trunk, they cannot be seen. Notice how the green and brown measuring-worm moth of New Guinea blends in with the tree trunk. It becomes almost invisible when it rests on the green lichen that grows on the tree.

Do you know why predators avoid the beautiful monarch butterfly? It is because of its bitter taste. The viceroy looks very much like the monarch butterfly but is a little smaller. You can tell them apart by looking at the hind wings. The hind wings of both butterflies have black borders, but the viceroy's have an extra black line going across the middle. You can tell the difference, but birds cannot. The viceroy resembles the monarch butterfly so closely that birds avoid eating either butterfly. When *one animal looks like another animal for protection*, it is called **mimicry**. The viceroy butterfly mimics the monarch.

Comprehension Check 5.11

True/False: *If the statement is true, write* true. *If the statement is false, replace the underlined word(s) with a word or phrase that will make the statement true. Do not write* false *in any blank.*

__Trνe__ 1. A grasshopper <u>plays dead</u> for protection.

__Trνe__ 2. A bombardier beetle uses a <u>chemical defense</u> for protection.

__bad taste__ 3. A monarch butterfly has a <u>stinger</u> for protection.

__Trνe__ 4. An animal that can blend in with its surroundings is using <u>camouflage</u>.

Think and Conclude.

5. The tiger-swallowtail caterpillar looks very much like a snake. What kind of protective design is this? How does it help the caterpillar?

____

____

5.12 Social Insects

Many insects live alone, such as most beetles, flies, butterflies, moths, and parasitic wasps. Other insects, such as honeybees, ants, termites, and paper wasps, live in communities. Insects that live in communities are called **social insects** because *they live and work together.* Their colonies are like small cities with thousands of inhabitants.

Ant Nests

Ants can live anywhere on Earth, but most ants prefer warm climates. Some ants live in trees, but most burrow underground. Some ants are "cowboys," herding and "milking" aphids for the sweet honeydew they give. Amazon ants capture baby ants from enemy ant nests and use them as workers. In Africa, driver ants travel in large droves and are so fierce that even elephants avoid them.

If you have ever observed ants going in and out of their home, you have probably noticed that they seem to be tireless workers. They are fascinating to observe because of the way they work as a team. If you have ever watched an ant struggle to carry a piece of food twenty times its size, you may have noticed that God gives ants enormous strength. An ant will not give up on the job it has to do. Other ants will often come along to help bring the food down into their tunnels.

God gives social insects, like ants, instincts so that they know what their purpose is and how to accomplish it.

ant nest in tree

anthill

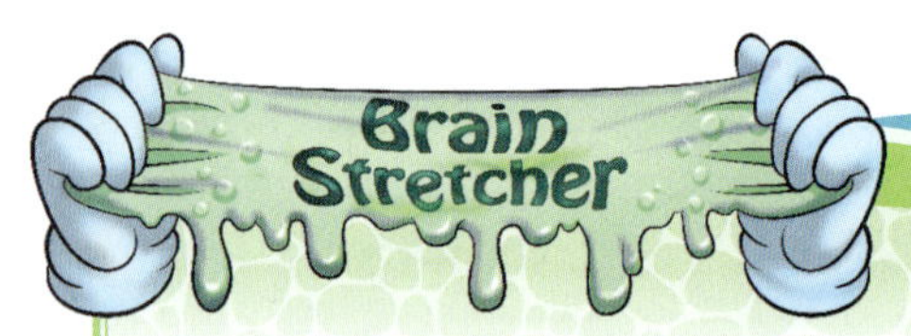

"Go to the ant, thou sluggard; consider her ways, and be wise: which having no guide, overseer, or ruler, provideth her meat in the summer, and gathereth her food in the harvest."

Proverbs 6:6–8

Teamwork: Have you ever been part of a sports team? If so, you learned that one person cannot play the game alone. Each individual member supports the other members. Everyone has the same goal of winning the game. In learning how to work together, the team becomes stronger.

The Bible instructs us to learn from the ant—to work hard without being reminded and to be prepared for the future. Ants were not created to live alone. They live and work in a community with other ants; they are a team. When ants have a job to do, their teamwork instincts tell them to work together until the task is completed. It is what God designed them to do.

How can this apply to you? You are a part of a larger group, or team, whether it is your family, your friends, your class, your church, or even your country. Working together, a group of people is stronger than one person alone. How can you help your "team"? Whose job can you help to finish? Whom can you encourage to keep trying? Do not be the one who stands aside and watches others do the work; be a team player. Consider the ant, and be wise.

Try This! Create an ant farm.

Materials needed:

- ✓ shovel
- ✓ large spoon
- ✓ old newspaper
- ✓ plastic bags
- ✓ jar with holes in lid
- ✓ dark paper
- ✓ tape
- ✓ water
- ✓ hand lens
- ✓ ant food (See below.)
- ✓ fabric piece slightly larger than mouth of jar

1. Dig up some soil and place it on an open newspaper. Look for adult ants, small white ant eggs, and small larvae. Try to find the queen, which you can recognize by her large size.

Because ants are social insects, you will want to gather many of them from the same location and keep them together in the same container. You can find them under rotting logs or rocks.

2. Spoon the ants into plastic bags, and later transfer them to a large jar with soil in it. Cover the jar opening with fabric; then screw on the lid. The fabric will give the ants oxygen without allowing them to escape.

3. Tape dark paper around the jar so the ants will build tunnels close to the glass. Remove the paper when you want to watch them.

4. Feed your ants bits of fruit, vegetables, bread crumbs, and meat. Watch how they use their strong jaws to cut, crush, and chew food. Remove spoiled food from the jar, and remember to keep the soil slightly moist.

5. You will enjoy watching your ants work together to make tunnels. Use a hand lens to observe their three main body parts, six legs, and antennae. Notice their large compound eyes.

Beehives

Because bees are vital, or important, pollinators, farmers and beekeepers usually provide wooden hives as shelter to encourage a variety of species to colonize and pollinate nearby crops.

When you think of bees, you probably think of a honeybee. Honeybees live together in a *hive which is made of many small wax cells.* The queen bee lays her eggs in some of these cells; honey and pollen are stored in the others. Bees take good care of their hive, repairing it and placing guards at its entrance to defend it from enemies.

Paper Nests

Social wasps, commonly called hornets, yellow jackets, or paper wasps, make paper nests. Some people think that the Chinese invented paper after observing the habits of these amazing insects. A social-wasp queen gathers *wood fiber from dead trees or posts, mixes it with water by chewing it well*, and forms it into sheets for her nest. She then fills the nest with six-sided cells that she builds from the same material. In these cells, she lays her eggs, which hatch into grubs, or larvae. The grubs spin cocoons where they grow and develop into adults. Do you remember what kind of metamorphosis this is? Because a larva hatches from an egg, this is complete metamorphosis.

With more workers, the queen is able to lay more eggs and raise more offspring. The workers gather wood fiber and water to build more paper walls and cells so the nest will continue to grow. Long before people learned how to air-condition houses and cars, wasps had "air-conditioned" nests. The wasps beat their wings at high speed to increase evaporation of the water that was used to build the nest. This keeps the nest cool. Some tropical wasps put small pieces of mica, a transparent mineral, in their walls as "windows" to allow light into the hive.

bald-faced hornet and nest

Termite Mounds and Tunnels

It is amazing what God's tiny insects can do when they all work together in harmony. Consider the hard-working termites. Their homes are usually found in dry wood above ground or in moist, rotting wood underground. Rarely venturing into daylight, termites build their houses from the inside. Whenever possible, they tunnel their way through rotten stumps or logs into wooden buildings or fences, eating as they go. Wood must be as delicious to termites as chocolate ice cream is to you! The termites cannot digest the wood alone; they have tiny organisms inside their intestines that do this job for them. This symbiotic relationship is a mystery to many scientists. Evolutionists cannot explain how the tiny organisms could have survived before the termite would have "evolved." We know, of course, that God made and planned each of these animals. Each animal was ready to accomplish its purpose on the day God created it.

Some species of African termites build waterproof mounds, or nests, of earth, debris, and saliva. The mounds can be twenty-five to thirty feet tall and as hard as rock. Some large mounds have been used in building roads, tennis courts, and bricks. The large queen stays in her own "bedroom" in this termite nest where she lays about 30,000 eggs per day for up to 50 years!

The nursery in these mounds has just the right amount of moisture and temperature whether in the cool of the night or the blazing African sun.

What do the diverse designs found in the animal world show you about their Engineer? Every creature, great and small, glorifies God. From a tiny insect buzzing on your window pane to a bird flying across the sky—each one shows it was miraculously engineered by God. As we understand each animal, we see that their intricate designs prepare them for the purpose and life God has for them.

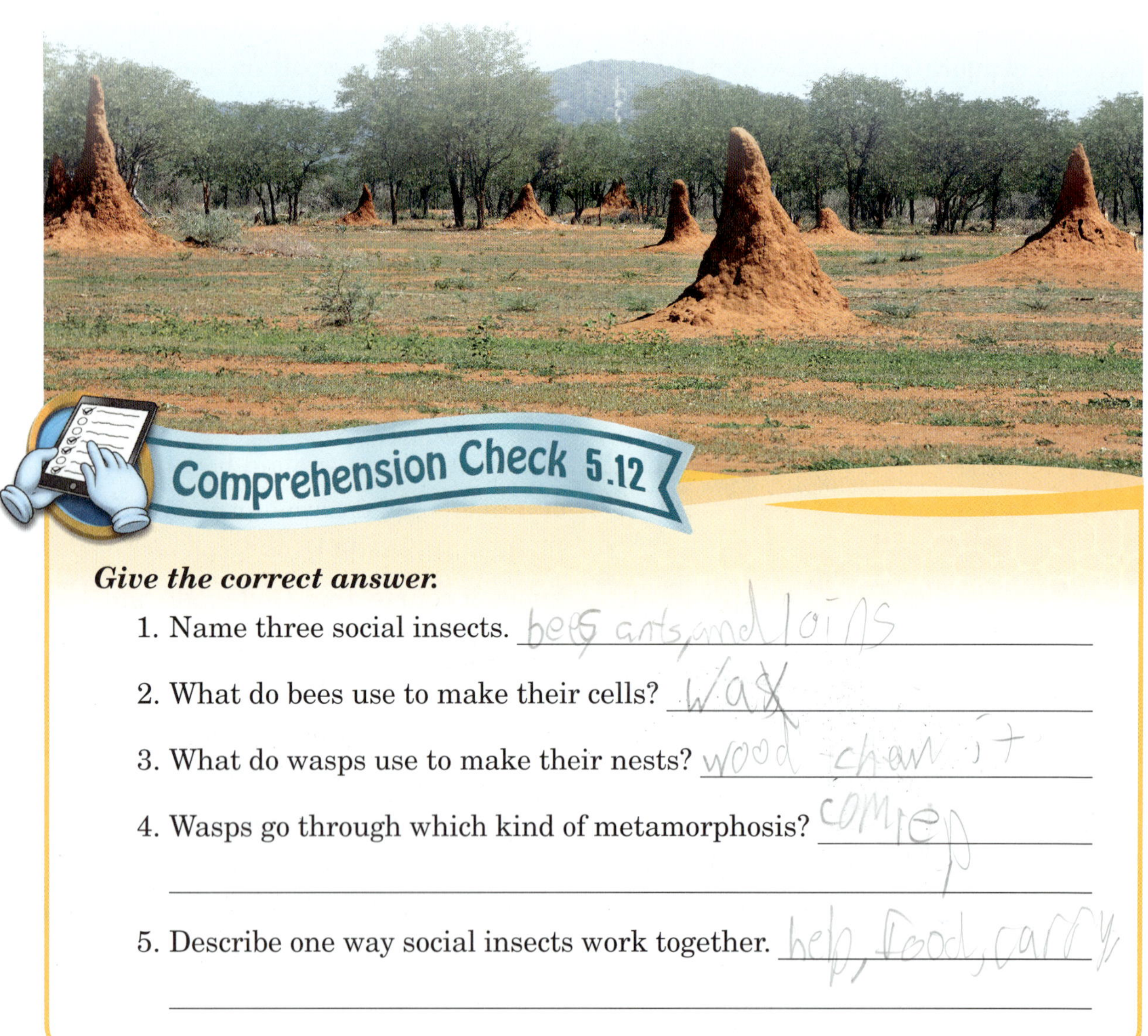

Comprehension Check 5.12

Give the correct answer.

1. Name three social insects. ______________________
2. What do bees use to make their cells? ______________________
3. What do wasps use to make their nests? ______________________
4. Wasps go through which kind of metamorphosis? ______________________
5. Describe one way social insects work together. ______________________

Jean-Henri Fabre:

The World's Greatest Entomologist

Jean-Henri Fabre grew up in France about 200 years ago. At that time, scientists knew very little about the lives and habits of insects. Many people had collections of dead insects, but Fabre was the first to devote his life to observing and describing the behavior of living insects in their natural habitats. He wrote ten books for adults describing what he saw in the fields and gardens around his home. Fabre delighted French schoolchildren with his famous science textbooks, which combined careful descriptions of nature with testimonies of his own faith in God Who designed and created the natural world.

Jean-Henri Fabre was born to a poor peasant family in a small French village. From his earliest childhood little Henri was a naturalist, a person who studies nature. He keenly observed the world of grasshoppers and sparrows, snails and lizards, and was always on the lookout for beetles, waterweeds, and bits of fossils. In the icy winter, he and his family had to live in the barn keeping warm by the heat of the animals. In the hot summer, it was Henri's job to drive the ducks to the pond each day, giving him time to study nature along the way. His one-room, one-window, country schoolhouse was often filled with pigs and chickens that wandered in, but the boy learned reading, writing, and arithmetic, and he even began studying Latin. Henri worked his way through school and then taught for many years in French schools where his students loved him. He was fired from one teaching job because he allowed girls to attend his science classes!

Throughout his life, Jean-Henri Fabre spent every spare moment with the insects that he found so fascinating. He researched bees, wasps, beetles, grasshoppers, and crickets. He discovered the importance of insect instincts and carefully described how insects behave together. Not until he was almost eighty years old, however, did others recognize the importance of his scientific work. *He is now known as the world's greatest entomologist.*

When Fabre was about thirty-five years old, an English naturalist named Charles Darwin published a book about evolution. Darwin said that God did not create the

living world as we know it and as the Bible says. Darwin believed that all living things somehow developed over long periods of time. He said that invertebrates like insects and jellyfish slowly turned into vertebrates like fish, lizards, and horses. He also said that there were no people at first, but then some tree-dwelling, monkey-like creatures with tails, hairy bodies, and pointed ears gradually developed into people.

Fabre knew that this was not what the Bible says. God created everything in the world, and He made the first man, Adam, from the dust of the ground, not from a creature like a monkey. We know that people are God's most special creation because the Bible says we were made in His image.

When many other people started believing Charles Darwin, Jean-Henri Fabre chose to believe God instead. Fabre had observed the beautiful design of nature much too closely to believe that it could all be just an accident. He did not think that Darwin's hypothesis was a reasonable or sensible explanation of the wonders we see in the world around us. He declared, "I observe, I experiment, and I let the facts speak for themselves.... The facts that I observe are of such a kind that they force me to dissent from [disagree with] Darwin's theories." Although many people of Fabre's day turned away from God, thinking that evolution was true, Jean-Henri Fabre did not. He testified, "Without [God] I understand nothing; without Him all is darkness.... You could take my skin from me more easily than my faith in God."

Chapter 5 Concepts Review

1 Vertebrate Classification Concepts 5.1

A. Remember

Short Answer: *Name the five classes of vertebrates. Write two characteristics for each.*

vertebrate class	characteristics	

B. Think Like a Scientist

Label: *Write* W *for warm-blooded and* C *for cold-blooded. Give an example of each description.*

_____ __________ 1. temperature stays the same no matter the environmental temperature

_____ __________ 2. fish

_____ __________ 3. babies depend on instincts instead of parent for survival

_____ __________ 4. reptile

_____ __________ 5. bird

_____ __________ 6. babies depend on parent for survival

_____ __________ 7. mammal

_____ __________ 8. temperature changes with the environmental temperature

_____ __________ 9. amphibian

continued

C. Fun with Terms

Puzzle: *Fill in the answers to find the word in the starred column.*

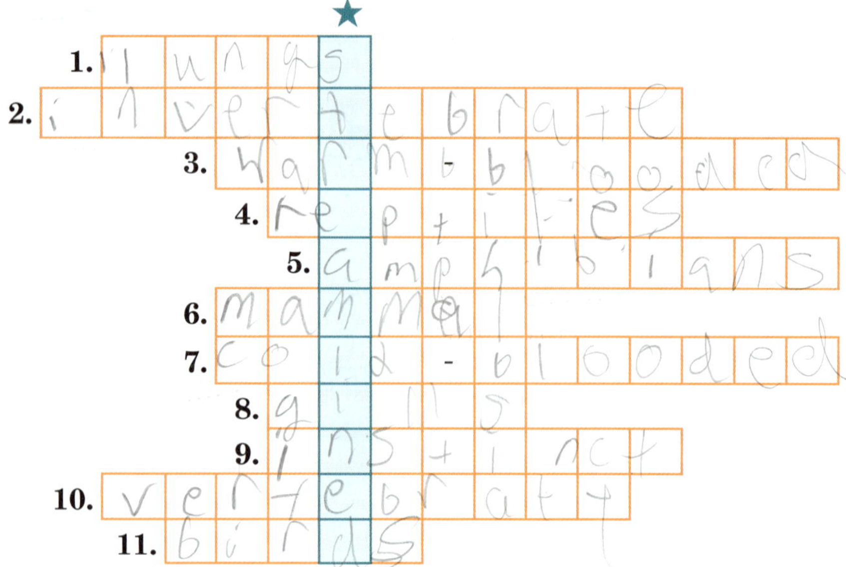

amphibians	**instinct**	**reptiles**
birds	**invertebrate**	**vertebrate**
cold-blooded	**lungs**	**warm-blooded**
gills	**mammal**	

1. organs that use oxygen from air for breathing
2. an animal without a backbone
3. having a body temperature that stays the same no matter the environmental temperature
4. vertebrate group with dry, scaly skin
5. vertebrate group with smooth, moist skin
6. an animal that gives its babies milk and teaches them to survive
7. having a body temperature that changes with the environmental temperature
8. organs that filter dissolved oxygen from water for breathing
9. a God-given ability or behavior that is inherited rather than learned
10. an animal with a backbone
11. only vertebrate group with feathers

★ designed to move easily through water or air

Bird Design Concepts 5.2–5.6

A. Remember

True/False: *If the statement is true, write* true *in the blank. If the statement is false, replace the underlined word(s) with a word or phrase that will make the statement true. Do not write* false *in any blank.*

_______________ 1. A bird that hunts other animals for food is called a bird of <u>predator</u>.

_______________ 2. A bird's body is <u>streamlined</u> so that it can move easily through the air.

_______________ 3. <u>Inventors</u> designed the force of lift.

_______________ 4. <u>Lift</u> can keep an airplane in the air.

_______________ 5. <u>Feathers</u> are very important to all birds.

_______________ 6. A layer of air between its feathers and skin <u>insulates</u> a bird.

_______________ 7. A bird's small, fluffy feathers are called <u>gizzards</u>.

_______________ 8. A bird's ears are <u>on top of</u> its head.

B. Think Like a Scientist

Identify: *Draw lines to match each bird's mouth structure with the food it probably eats.*

continued

Identify: *Draw lines to match each bird's foot structure to life in its habitat.*

C. Fun with Terms

Look It Up: *What do you know about these words? If you are not sure, look up the definitions and study them. Fill in the missing letters; then match each word to a fact about it.*

animal evolution	**extinct**	**incubate**
	gizzard	**insulate**

1. g__zz__rd
2. __xt__nct
3. __v__l__t____n
4. __nc__b__t__
5. __ns__l__t__

- no longer exists
- to keep an egg warm until it hatches
- part of a bird's stomach
- idea that animals slowly turned into other animals
- to keep warm by preventing heat loss

3 Invertebrate Design Concepts 5.7–5.8

A. Remember

***Fill in the Blank:** Choose the correct word from the box below and write it in the blank.*

1. Some invertebrates have an outside skeleton called a(n) ______________________________.
2. Sense organs, called ______________, help some invertebrates respond to their environment.
3. An earthworm uses ____________ to hold tightly to the wall of its burrow.
4. Corals live as a group called a ______________________.
5. A(n) ______________ is an invertebrate with jointed legs.

antennae	**colony**	**segments**
arthropod	**exoskeleton**	**setae**

B. Think Like a Scientist

***Label:** Write I for invertebrate and V for vertebrate. Circle the arthropods. Be prepared to explain your answers.*

_____ 1. shark
_____ 2. leech
_____ 3. lobster
_____ 4. clownfish
_____ 5. owl
_____ 6. sea anemone
_____ 7. whale
_____ 8. ant
_____ 9. sea urchin
_____ 10. spider

***Short Answer:** List 2 characteristics of invertebrates and vertebrates.*

Vertebrates	Invertebrates
11. ______________	13. ______________
12. ______________	14. ______________

C. Fun with Terms

Word Scramble: *Unscramble the words; then write the correct term in the blank.*

1. **gnolmit** — the shedding of an exoskeleton, skin, or other body covering ______________________

2. **stemeng** — a body section ______________________

3. **combisity trailinshope** — two different organisms that live closely together and benefit from each other ______________________

4 Insect Design Concepts 5.9–5.12

A. Remember

Identify: *Label the parts of this insect. Circle the names of the three main body parts.*

Identify: *Label the stages of incomplete metamorphosis. Begin at number 1.*

Identify: *Label the stages of complete metamorphosis. Begin at number 1.*

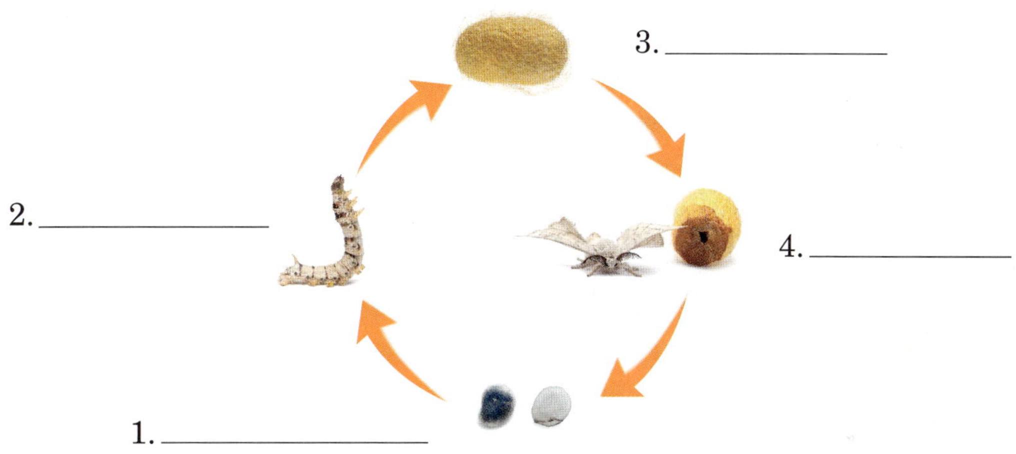

B. Think Like a Scientist

Label: *Write* I *for insect and* N *for not an insect. Be prepared to explain your answers.*

______ 1. bee	______ 4. mosquito	______ 7. butterfly
______ 2. cricket	______ 5. starfish	______ 8. lice
______ 3. crab	______ 6. spider	______ 9. ladybug

C. Fun with Terms

Look It Up: *What do you know about these words? If you are not sure, look up the definitions and study them. Give an example of each word.*

parasite	**host**	**mimicry**	**camouflage**
______________	______________	______________	______________

Chapter 6 Understanding Ecosystems at Work

TERMS

habitat: the natural home of an animal or plant

climate: the weather conditions an area receives over time

altitude: the height above sea level

salinity: saltiness

6.1 What Makes a Habitat?

A **habitat** *is the natural home of an animal or plant.* These natural homes for living things contain all the space, shelter, water, and food that each organism needs. Habitats are fascinating to study! Wherever you live, you can be sure there is a variety of habitats to learn about. *A habitat's location determines its living and nonliving factors.* Some of the nonliving factors come from a location's climate, such as the amount of sunshine and rainfall it receives, the type of water it contains, and the amounts of gases in the air. Other factors, such as types of soil, rocks, and minerals, come from the earth.

Climate Components

The Sun's Light and Heat Energy

Different habitats receive different amounts of sunshine. A habitat is most affected by its location on Earth, giving it its unique climate. **Climate** *refers to the weather conditions an area receives over time.* As Earth revolves around the sun, it rotates on its axis, causing some places to receive more of the sun's heat energy than others. *Earth's tilt as it revolves and rotates gives some places four seasons, while other places have only two.* Along the equator, tropical habitats receive a consistent amount of heat energy from the sun. Polar climates are cold and frigid because of where they are located on Earth. For half of the year, polar habitats receive little to no sunlight at all, making their winter season very cold. In the summer, the sun shines constantly, but the subsoil is still frozen. Between the equator and the earth's poles, we find habitats in more temperate, or balanced, climates. These habitats have four seasons: winter, spring, summer, and autumn.

Water

All habitats contain water. Rainforests are the wettest land habitats. They are usually found in tropical climates because of the amount of rain they receive and high level of humidity, or moisture, in the air. Deserts are the driest habitats because it is difficult for rain and moist air to reach them. Sometimes, this dryness is a result of a desert being located in the middle of a continent or a mountain chain that blocks rain. Did you know that not all deserts are hot? Antarctica is considered a desert continent because it receives fewer than ten inches of rain each year.

Some habitats are made of water. Salt water has a high level of dissolved salt; that is why the ocean tastes salty. The ocean is filled with saltwater habitats, such as trenches, coral reefs, kelp forests, and salt marshes. Freshwater habitats have little or no salt. Lakes, ponds, rivers, and streams can be found all over the world.

Air and Its Gases

Not every place on Earth is at the same **altitude**, or *height above sea level*. The top of a mountain is at a very high altitude while the bottom of a valley or canyon is at a much lower altitude. The higher in altitude a person is, the less oxygen there is to breathe. Gravity is what holds atmospheric gases to the earth, making gases less dense on the top of a tall mountain than gases near sea level. Organisms in mountainous habitats are well suited for the air they live in.

Earth Components

Soil Type

Plants absorb water, minerals, and nutrients from the soil. Different kinds of soil help different kinds of plants to grow. Some soil, called clay, can quickly turn to mud because it holds water. Some soil is sandy, and water can easily flow through it. Desert soil, for example, is usually drier and sandier than the fertile, or rich, soil of a forest or riverbank.

Earth's soil contains layers. The *topsoil* is the top layer of soil. It is *usually dark and rich in humus and nutrients.* It is soft and absorbent enough to hold water.

The *subsoil* is the layer of soil below the topsoil. It is harder and more packed down than topsoil. It is also lighter in color because it does not have as much humus. Large plants have roots that grow down into the subsoil in search of more minerals.

Soil Layers

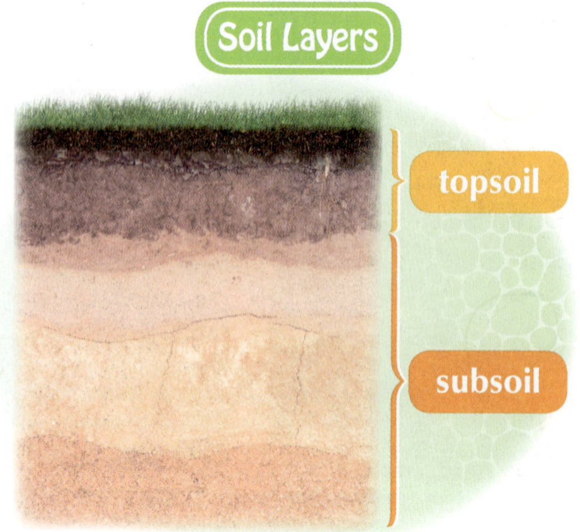

Rocks and Minerals

Some living things use rocks for homes. Algae and mosses can grow on top of rocks. You can often find decomposers by simply lifting a rock. Try it sometime and watch them scurry for cover! You will see that an entire community is living under the rock.

Did you know some animals make the mountains their habitat? God designed these animals to be sure-footed. They can climb in steep, rocky areas where it would be difficult for you to travel. Though trees cannot grow beyond a certain height in some mountainous areas, other plants can.

Rocks are made of minerals. The soil is filled with important minerals that help plants to grow. Some plants need certain kinds of minerals more than others. You may remember that *salt is a very common mineral.* Some organisms thrive where the **salinity**, or *saltiness*, is very high.

Living things in a habitat respond to the nonliving things of their habitat. No matter how harsh or challenging a habitat's environment is, our caring Creator equips the living things with what they need to thrive.

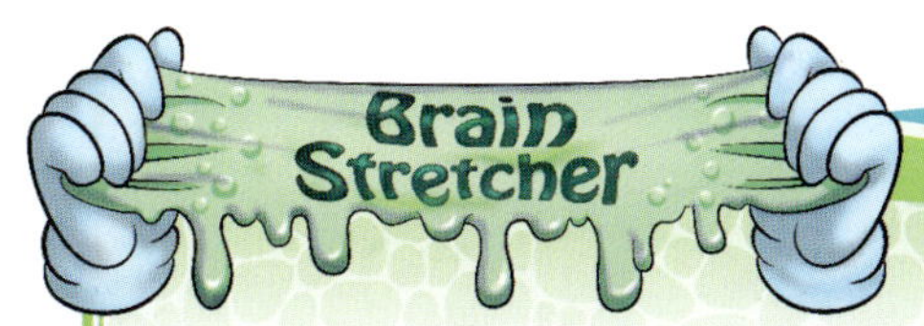

Biomes: God created the world with a variety of life to live in a variety of habitats. Each unique design shows God's creativity and provision. Habitats are a part of larger areas called biomes. *Bio* means "life." All of life takes place in the biomes of the world. Biomes are huge ecosystems with similar climates, soil types, plants, and animals. Land biomes include tundra, coniferous forests, temperate forests, deserts, grasslands, and rainforests. Water biomes include oceans, lakes, rivers, streams, ponds, and wetlands.

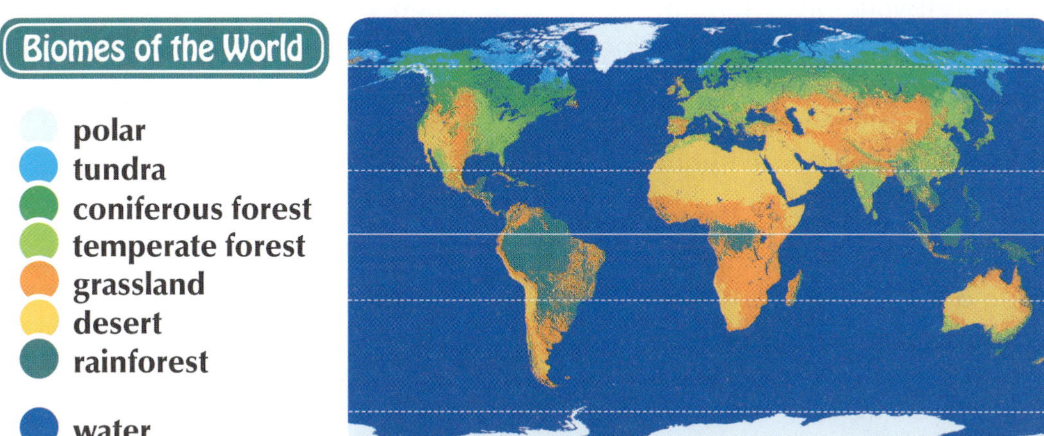

Within a biome are many habitats—some are very large while others are very small. For example, the temperate-forest biome not only contains forest habitats, but also rivers, streams, and ponds. We can find many microhabitats, or small habitats, in this biome as well, such as living trees, fallen trees, rotting tree trunks, and stagnant mud puddles.

Within each biome, we find ecosystems broken down further into levels. Each ecosystem contains a community of living things. Within that community are populations of each species; then each species contains individual living things, or organisms.

Try This! Create a shoebox diorama.

Materials needed:

- research materials on your favorite habitat
- shoebox
- tape
- construction paper
- colored pencils
- scissors
- glue
- *small plastic animals
- *artificial plants
- *twigs
- *sand
- *gravel
- *cotton

1. Research your favorite habitat using a variety of sources. Find out what kinds of living things and nonliving things can be found in that specific habitat.
2. A diorama is a three-dimensional miniature model for display. Plan a way to illustrate your habitat using your shoe-box and gathered materials.
3. Open your shoebox and nest both pieces together so that the lid becomes your habitat "ground" and the bottom piece becomes the background. Tape both pieces together.
4. Decorate and assemble your diorama.

*optional

Comprehension Check 6.1

***Multiple Choice:** Circle the correct answer.*

1. A habitat's __?__ determines its nonliving factors.
   a. animals b. location c. plants
2. The tilt of the earth gives it its __?__.
   a. air b. seasons c. soil type
3. __?__ is rich in nutrients and humus.
   a. Sand b. Subsoil c. Topsoil
4. Salt is a common __?__.
   a. decomposer b. mineral c. organism

***Think and Predict:** Have a friend name an animal; then find a biome where that animal may live. Refer to the map on p. 259.*

ecosystem: a community of organisms, such as plants and animals, interacting with their physical environment

niche: the special job of an organism in its environment

6.2 What Is an Ecosystem?

Think of a time when many animals were in one place. Perhaps you are thinking of the perfect Garden of Eden. Here, God entrusted Adam with the stewardship, or care, of the plants and the animals that lived there. Maybe you are thinking of Noah's Ark. God rescued His creation by sending Noah's family and every kind of land animal into the ark for refuge, or safety, from the judgment of sin. Have you ever seen a great variety of animals in one place?

Zoos and aquariums are places where animals from many habitats are cared for. They are wonderful places to observe and enjoy animals that do not normally live where you do. While we can observe some things about animals living there, we are not able to observe everything because the animals are not in their natural habitat. Fences, glass tanks, buildings, and paved sidewalks help keep both the animals and visitors safe, but they are not natural parts of an animal's environment. Natural habitats are filled with things God made specifically for the animals that live there.

Ecosystem Communities

Just as God intended, the earth is filled with animals and plants. We find them in a variety of natural habitats, such as ocean reefs, ponds, rivers, prairies, mountains, forests, deserts, and frozen tundra. How did God plan to take care of His creation in their habitats? God cares for animals and plants through His design of ecosystems. *An* **ecosystem** *is a community of organisms, such as plants and animals, interacting with each other and their physical environment.*

Because of God's plan, an ecosystem does not need a zookeeper or farmer to care for it. Each plant and animal has its own **niche** [nĭch], or *special job,* in

an ecosystem that *helps the other members of its community*. Each organism is fully equipped for the purpose God has for it.

Living Parts of an Ecosystem

The living parts of an ecosystem are its organisms—animals, plants, bacteria, and fungi. Living things depend on each other for food in an ecosystem. Sometimes, they also use each other for shelter. A tree, for example, is home to many organisms while others use it for food. Some living things have symbiotic relationships, mutually depending on each other. Lichens protect trees by absorbing toxic air pollution. Trees help lichens by giving them a place to grow and reproduce.

Nonliving Parts of an Ecosystem

God designed the living parts of ecosystems to use the *nonliving parts like water, sunlight, air, temperature, soil, rocks, and minerals.* Without nonliving parts, an ecosystem could not exist.

Water

Living things need water. Without water, seeds could not germinate; plants could not grow. Plants need water to complete photosynthesis.

When it rains, plants get the water they need. Water sources fill up so that insects, birds, and other animals are able to drink. Some plants and animals need water every day; others can live without rain or a water source for weeks. You will find that organisms are designed to use the amount and kind of water found in their natural habitat.

Sunlight

Why do all living things depend on the sun? Sunlight energy is needed for the process of photosynthesis. Without photosynthesis, plants could not make food. Without plant producers, consumers could not live. As the sun shines, plants use its energy to grow and reproduce. Different plants need different amounts of sunlight. Organisms living on the forest floor were designed to live where they are shaded from sunlight. Fungi, bacteria, and spore-producing plants are healthy in this kind of environment. Meanwhile, as trees grow, they reach for the sunlight they need. Plants growing in open grasslands or deserts also thrive in the hot sun.

Air and Its Gases

The *atmosphere* is the layer of air that surrounds the earth. It is made of many gases. Do you remember what gases are made of? Gases are made of matter molecules. Different living things need different types of gases for life. For example, people and animals need oxygen which plants release into the air. When people and animals breathe in oxygen, they exhale carbon dioxide as a waste gas. Plants use the carbon dioxide to make more oxygen. Even water contains dissolved oxygen for fish and other water animals to breathe. Water animals give off carbon dioxide for plants to use. Nitrogen is another important gas in our air. In fact, there is more nitrogen than any other gas in our atmosphere.

Temperature

Some plants and animals were designed to live in extreme temperatures, while others need a temperate climate to survive. A penguin is at home in Antarctic conditions because its body was designed with warm insulation and thick feathers. A housefly, however, could not survive in frigid climates because of its cold-blooded body. Every living thing tolerates temperature changes differently.

Soil

Not all soil is the same. Soil is made of different components, such as sand, clay, and organic matter, or humus. Different habitats have different amounts of each soil component. In His wisdom, God designed a variety of plants to grow in a variety of soil types. Desert plants grow well in dry, sandy soil. The plants of a riverbank, however, depend on the rich, moist soil where they grow.

Rocks and Minerals

A *mineral* is a substance found naturally in the earth. Rocks are made of minerals. Soil is full of tiny pieces of minerals that have been worn away from rocks. Each ecosystem has different types of rocks and minerals that living things can use.

The Ecosystem in Your Backyard

Did you know your backyard is its own ecosystem? If you observe closely throughout the year, you will see that it is full of life. By now, you may recognize a few birds that feed in your backyard daily. Although they may be shy, other vertebrates like squirrels, rabbits, mice, frogs, snakes, and the occasional raccoon or shrew visit your backyard—especially during spring and summer. Look even closer, and you will see that many invertebrates depend on your backyard ecosystem. Besides pollinators, such as butterflies and bees, underground decomposers are hard at work. Earthworms, beetles, termites, and insect larvae are all doing the job God created them to do. They break down dead and decaying things and give nutrients to the soil. Tiny snails and slugs munch away at the plants in your flower bed or garden.

Do you have a backyard tree that you have been observing? As a tree, its job in your backyard ecosystem is very important. Trees give off plenty of oxygen during photosynthesis. Many animals make their homes in your backyard tree and depend on it for food.

By late summer, most of the flowers from your backyard plants are finished blooming and are ready to rest. What will happen to the seeds after their dormant season? In the warm spring, seeds of grasses, small plants, and flowers will germinate and grow. New flowers will bloom and be pollinated. In the hot, sunny summer, flowering plants will be able to make new seeds and fruit.

If you live in a temperate climate, your backyard ecosystem may already have entered its winter dormancy. Temperatures have dropped, the hours of daylight have decreased, and instead of rain, you might be experiencing snow. During winter, it may seem that your backyard ecosystem is taking a break, but even during this time, there is plenty to observe and enjoy. Look for animal tracks in freshly fallen snow. Notice what backyard animals are missing from your ecosystem this time of year. Which ones are missing because they have traveled south for the winter? Which ones are only hiding during their winter sleep? Remember that water, temperature, and sunlight are all nonliving parts of your backyard ecosystem. Each living thing responds to the changes in different ways.

Observe to Understand: An Ecosystem

Throughout your study of this chapter, observe an ecosystem at work. Choose an ecosystem you can observe often, such as your own backyard or a local park. Look closely and gather information.

What kind of climate is it in? Does it experience two or four seasons? Write down what you notice about the nonliving parts of this ecosystem. What does the soil look like? How much rain does it receive? What rocks can you find? List the living things that you see. Write down one or two sentences to describe how some of them interact with each other. Try to observe how plants and animals interact with the nonliving parts of their ecosystem as well.

Comprehension Check 6.2

Fill in the Blank: *Choose the correct answer from the box below and write it in the blank.*

1. An ________________ is a community of organisms interacting with each other and their physical environment.
2. An organism's ____________ is its special job that helps the other members of its community.
3. Animals are an example of a ____________ part of an environment.
4. Sunlight is an example of a ________________ part of an environment.

ecosystem	**living**	**niche**	**nonliving**

Think and Conclude: *Study this picture. List the living and the nonliving parts of this ecosystem.*

Living Parts

5. ________________
6. ________________
7. ________________
8. ________________
9. ________________

Nonliving Parts

10. ________________
11. ________________
12. ________________
13. ________________
14. ________________

food chain: the transfer of energy from one living thing to another for survival

herbivore: an animal that eats only plants

omnivore: an animal that eats both plants and meat

carnivore: an animal that eats meat

apex predator: a predator at the top of its food chain with no natural enemies

6.3 What Is a Food Chain?

As a Junior Scientist, you are learning to be more observant of the world around you. Have you observed your backyard ecosystem yet? Perhaps you have another favorite place to observe nature. You are probably becoming more familiar with the living things in that ecosystem. Have you learned some of the animals' names? Living things interact with other living things in an ecosystem through food chains. Food chains are how every member of a community is fed. *A* **food chain** *is the transfer of energy from one living thing to another for survival.* All living things need energy to grow and reproduce. Sunlight energy is where food chains begin.

A Flow of Energy

Plant producers get their energy from the sun; this energy is what helps them make and store food. Plants produce food energy for the animals that consume them.

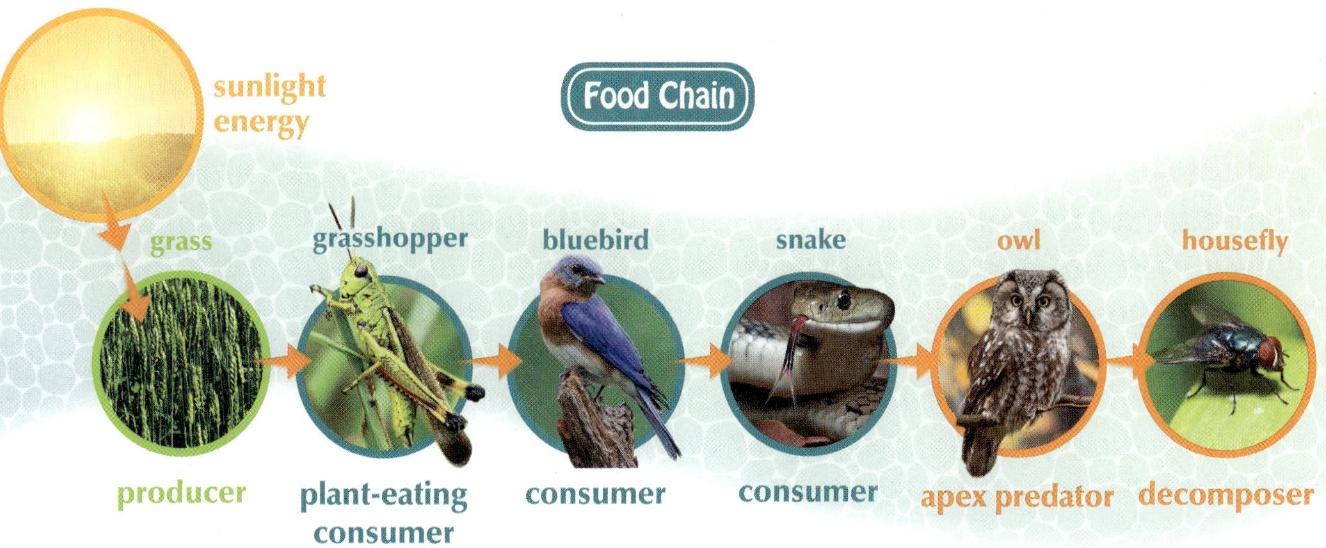

Every consumer gets energy by eating, or consuming, something for food. A *plant-eating animal*, or **herbivore**, continues the food chain by eating, or consuming, a plant for energy. Another animal, sometimes an omnivore, continues the food chain. **Omnivores** *eat both plants and meat.* Herbivores and omnivores are usually in the middle of a food chain. A *meat eater*, or **carnivore**, continues the food chain after that.

An Ecosystem Balancer

Some carnivores have no natural enemies. *If an animal has no natural enemies, it is at the top of its food chain*, making it an **apex predator**. Most ecosystems have one or two apex predators. *Their niche in the ecosystem is very important because they eat extra plant-eating consumers.* What could happen to an ecosystem if there were no apex predators? Think about your backyard ecosystem food chain. If there were no apex predators, such as hawks, owls, snakes, and cats, the mice population would flourish. The extra mice might spread disease and cause a lack of food. Soon, your backyard ecosystem would not be a healthy place for living things—especially you!

A healthy ecosystem is balanced, meaning there is not too much of one kind of plant or animal living there. Too much of one thing or not enough of another can disrupt this balance, which affects the whole ecosystem community.

Every bit of energy is used or transferred in a food chain. Whatever is not eaten is broken down and used up by decomposers. These bacteria, fungi, and tiny animals are habitat cleaners. Without them, habitats would be filled with dead things instead of living things. Decomposers give life to their ecosystem by adding nutrients back into the soil. Nutrient-rich soil helps plants grow and produce more food.

Brain Stretcher
Habitats of Apex Predators: What type of ecosystem, or habitat, might these apex predators belong to?
shark
lion
orca
crocodile
tiger
wolf
A.
B.
C.
D.
E.
F.

Backyard Food Chains

Think about what happens in your backyard ecosystem throughout the year. How would living things interact with each other and their environment? What food chains would you notice?

Spring showers awaken plants from their winter dormancy. Grass and other plant seeds begin to germinate. Instead of a brown, straw-like lawn, your backyard becomes greener and greener. The bare branches of an apple tree begin to bud. Soon, apple blossoms will attract pollinators. Tiny insects pollinate little grass and weed flowers as well. Frogs, rodents, and birds visit your backyard to hunt insects or eat plants.

During the summer, your backyard ecosystem is flourishing. On the weekends, your family is busy mowing the grass, uprooting weeds, and picking flowers or vegetables. More consumers depend on your yard for food. Squirrels, rabbits, and mice sneak in to munch on plants. Birds stop by for earthworms, insects, and seeds. Apex predators, like cats, snakes, and birds of prey, help the ecosystem by eating the extra mice.

In autumn, the days become shorter and the temperature becomes cooler. The plants of your backyard ecosystem respond by entering dormancy. As the days become shorter, there is less sunlight. The leaves of your backyard apple tree respond by shutting down. Their veins become blocked, and water stops traveling to each leaf. Photosynthesis stops happening, and the chlorophyll breaks down. Leaves turn from green to golden and eventually to brown. Because your apple tree is a deciduous tree, the brown leaves fall to the ground and the branches become bare.

Apples also drop from the tree in autumn. Some of the apples are harvested by your family; others are munched on by animal consumers. What is left on the ground begins to break down with the help of fungi and bacteria. Little creature decomposers, such as insect larvae, eat the rotting-apple pieces. Woodlice and beetles nibble on the decaying leaves around it. Soon, the apples and leaves become part of the soil. Earthworms turn this material into humus by digesting it. Humus enriches the soil, providing nutrients for new plants to grow in the springtime.

woodlouse

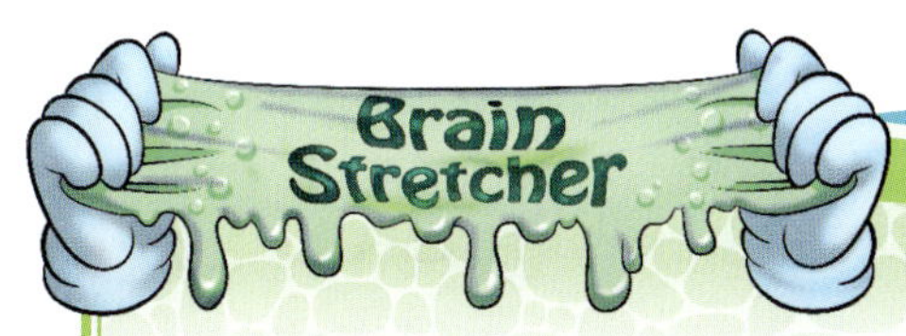

Food Chains: Food chains were much simpler when God first created the earth. Adam, Eve, and all the animals were herbivores; they ate only plants.

> *And God said, Behold, I have given you every herb bearing seed, which is upon the face of all the earth, and every tree, in the which is the fruit of a tree yielding seed; to you it shall be for meat. And to every beast of the earth, and to every fowl of the air, and to every thing that creepeth upon the earth, wherein there is life, I have given every green herb for meat: and it was so. Genesis 1:29–30*

The panda bear gives us a picture of what food chains could have looked like. Even though a panda has sharp teeth like any other bear, it is not a predator. It usually eats plants, bamboo being its favorite food.

Many years after Creation, when the floodwaters of Noah's day were gone, God gave Noah a promise that He would provide for him and his family, just as He had provided for Adam, Eve, and the animals.

> *Every moving thing that liveth shall be meat for you; even as the green herb have I given you all things. Genesis 9:3*

Some people still choose to eat only plants and plant-based foods. Others eat meat along with plants and plant-based foods. Animals in their ecosystems live as God designed them to live. In His wisdom, He gives His creation what it needs to survive.

Producers and Consumers

Observe your backyard or park ecosystem. Make a list of the producers and consumers you see. Use field guides to form a hypothesis about what kind of apex predators might be part of this ecosystem. You may choose to sketch a possible food chain.

True/False: *If the statement is true, write* true. *If the statement is false, replace the underlined word(s) with a word or phrase that will make the statement true. Do not write* false *in any blank.*

_______________ 1. In a food chain, energy is <u>transferred</u> from one living thing to another.

_______________ 2. Plant producers get energy from the <u>soil</u>.

_______________ 3. Animal consumers get energy from the <u>food</u> they eat.

_______________ 4. A(n) <u>omnivore</u> eats both plants and animals.

_______________ 5. A(n) <u>carnivore</u> is most likely the first consumer of a food chain.

_______________ 6. An apex predator has <u>many</u> natural enemies.

_______________ 7. An apex predator is <u>harmful</u> to its ecosystem.

Think and Predict: *Study this food chain; then answer the question.*

8. What could change in this ecosystem if all the wolves were removed?
   a. There would be more plants because there would be fewer deer.
   b. There would be fewer plants because there would be more deer.
   c. There would be more deer because there would be more plants.

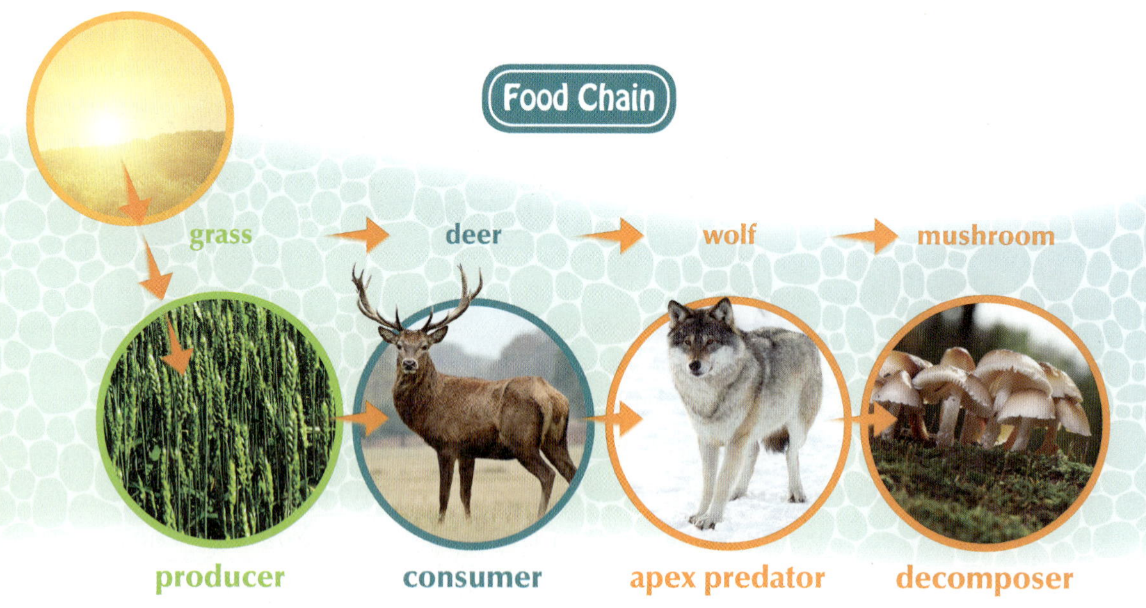

permafrost: permanently frozen subsoil

tundra: treeless land because of permafrost

tree line: the altitude at which trees do not grow

biodiversity: variety and amount of life within an ecosystem

migration: the movement from one place to another with the change of seasons

hibernation: a deep winter sleep to conserve energy

6.4 Forest Ecosystems

Picture yourself walking through a forest. What are you imagining? Are you in a quiet forest, surrounded by pine trees and snow? Are you in a noisy forest with dense vegetation and hot, sticky air? Perhaps you are in a peaceful, cool forest. As wind rustles the leaves, birds chirp, and insects hum about from plant to plant. There are many forests all over the world. Each forest ecosystem is unique in its own way. We can recognize three main kinds of forests: *coniferous forest*, *temperate forest*, and *rainforest*.

The Forest Environment

Usually we find certain kinds of forests in certain climates. Forests grow in places where tree roots can grow deeply. In polar climates, tree roots cannot grow into the permafrost. **Permafrost** *is subsoil that is frozen year-round*. This kind of habitat is called tundra. **Tundra** *is a treeless land because of permafrost*. If you were to look at a mountain from above, you would see that trees stop growing at a certain point. *The altitude at which trees do not grow is called the* **tree line**.

Besides needing better soil conditions for growing, the air is too thin this high in the atmosphere. Trees need more heat energy from the sun than mountainous temperatures can provide.

Coniferous Forests

Coniferous forests can be found in most climates because *conifer trees thrive well in differing temperatures. Coniferous forests are most common where it is cold and there is less sunlight or the altitude is higher.* Very cold climates near the tundra or on a mountainside cannot support deciduous trees.

Temperate Forests

Temperate forests grow in temperate climates that have four seasons. The trees of a temperate forest were created to respond to the change of season and less sunlight in winter by becoming dormant. Many deciduous trees grow in temperate forests.

Tropical Rainforests

Tropical rainforests grow in tropical climates where there is much rain year-round. Remember, tropical climates also receive the most sun because they are the closest to the equator. Tropical climates feel hot and humid. Many plants can grow best in the wet, humid conditions the rainforest provides. *Rainforests have the highest level of* **biodiversity**, or *variety of life*, of all land habitats. Because the rainforest is filled with so many plant producers, many animal consumers depend on it for food and shelter. A rainforest supports much more life than any other type of land habitat.

coniferous

temperate

tropical rainforest

Different forests have different soil quality and composition. A temperate forest has very fertile soil. Its deciduous trees produce leaf litter every year. These leaves decompose, and nutrients are recycled. More humus in the soil soaks up more water and oxygen. Coniferous- and tropical-forest soils are not as fertile. They have trees that do not shed their leaves when the seasons change. Surprisingly, tropical forests have the poorest soil of any forest type. The amount of rainfall and humid conditions affect how quickly nutrients in the soil are recycled.

Try This! Create a rainforest terrarium.

Materials needed:

- ✓ large jar with lid
- ✓ pebbles
- ✓ activated carbon charcoal
- ✓ dried moss
- ✓ potting soil
- ✓ living moss
- ✓ small rainforest plants (ferns or palms)
- ✓ spray bottle of water
- ✓ *toy rainforest animals

1. Place a layer of pebbles at the bottom of your jar for drainage; then add the charcoal to act as a filter; then a layer of dried moss to act as a barrier.
2. Add a layer of soil for planting the living moss and plants. Spray your plants with plenty of water. You may choose to add a few toy rainforest animals to enhance your terrarium. Seal your jar and place it in indirect sunlight. Observe at least once a day.

A rainforest has a high level of moisture in the air, or humidity. By placing a lid on the top of your terrarium, it can water itself because of transpiration, which is water released from the leaves of plants. Moisture droplets will condense on the plants and the glass inside your jar. Some droplets will "rain" back down on the plants. Every few days, open the lid to your terrarium so that this miniature ecosystem can receive fresh air. Spray a little water inside before resealing it.

*optional

Life in the Forest

Though each kind of forest has varying amounts of vegetation and types of life, all forests have some things in common. Forests are made of many plants; chiefly the trees that help give them their names. Forests have three main layers: the *canopy*, *understory*, and *forest floor*. Each layer has its own ecosystem of plants and animals because each layer receives different amounts of rain and sunlight. Depending on the types of plants, a forest could even have two canopy layers or two understory layers. Sometimes, a forest has *very tall, emergent trees that rise above the canopy*. In the rainforest, this is called an **emergent layer**.

The **forest floor** *is the very bottom layer of the forest*. Most of the time, the forest floor is very dark because trees are blocking the sunlight. The lack of sunlight makes the forest floor an ideal place for fungi, like mushrooms, to grow. Fungi, bacteria, and other decomposers work together to break down leaf litter, fallen trees, and other dead and decaying things.

In the **understory layer**, we find *plants growing close to the ground*. The dark, damp environment helps spore plants, such as mosses and ferns, to grow. Young trees, shrubs, and flowering plants also grow in the understory.

The **canopy layer** *is made of the crowns of large trees*, as well as vines and mosses. The canopy *acts like an umbrella*, keeping much of the sunlight and some of the rain from reaching the forest floor.

Food is abundant in the forest. Every forest has birds that depend on its insects, seeds, berries, and nuts.

Forests attract plenty of herbivores, such as rodents and other small mammals. Amphibian and reptile omnivores enjoy the cooler temperatures and plenteous plants and insects the forest provides. If herbivores and omnivores live in the forest, you can be sure carnivores live there! Every forest ecosystem has at least one kind of apex predator.

How do animals respond to the change of seasons in a temperate forest? During winter, the temperature drops, and food is scarce. Some animals migrate to warmer climates in autumn and return in the spring. **Migration** *is the movement from one place to another with the change of seasons.*

Migration is a mystery to many scientists. How do animals know when to travel? How do animals navigate? God-given instincts cause some animals to migrate. Though no one knows exactly how migration occurs, scientists have some ideas. It could be that some animals use the earth's magnetic field to help them know when to travel and where to go. Certain insects, such as butterflies and bees, use the sun as a compass. Animals are able to sense changes in their environment, such as a change in weather or less sunlight. Have you ever observed an animal behaving differently before a storm? Animals have been known to sense a catastrophic event coming, such as a volcanic eruption. Creation scientists believe that God allowed certain animals to know where and when to travel to Noah's Ark before the Flood. No matter the reason or method, animal migration points to a wise and caring Creator.

Some animals do not need to migrate because they can make changes with their surroundings. Some grow thick winter coats or a layer of fat, called *blubber,* for insulation. Others hibernate when the temperature gets too cold. *To* **hibernate** *is to fall into a deep winter sleep*. The main reason animals hibernate is to *conserve energy* when food is scarce. Dormice and other small animals hibernate during winter. Their body temperature lowers and their organs slow down so that they can fall into a winter sleep. Bears have a different type of hibernation. Their body temperature does not change as much so that they can wake up to eat food or protect their cubs from danger.

hibernating dormouse

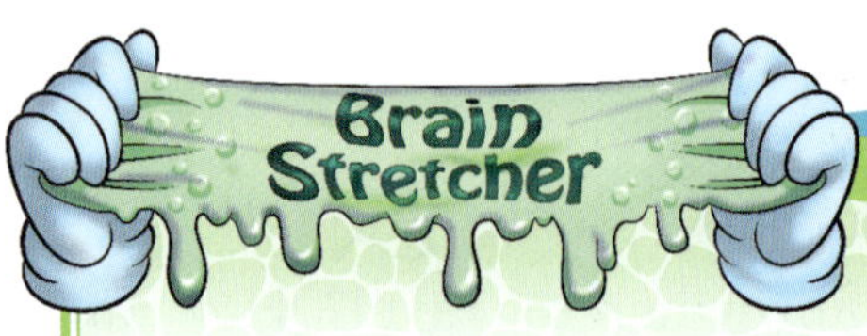

Forest Fires: Some forests, such as coniferous forests that contain many pine trees, are especially vulnerable to forest fires. Sometimes, a forest fire occurs as a natural event, like a lightning strike; other times fires happen unnaturally. An entire ecosystem can be destroyed in a forest fire. Many living things are killed while others escape to find a new home. God gives animals instincts that tell them when they are in danger. Surprisingly, it seems that the forest ecosystem benefits from naturally caused fires. Some pinecones only release seeds in extreme heat. Ashes contain minerals that make the soil very fertile for regrowth. God's plan is for the forest to replant itself and grow back even better than before.

When parts of the forest do not grow back, those parts become fertile, open land for other kinds of plants to grow. If God had not designed forests to regrow, some forest animals would not have a place to live. Without any forests on our planet, the air would not have enough oxygen for people and animals to breathe.

We would never want to put a forest at risk for an unnatural forest fire. If your family makes a campfire or has a cookout in or near a forest, be sure to follow correct procedures for completely putting out the fire afterwards. Do your part to be a caretaker of the forest.

Matching: *Write the letter of the correct answer in the blank.*

______1. permanently frozen subsoil

______2. altitude at which trees do not grow

______3. forest that can thrive in very cold climates

______4. forest habitat with the highest biodiversity

______5. lowest level of a forest

______6. forest layer that can act as an umbrella

______7. to move from one place to another when seasons change

______8. deep winter sleep to conserve energy

A. canopy
B. coniferous
C. forest floor
D. hibernate
E. migrate
F. permafrost
G. rainforest
H. tree line

Think and Predict.

9. After a forest fire, what order of regrowth might you expect? Could the forest flourish again someday? ______________________________

____

____

____

____

____

10. During the winter in temperate climates, cold-blooded animals are dormant. This dormancy is usually different from true hibernation, depending on the species. How might dormancy affect some forest consumers? ______________________________

____

____

____

____

grassland: a large, flat, open area of grasses

savanna: a tropical grassland

burrow: to tunnel

grazer: an animal that feeds on grasses

browser: an animal that feeds on trees and shrubs

scavenger: an animal that eats other animals that have already died

6.5 Grassland Ecosystems

Almost every continent has them—"seas of grass" as far as the eye can see. *A* **grassland** *is a large, flat, open area of grasses.* Most grasslands have few trees. Grasslands are usually found between a desert and a forest area in the innermost part of a continent. In North America, grasslands are called *prairies*; in South America, they are called *pampas*. Australia calls their grasslands *rangelands*. The largest grasslands stretch from Europe to China, where they are called *steppes*. The *savannas* in Africa cover over half the continent.

The Savanna Environment

Without mountain chains and trees to block wind, most grassland habitats are windy and dry. Sometimes, grasslands grow larger because of natural fires that burn through the dry prairie grasses and into the forest. Wild game, such as deer, antelope, buffalo, and other grazers, increase in population when there is plenty of grassland for food. A long time ago, it was common to burn a forest on purpose to create hunting land. Hunting was a way of

life in the history of the American prairie. While the prairie is a temperate grassland, *the African* **savanna** *is considered a tropical grassland* because it grows in a tropical climate. It is hot all year long and has *two seasons:* a *long, dry season* and a *short, wet season.*

The soil in the African savanna is sandy and dusty; people call it "red earth." *It is reddish in color because it is high in a mineral called iron.* Because of the dry, then rainy climate, humus does not have a good chance to develop in the savanna soil. This lack of humus means the savanna does not have as much fertile soil as the North American prairie.

The rainy season fills rivers, flood plains, and waterholes. Land animals depend on these water sources all year long.

Life in the Savanna

In many ways, we can thank the termite for life in the savanna. Its niche is very important. Termites help the savanna soil by **burrowing**, or *tunneling underground.* Their burrows help air and water circulate. *Termites are valuable decomposers of the savanna.* What they cannot digest is broken down by special fungi that grow on their mounds. Termites then redigest the broken-down matter, and the soil is enriched. You might remember the organisms that live inside the termites' intestines. This symbiotic relationship helps the termite do its job.

Savanna elephant grasses reach ten feet down into the soil. Their niche is

to prevent soil erosion and feed many consumers. Most savanna plants receive nutrients, water, and even oxygen from what they can find deep underground. Can you guess when the growing season begins in the African savanna? The rainy season saturates the soil and all the seeds waiting to be germinated. The rain causes tough-shelled seeds to germinate and sprout. A new growth of plant producers will be ready for animal consumers to eat. Grasses that have been cut by grazing flourish and grow tall again.

Grazing animals, such as zebra, antelope, gazelles, and wildebeests time their migration according to when and where plants are growing. **Grazers** *are important herbivores in grassland habitats.* As they roam, their hooves prepare the soil for plant growth. Seeds get pressed into the soil, and the broken soil absorbs more rainwater. Grazing animals also help the grasslands by cutting the grass with their teeth. This helps the grass plants to grow stronger and gives the roots a chance to make new soil.

There are more trees in the African savanna than any other grassland habitat. Some common trees are the baobab [bā′ō·băb] and acacia [ə·kā′shə]. The baobab has shallow roots compared to its height. Water is stored in its trunk so that it can survive the dry season. Giraffes and elephants *feed on the leaves and bark of trees and shrubs. These animals are called* **browsers**. Their niche in the savanna is to keep trees and shrubs from competing with the grasses.

Lions and cheetahs are apex predators of the savanna, along with hyenas, African wild dogs, and leopards. They have an important niche because they hunt extra grazers and browsers. If the savanna did not have these predators, too many herbivores would eat all of the plant life. Soon the savanna would become a wasteland.

The lion's size makes it the most fearsome predator, but it is not always successful. Lions often steal the catch of smaller predators such as wild dogs and hyenas, which hunt in large packs. The cheetah carefully stalks its prey before it races to attack, running faster than any other animal on Earth. Leopards, like cheetahs, hunt alone and often drag their prey into the high branches of a tree to keep it from being stolen by other apex predators.

Jackals and vultures feed on meat left over from a predator's meal. Predators like hyenas often scavenge food for an easy meal. *A* **scavenger** *is an animal that eats other animals that have already died; they do not need to hunt their food.* Their niche is important because they quickly break apart carcasses that would take years to decompose. Scavengers protect ecosystems from the spread of disease. Termites, worms, fungi, and bacteria continue the process of decomposition.

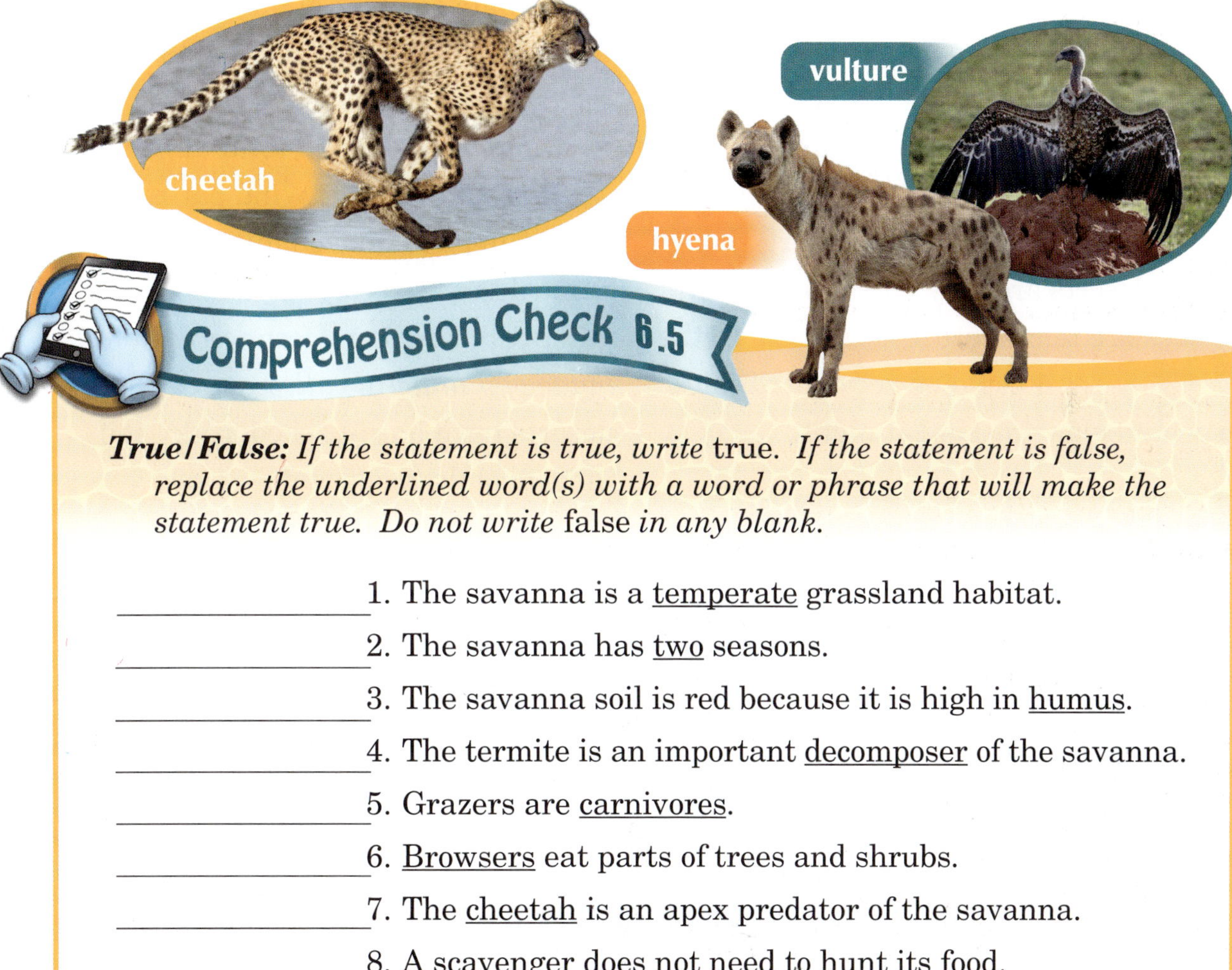

Comprehension Check 6.5

True/False: *If the statement is true, write* true. *If the statement is false, replace the underlined word(s) with a word or phrase that will make the statement true. Do not write* false *in any blank.*

_______________ 1. The savanna is a temperate grassland habitat.

_______________ 2. The savanna has two seasons.

_______________ 3. The savanna soil is red because it is high in humus.

_______________ 4. The termite is an important decomposer of the savanna.

_______________ 5. Grazers are carnivores.

_______________ 6. Browsers eat parts of trees and shrubs.

_______________ 7. The cheetah is an apex predator of the savanna.

_______________ 8. A scavenger does not need to hunt its food.

desert: a dry land with little plant growth

oasis: fertile place in the desert where there is water

nocturnal: to be active during the night

6.6 Desert Ecosystems

The Desert Environment

A **desert** *is a dry, or arid, land with little plant growth.* Deserts are arid because they have little rainfall—less than 10 inches per year. Even the air of the desert feels dry. It is the dry air and lack of precipitation, not necessarily the temperature, that makes a desert, though many deserts are very hot. Some deserts experience temperature extremes. In the same desert, it could be very hot at mid-day but below freezing at night.

Desert soil is coarse and filled with minerals but very little humus. Some desert soil has a very high salt content. How would the lack of humus affect the roots of desert plants? Without root systems to hold sand in place, *the force of wind can push sand dunes from one place to another.* Wind patterns cause dunes to take on different shapes. Some dunes are very large and tall; others are smaller. *Most desert sand is made of a mineral called quartz.* Many deserts have hard-packed, rocky soil instead of drifting sand.

Sahara Desert

The Gobi Desert: A Cold Desert

The Gobi Desert borders China and Mongolia and is surrounded by mountains. The desert itself is a very flat plain, and the soil is very rocky. If you were to walk across the Gobi, you could see for miles and miles in front of you because the land is so flat. Bactrian camels, snow leopards, golden eagles, and a curious rodent called the jerboa are among the animals that make their home here.

The Sahara Desert: A Hot Desert

The Sahara Desert of Africa is the hottest desert in the world. It has mountains, sand dunes, and gravelly soil. Gazelles, foxes, baboons, hyenas, mongooses, and venomous reptiles are among its native animals. People living in the Sahara have camels, sheep, and goats. Most of the people living in the Sahara live at its oases. *An* ***oasis*** *is a fertile place in the desert where there is water.*

The Antarctic Desert: A Special Case

Antarctica is unique because most of its land is covered in ice, keeping it from being considered tundra. Though Antarctica is a polar region, its precipitation level is so low that it is also considered a desert—the most unique desert in the world.

Life in the Desert

Deserts have the lowest level of biodiversity, or variety of life, of all habitats. Why does a rainforest have the highest level of biodiversity while a desert has the lowest? Think of how the food chain of an ecosystem works. Most plant producers need plenty of water to complete the process of photosynthesis. Photosynthesis is what gives plants the ability to make, or produce, food. Animals thrive where there is plenty of food. A rainforest supports more life than a desert.

Though its level of biodiversity is low, a desert ecosystem can support more life than you may think. God gives the plants and animals living there the ability to thrive in their environment. Birds, reptiles, mammals, and arthropods all make their home in deserts. Specially designed amphibians can live in the desert, too.

How do plant producers survive without much water? Cactus plants have a unique system of photosynthesis. In most plants, this food-making process happens in leaves. In cactus plants, however, photosynthesis happens in the stem and trunk. A cactus plant's leaves are its spines, which help it retain moisture. The fully grown saguaro [sə·gwär′ō] cactus can stand between forty and sixty feet tall and can hold between one and two tons of water. Its root systems spread out to collect and store up to two hundred gallons of water.

saguaro cactus

Many desert plants have spines or thorns for protection. *Thorns and spines can protect a plant from being completely eaten.* Cactus plants need this extra protection in the desert where food is scarce. Without its spines, all of the cactus plant would be eaten. How might the loss of cactus plants affect the desert ecosystem? After some time, the consumers who depend on cactus for food might starve. If the population of a certain species becomes too low, other animals in that food chain may also be affected. Underpopulation and overpopulation can disrupt the balance of an entire ecosystem.

The desert only receives rainfall once or twice a year. Desert plants and animals take advantage of this water. Desert plants absorb and store water for photosynthesis. Some plants, such as grasses, spread out their roots to get every drop. Cactus plants become thick and swollen with water, storing it to use until the next rainfall. Dormant seeds are germinated, and flowers bloom for pollination.

The rainwater forms small pools. Desert amphibians and insects lay eggs in these pools. If you look carefully, there will soon be new life all over the desert.

Animals of the desert have two main ways of managing in the extreme temperatures of their habitat. Some rodents and amphibians find shelter by burrowing underground. Burrows stay cool away from the sun's rays, and the soil helps insulate the animals from the heat. The fennec fox is a burrowing animal. God gave this fox extremely large ears to keep it cool and to hear its prey underground. This fox and other animals are **nocturnal**, or *active at night* when it is cool instead of during the day when it is hot. When the sun sets, they come out to look for food; when the sun rises, they find shelter and go to sleep.

If you were a desert animal, where would you find shelter? Some desert animals, such as reptiles, live in the crevices between rocks or rock formations. Some birds of the desert live in a hole made in a cactus plant. Desert arthropods are sheltered in their strong exoskeletons. This armor not only protects them from heat, but it also keeps them from drying out.

Birds can thrive in desert habitats because their normal body temperature is between 103 and 106 degrees Fahrenheit. As you know, birds are well insulated, keeping the heat outside their bodies. On very hot days, desert birds can move their feathers to let cool breezes flow through them.

fennec fox by burrow

lizard

Gila woodpecker

blue scorpion

stink beetle

tarantula

Soaring birds, such as eagles and vultures, can find cooler temperatures by flying higher into the sky.

God designed desert animals to thrive where water is scarce. Many desert animals do not actually drink water. Instead, they get water from the plants they eat. Some animals get water from seeds and other plants; other animals get water by licking the morning dew off plants. Still others can go months without drinking anything. These animals have body systems that give them water from the stored fat in their bodies. What do the special designs of desert plants and animals tell you about their Maker?

dew on succulent

God equipped the plants and animals He made to do their jobs, wherever those jobs may be.

> *Thy righteousness is like the great mountains; Thy judgments are a great deep: O Lord, Thou preservest man and beast.* Psalm 36:6

Comprehension Check 6.6

Give the correct answer.

1. What does *arid* mean? ______________________
2. What mineral is most desert sand made of? ______________________
3. What causes sand dunes to move from place to place? ______________
4. What cold desert is found near China and Mongolia? ______________
5. Name the hottest desert. ______________________
6. What life-giving substance is found at an oasis? ______________
7. What structures protect some desert plants from being completely eaten? ______________________
8. Why are some desert animals nocturnal? ______________________

____

ice cap: a year-round covering of ice and snow over polar land

Ice Age: a period of cold temperatures bringing ice and snow after the Flood

glacier: a large, moving body of ice on land

iceberg: a large piece of ice that has broken from a glacier

ice shelf: frozen, floating seawater connected to land

plankton: microscopic, floating water plants and animals

phytoplankton: single-celled algae; able to use the sun's energy for photosynthesis

zooplankton: small water animals

Polar Ecosystems

As you learn about ecosystems, you may find that the habitats with the most extreme conditions are the most fascinating. Perhaps nothing is as extreme as Antarctica—the coldest place on Earth. Antarctica is the largest, driest, coldest desert. Because it is located in a polar climate, its situation is unique; it is often called a polar desert.

The Polar Environment

While Antarctica is considered a desert because of its very low precipitation, it is also a frigid, polar habitat. Polar climates are unique. Seasons work differently in polar climates. Instead of four seasons, Antarctica has two—summer and winter. For six months it is always daylight. This is when the South Pole is tilted toward the sun, giving Antarctica its summer. For the other six months, when the South Pole is facing away from the sun, it is always dark. This is Antarctica's winter. The Arctic North Pole is also a polar climate, but its seasons occur opposite of the Antarctic South Pole. While the South Pole has summer, the North Pole has winter.

Most of Antarctica is always covered with a polar **ice cap**, *a year-round covering of ice and snow*. Both the North Pole and the South Pole have polar ice cap areas, but the Antarctic has more. Creation scientists believe the ice caps were formed after the Flood. A worldwide Flood would have triggered a series of major changes in the earth's climate. As the "fountains of the deep" (Genesis 7:11) were opened, volcanic eruptions could have filled the atmosphere with ash. Volcanic ash in the atmosphere could have caused the sun's rays to be reflected back into space instead of being absorbed by the earth. *A period of worldwide cold temperatures bringing ice and snow* would have followed, called the **Ice Age**. Evolutionists believe there were *many* climate changes and *many* ice ages over millions of years. Creation scientists see the polar ice caps of the North and South Poles as evidence of a world recovering from *one* Ice Age caused by the Flood. They believe that over the several hundred years of the Ice Age, layers of snow and ice caused this deep layer of ice at the earth's poles.

> *The waters are hid as with a stone, and the face of the deep is frozen.*
> *Job 38:30*

Looking out at the Antarctic landscape, you would see glaciers, icebergs, and ice shelves. *A* **glacier** *is a large body of ice moving over land.* **Icebergs** are mountainous *pieces of glacier that break off* and float out into the ocean. The ice cap, glaciers, and icebergs are frozen fresh water. An **ice shelf**, however, is *frozen, floating seawater connected to land*. The Antarctic ice shelf forms a giant ring around the entire continent.

Observe iceberg buoyancy concepts.

Materials needed: ✓ a balloon ✓ a freezer ✓ large container ✓ water ✓ scissors ✓ ruler

1. Fill a balloon with water and place it in the freezer until it is solid. Carefully cut the balloon away from your "iceberg."
2. Place your "iceberg" into a large container filled with water. Measure the amount of iceberg in the water and the amount protruding out of the water.

Icebergs float because of their density and buoyancy. Freshwater molecules expand by 10% when they freeze. An iceberg is less dense than the water it is in. Buoyancy is the upward force from the water to push an object to the surface. The surface area of an iceberg displaces enough water for the iceberg to float. About 90% of the iceberg is always submerged; 10% always protrudes out of the water, balancing its buoyant force. Ships that travel in polar regions must be very careful as they navigate around icebergs because they can only see 10% of what they could collide with.

Life in Antarctica

The only people who live on Antarctica are scientists who live there temporarily. They study the ice and organisms living there. You might wonder, "What organisms could possibly live in Antarctica?" The Antarctic tundra does not have any trees or large plants, but types of mosses, lichens, and algae can grow even in these extreme conditions. Bacteria and fungi also help the Antarctic ecosystem. Scientists have found types of bacteria under the ice that decompose certain gases that could disrupt the balance of gases in the atmosphere.

Why do animals migrate to the Antarctic every year? Water currents bring rich nutrients to the Southern Ocean, causing plankton to flourish. **Plankton** *are mostly small, floating water plants and animals. The main producers of all Antarctic food chains are single-celled algae called* **phytoplankton**. These *tiny plants* grow near the ocean's surface and are able to use the sun's energy for photosynthesis. *Small water animals, called* **zooplankton**, feed on phytoplankton. Krill, large zooplankton, feed on other zooplankton. Whales, seals, fish, and penguins all enjoy the plenteous krill living in the Southern Ocean during the Antarctic summer. The blue whale can eat four tons of krill in a single day!

Toothed whales are the apex predators of the Southern Ocean. Sperm whales eat fish and invertebrates like the colossal squid. Orcas, or killer whales, are the largest members of the dolphin kind, or family. Orcas hunt many animals including seals, penguins, and whales. They hunt together in a group called a pod. Many generations of an orca family often make up a pod. These whales can live between fifty to eighty years and are very intelligent.

God designed Antarctic animals with bodies that can thrive in extremely cold temperatures. Most of the fish in the Antarctic waters have special blood that acts like "antifreeze" for their cold-blooded bodies. Penguins have dense feathers that keep them warm. Whales and seals have a thick layer of blubber to insulate them. God has given Antarctic animals instincts to help them know when to feed at the South Pole and when to migrate to warmer waters to have babies.

The arctic tern has the longest migration of any animal. It travels from the Arctic to the Antarctic to take advantage of both summer seasons. In its lifetime, the arctic tern will have traveled enough miles to make three round trips to the moon. Who gave the arctic tern these amazing flight capabilities? God engineered each bird for the life it has. What does His care for Antarctic animals show you about His care for you? Since God found a way to care for animals that live in challenging conditions, how much more will He care for you when you are challenged? God equips His children with His strength; trust in Him to help you!

Behold the fowls of the air: for they sow not, neither do they reap, nor gather into barns; yet your Heavenly Father feedeth them. Are ye not much better than they?

Matthew 6:26

Comprehension Check 6.7

Matching: *Write the letter of the correct answer in the blank.*

C 1. year-round covering of ice and snow
B 2. period of cold temperatures after the Flood
A 3. large, moving body of ice
D 4. frozen, floating seawater connected to land
F 5. single-celled water algae
G 6. small water animals

A. glacier
B. Ice Age
C. ice cap
D. ice shelf
E. iceberg
F. phytoplankton
G. zooplankton

TERMS

ocean basin: the surface of the earth covered by an ocean

continental shelf: the underwater land along the coast of a continent

trenches: the deepest places of the ocean floor

abyssal plain: the valley-like floor of the ocean between continents

sunlight zone: the ocean depth with the most light; has the most ocean life because of photosynthesis

twilight zone: the ocean depth with only blue light; does not allow photosynthesis

midnight zone: the dark depth of the ocean after the twilight zone

abyss: the deepest, darkest zone of the ocean

6.8 Saltwater Ecosystems

The main reason Earth is habitable is that it is the only planet with liquid water. The ocean covers just over 70% of the earth's surface. There is so much ocean that 97% of the earth's water is salt water. Though this water is not the kind we can drink, we depend on it for many reasons. It is an important part of the water cycle. Evaporation of the ocean gives rainwater to land habitats, giving us fresh water to drink. Plant producers could not survive if it were not for water. The ocean also helps regulate the earth's temperature. In summer, ocean water stores heat energy from the sun; in winter, the ocean releases the heat into the atmosphere.

The Ocean Environment

The surface of the earth that is covered by an ocean is called an **ocean basin**. The ocean basins contain the five oceans. They support much life in their many different saltwater habitats. Each ocean is unique because of the nonliving components there. Different ocean depths are able to receive different amounts of sunlight and water pressure. For example, *underwater land along the coast of a continent is the* **continental shelf**. Here the water is shallower. Ocean **trenches** are very deep and dark. They are *the deepest places in the ocean floor*. Here the water pressure is too heavy for most living things—except the animals that God designed to live there. *Between continents, the ocean spreads out like a giant bowl. This valley-like floor is called the* **abyssal plain**. If we were to take a soil or sand sample from a trench and compare it with the sand of a continental shelf, we would find differences there, too.

The temperature of the ocean and its habitats depends on its depth and location on Earth. Polar waters are very cold while tropical seas are warm. What gives the ocean its heat energy? Energy from the sun warms the ocean. Deeper parts of the ocean do not receive heat energy from the sun like shallower depths do.

Water in the ocean is always moving. Gravity from the sun and moon pulls the water so that it bulges out, causing ocean tides. The earth's rotation, wind, water temperatures, the shape of the ocean's floor, and even the ocean's salinity affect the water's movement. Living things depend on this movement in many ways.

Life in the Ocean

Ocean Zones

The various ecosystems and habitats found in the ocean are usually divided by depths. *The* **sunlight zone** *has the most life because the most photosynthesis happens here.* Because of the plenteous plant producers, many animal consumers live in the sunlight zone. *The* **twilight zone** *has only blue light that does not give plants energy for photosynthesis.* Plants do not live in the twilight zone, but animals do. Some of these animals are luminous, giving off their own light. At lower depths is the **midnight zone**. There is *no light in the midnight zone*, but some sea animals do live there. Sperm whales can dive to these depths in search of food. Their ability to dive this deep and surface again is an incredible design. People need special submarines and equipment to survive the water pressure here. *The deepest, darkest zone of the ocean is the* **abyss**. Here the water pressure is too great for many living things. Because this zone is so far from plants, not much oxygen is dissolved in water at this depth. God designed the animals that live in the abyss to thrive without much oxygen and in extreme water pressure. Even deeper than the abyss are ocean trenches. *Earthquakes often occur near ocean trenches.*

Ocean Food Chains

Phytoplankton are the main producers of the ocean because they receive sunlight energy for photosynthesis. *Kelp is a type of alga because it does not have roots.* Kelp can grow to be so large that the plants form kelp forests. Zooplankton continue ocean food chains by feeding on phytoplankton. Little fish, ocean invertebrates, and even great whales feed on plankton, and ocean food chains continue.

You may have heard someone say, "He swims like a fish." A person can learn to swim, but fish were born knowing how to swim. Do you remember what we call the inherited abilities of animals? God gives fish and other water animals the instincts and equipment that help them swim.

A fish needs oxygen to breathe as much as you do. How does it breathe this important gas while it lives in water? The oxygen that fish breathe is given off by the plants of the ocean and dissolved in the water. When fish breathe, they take in water through their mouths. The water passes over their gills. The *gills remove the oxygen* for the fish and send the water and carbon dioxide back into the ocean or wherever the fish is swimming. The carbon dioxide is then used by algae.

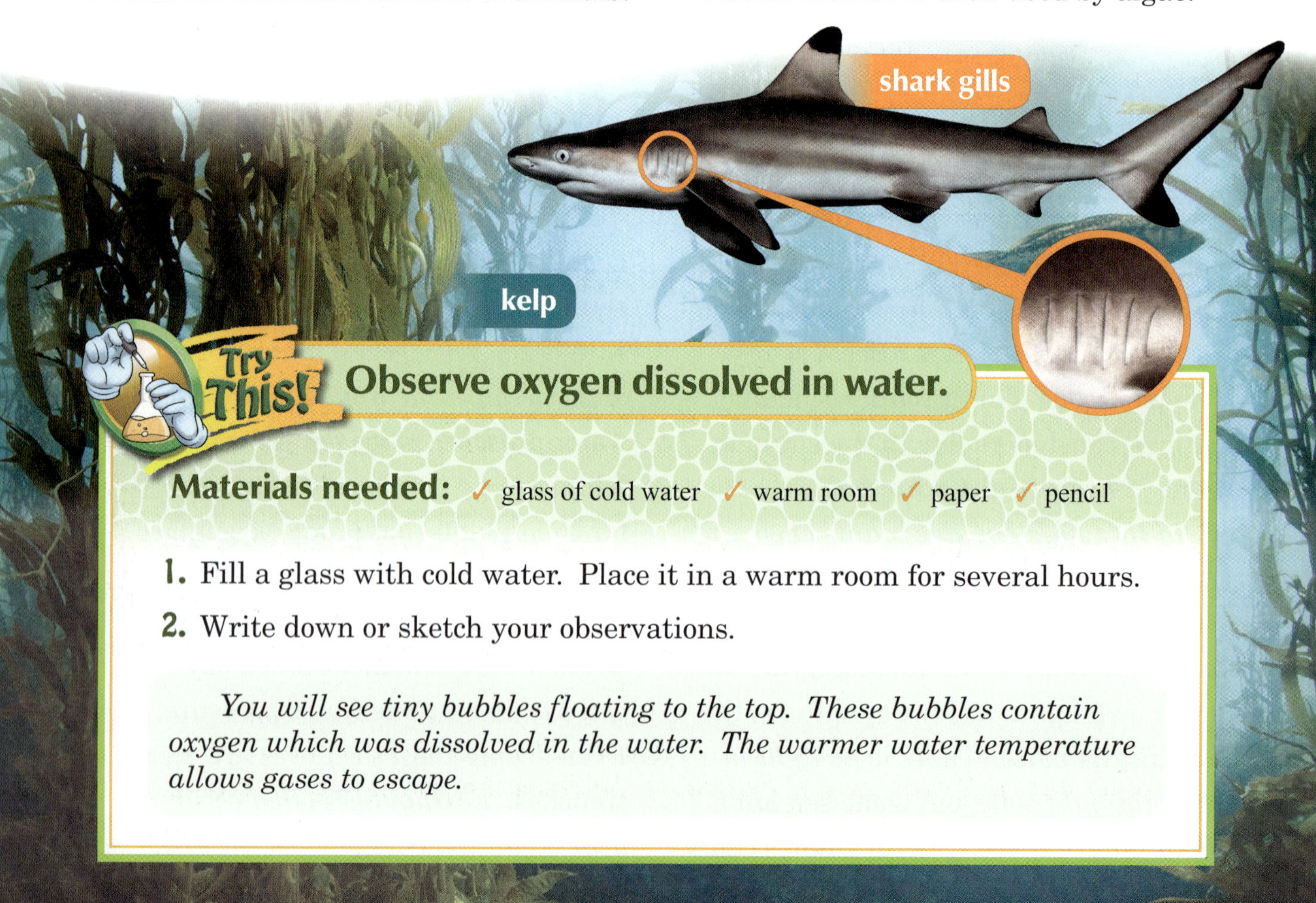

Try This! Observe oxygen dissolved in water.

Materials needed: ✓ glass of cold water ✓ warm room ✓ paper ✓ pencil

1. Fill a glass with cold water. Place it in a warm room for several hours.
2. Write down or sketch your observations.

You will see tiny bubbles floating to the top. These bubbles contain oxygen which was dissolved in the water. The warmer water temperature allows gases to escape.

Try This! Construct a model of the ocean's zones.

Materials needed:

- ✓ jar
- ✓ beaker
- ✓ spoon
- ✓ funnel
- ✓ corn syrup
- ✓ blue liquid dish soap
- ✓ water
- ✓ liquid coconut oil (or another liquid oil)
- ✓ rubbing alcohol
- ✓ black, purple, green, blue food coloring
- ✓ *blue oil-based food coloring
- ✓ tape
- ✓ pen

1. Make a jar of ocean zones. Begin with an ocean trench. Mix 150 milliliters of corn syrup with black food coloring. Using a funnel, carefully pour your mixture into the jar. Add the remaining layers using these materials:
   - ✓ Abyss—mix dish soap with purple food coloring.
   - ✓ Midnight—mix water with dark green food coloring.
   - ✓ Twilight—mix coconut oil with dark blue food coloring. (Oil-based food coloring works best.)
   - ✓ Sunlight—mix rubbing alcohol with light blue food coloring.
2. Label each zone and place a lid on top of your jar.

Each liquid has a different density so the layers will remain separated unless shaken. While we used dyes to color each zone, the ocean zones have different "colors" because of the amount of sunlight they receive.

*optional

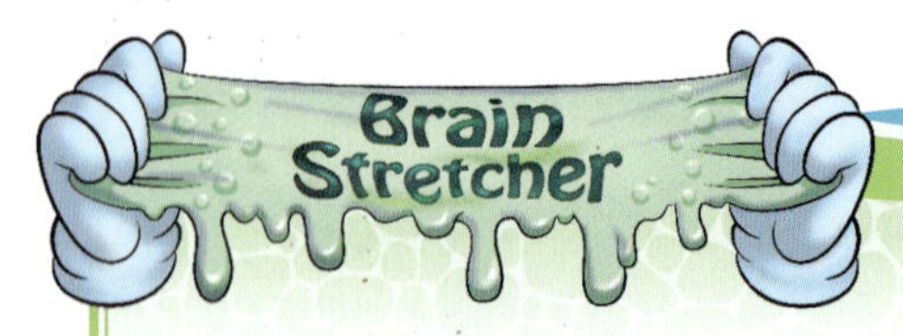

Tide Pools: Along the rocky shore of the sea, you will find pools of seawater. These pools are left in the hollows of rocks when the tide goes out. When the tide comes back in, it washes away what was in the pools and leaves small sea animals and seawater. The small sea animals cannot get back to the sea until the tide washes them back out. These rocky little hollows are called tide pools.

A tide pool is a wonderful place to study a microhabitat. Tiny plants and animals are trapped for a few hours before returning to the ocean. Some organisms are more vulnerable than others to changes in a tide pool's nonliving parts: the amount of sunlight, temperature changes, and salinity level. Other organisms are more resilient because of the way God designed their body systems to work. For example, clams can shut their shells tightly until the tide returns. This helps their bodies to continue to live in the right amount of salt water. Starfish, oysters, clams, and mussels are often washed into tide pools, as are lugworms, which live in the sand as an earthworm lives in the dirt. You may even find a sand dollar, crab, or seaweed.

Multiple Choice: *Circle the correct answer.*

1. The underwater land found along the coast is called the _?_ .
   a. abyssal plain b. continental shelf c. ocean basin

2. Between continents we find the _?_ .
   a. abyssal plain b. coast c. continental shelf

3. Photosynthesis can happen in the _?_ .
   a. abyss b. sunlight zone c. twilight zone

4. The _?_ has only blue light.
   a. abyss b. midnight zone c. twilight zone

5. The _?_ is the darkest ocean zone.
   a. abyss b. midnight zone c. twilight zone

6. Earthquakes occur near many ocean _?_ .
   a. coasts b. trenches c. zones

7. Kelp is a type of _?_ .
   a. alga b. plankton c. tree

8. Fish breathe _?_ in water using gills.
   a. air b. carbon dioxide c. oxygen

wetland: land covered with water for at least part of the year

mouth: the place where a river flows into the ocean (or other body of water)

delta: a triangular landform at the mouth of a river made from deposited soil

estuary: a body of water where the ocean tide meets a river or stream, causing salt water to mix with fresh water

6.9 Freshwater Ecosystems

Freshwater habitats, such as lakes, ponds, rivers, streams, and certain wetlands, have very little salt. Besides being an important water source for land animals, each of these habitats is its own ecosystem. How does God provide fresh water all over the world? It all starts with precipitation.

Precipitation is any form of water which falls from the sky. Because of the sun's heat energy, water vapor from the ocean rises and leaves the salt behind. Clouds form in the sky, cool down, and release water in the form of precipitation. *Freshwater sources are constantly being resupplied because of*

Everglades

precipitation, such as rain or snow. A snowy mountaintop can begin *water runoff*, which eventually trickles into streams. Streams join together and flow into a river.

As rivers flow, they often empty into lakes and ponds, which are also freshwater habitats. A lake is a large body of fresh or salt water, while a pond is a small body of fresh or salt water. Some freshwater lakes, such as the Great Lakes, are very large. Other freshwater habitats are the various wetlands we find inland. *A* **wetland** *is land covered in water for at least part of the year.* Freshwater wetlands include swamps, bogs, and flood plains and are usually near rivers or lakes.

God supplies plants and animals with life-giving water all over the world and in every kind of ecosystem. Fresh water continues to flow from one freshwater source to the next until a large river flows into the ocean. *The place where a river flows into the ocean is called the* **mouth**. *Soil and minerals carried by the river are deposited and gathered here, forming a* **delta**, or *triangular landform at the mouth of the river.* Other times, a river's mouth opens into a type of saltwater wetland called an estuary. *An* **estuary** *is a body of water where the ocean tide meets a river or stream, causing salt water to mix with fresh water.* This kind of body of water can become a type of saltwater wetland.

Wetlands protect the coast in some very important ways. As water flows through an estuary or wetland, *pollution is filtered out.* Wetlands also *protect land from erosion and flooding by slowing the flow of water.* They can *take on extra water when rivers overflow.* During a hurricane, wetlands act like a sponge by absorbing extra floodwater.

> *All the rivers run into the sea; yet the sea is not full; unto the place from whence the rivers come, thither they return again.* — *Ecclesiastes 1:7*

The Everglades Environment

The Florida Everglades is the largest freshwater habitat in the world. This complex tropical wetland is about 1.5 million acres of marshland, sloughs, swamps, and forests. At its southernmost tip is Everglades National Park. This part of the tropical wilderness is protected by the U.S. National Park Service. The Everglades feeds an important estuary at the southern tip of Florida, called Florida Bay, that is fed by the fresh water of Lake Okeechobee.

As a tropical ecosystem, the Everglades experiences two seasons: a wet season and a dry season. The wet season begins in April and is very hot and humid. Heavy rain, tropical storms, and even hurricanes bring almost 50 inches of rain per year. This precipitation makes the water level rise drastically, causing many important habitat changes. Sometimes, tropical storms and hurricanes open new areas for plants to grow.

November brings the dry season. The temperatures become more cool and mild and the heavy rains stop. Migratory birds and other animals depend on the dry season of the Everglades for a safe place to nest and raise their young.

The soil in the Everglades usually contains plenty of water, but little oxygen. Wetland soil is rich and fertile for plant growth. Because of water runoff, minerals are plenteous and rich in nutrients. Sometimes, decaying vegetation mixes with soil matter to form a dark, sticky, muddy soil.

Life in the Everglades

A long time ago, Native Americans lived in the Everglades. They called their home grassy waters. The marshland of the Everglades is filled with sawgrass, which gets its name because of its saw-like leaves. Many people today call it the River of Grass. Some water plants float on the top of the water; others grow underwater.

In the northwestern part of the Everglades is Big Cypress Swamp. Cypress trees are well equipped to live in watery environments. Their unique root systems reach above the water for oxygen since the thick, muddy soil they grow in contains little oxygen. The part of the roots that reaches above the water is called its knees. Only some plant roots can tolerate being wet for a long time without rotting.

Mangrove forests are important to the ecosystem because they prevent erosion and cushion the coast against storms. Mangrove trees are able to

get their oxygen from "pores" on their roots and the gaps of their bark. Some smaller wetland plants grow completely submerged in water. Other wetland plants like water lilies and duckweed float on the surface.

Wetland environments are home to many water insects, such as the whirligig beetle. Do you remember what is special about the whirligig beetle's eyes? It uses its double set of compound eyes to see above and below the water. With the help of a little air bubble under its abdomen, it can dive underwater for food. Its niche in the Everglades is to keep the water clean from insects that fall in. In many ways it is like a tiny wetlands scavenger.

Perhaps the most curious of all is the dragonfly. Dragonflies lay their eggs in a quiet freshwater habitat, such as a marshland or a swamp. The nymphs that hatch have gills and spend the next few years in the water, eating insect larvae, snails, tadpoles, and even small fish. Their niche is important as ecosystem balancers, for they have big appetites! When a dragonfly nymph is ready to leave the water, it climbs up onto a water plant, sheds its exoskeleton, and then flies away. As an adult, the dragonfly's niche changes. Now it hunts for flying insects, such as mosquitoes. Too many mosquitoes can carry diseases. As it hunts, the dragonfly uses its superior flight skills to sneak up on and surprise its prey. It can maneuver better than any helicopter, for it can quietly remain in one place during flight. It can even fly backward as skillfully as it flies forward!

In the Everglades, the most infamous animals are its reptiles. Turtles, water moccasins, rattlesnakes, and alligators live there. What niche do you suppose the alligator has in its ecosystem? The alligator is an apex predator. Besides being important ecosystem balancers, alligators help the Everglades by making shallow ponds with their feet and snouts, creating microhabitats called gator holes. Fish, birds, amphibians, insects, and other invertebrates depend on these microhabitats.

The birds of the Everglades are almost as famous as the alligators. Wetlands are essential to many types of birds. Long-legged wading birds, such as herons, egrets, spoonbills, and ibises, are found in the Everglades. Waterfowl and shorebirds need this habitat, too. During the dry season, many birds migrate to these wetlands to nest and raise their young. At that time, there are so many birds in the Everglades that they are everywhere you look. John James Audubon, a naturalist who was most famous for his study of birds, said of the Everglades birds, "We observed great flocks of wading birds flying overhead toward their evening roosts. . . . They appeared in such numbers to actually block out the light from the sun for some time."

Swimming mammals, such as otters and manatees, enjoy the shelter and plenteous water in which to swim, find food, and play. Some people call the manatee a sea cow because it "grazes" on water plants and has a gentle nature. Apex predators, such as black bears and cougars, hunt for deer or smaller mammals that hide among wetland water plants.

Try This! Watch a leaf "breathe."

Materials needed:
- ✓ large beaker (or glass bowl) of lukewarm water
- ✓ freshly cut leaf ✓ sunny place ✓ *hand lens

1. Submerge a freshly cut leaf into your beaker of water. Place the beaker in a sunny place.
2. After several hours, observe the leaf. You may choose to use a hand lens to get a better look. Do you see the tiny bubbles?

Underwater plants can still make food using photosynthesis if they receive sunlight. As a leaf in any plant makes food, it gets rid of oxygen as waste. What might need oxygen in water habitats? Fish and other aquatic animals use oxygen and give off carbon dioxide for water plants to use.

*optional

Comprehension Check 6.9

Give the correct answer.

1. What does water runoff begin as? ______________________________
____
2. Name a freshwater wetland. ______________________________
____
3. Name three ways wetlands help the coastal land. ______________
____
4. What type of habitat is the Florida Everglades? ______________
____
5. How many seasons occur in the Everglades? ____________________
6. Name a producer of the Everglades. ____________________________
7. Name a consumer of the Everglades. ____________________________

Think and Conclude.

8. Why is the biodiversity of life so much greater in a wetland than in a desert? ____
9. What is the difference between a delta and an estuary? ____________
____

Wetland Case Study 1

An Invasive Species

Burmese python

What can happen when a non-native plant or animal species is introduced to a new habitat? Most of the time, the ecosystem community makes adjustments and the newcomer becomes part of the habitat. For example, the ring-necked pheasant originally came from China. Now it is the state bird of South Dakota, a delightful sight in many prairie habitats. But sometimes, a non-native organism can become invasive. An *invasive species* disrupts the balance of an ecosystem and causes harm. Usually this disruption happens because the species can quickly reproduce, or have many babies, and has no enemies in its new habitat. An invasive species of the Florida Everglades is the Burmese python.

In the 1970s, people could purchase Burmese pythons as pets. This kind of snake, however, is not a good pet because it can grow too large to keep contained. Would you have room for a twenty-foot snake in your house? Pythons can also be dangerous to other pets and small children because they are predators. Instead of having a venomous bite, a python squeezes its prey to death and eats it whole. Scientists think that the Burmese python was introduced into the Everglades because people released their large pet snakes into the wild. Snakes could also have escaped captivity during Hurricane Andrew, a very severe hurricane in Florida in 1992.

In Southeast Asia where the Burmese python lives in its natural habitat, it is a helpful part of its own ecosystem. It hunts and eats extra plant-eating consumers, such as rats, and has more natural enemies to balance its population. But in the Florida Everglades, it has no enemies. The python has become an invasive apex predator, eating mammals, birds, and other reptiles—even alligators!

Park rangers and people who care for the Everglades have noticed a drastic difference in many animal populations. This is probably because there are more than 100,000 Burmese pythons living, growing, reproducing, and eating a lot of animals. Some people believe there may be a solution for this ecosystem problem, such as capturing or hunting these snakes. Though ecosystems have been known to make adjustments when an invasive species is introduced, people can make wise choices and attempt to correct the problem.

Think and Predict.

The Burmese python has drastically reduced the raccoon population. Turtle eggs are the raccoon's favorite food. What could happen to the turtle population? How might this affect the ecosystem?

Wetland Case Study 2

Stewardship and Conservation

Because of water runoff and the way water travels from source to source, people find ways to take special care of wetlands and other freshwater habitats. Anything that seeps into or is dumped into a river upstream can contaminate water sources downstream. Long ago, many gold miners used a liquid metal, called mercury, to help them purify gold. The mercury washed into streams and made fish very sick. People who ate the fish became ill because mercury is poisonous if touched or swallowed.

Other pollutants can wash into water systems, causing problems in ecosystems and hurting the people and animals that drink the water. God has some natural plans in place for water pollution. Some of the tiniest organisms in water habitats are bacteria that are able to decompose dead things and pollutants. Some bacteria can even decompose oil spills in the ocean. As caretakers of God's creation, we can look for ways to limit water pollution.

Have you noticed that God has many systems in place for recycling waste gases? Carbon dioxide is a waste gas that we breathe out. Plants use carbon dioxide to give people and animals oxygen; then as we use oxygen, we breathe out more carbon dioxide. Nutrients get recycled through food chains and decomposition. Even the water in our world is constantly being recycled because of the water cycle.

As people, we are the only living things who produce waste that cannot be naturally decomposed, such as plastics, metal scraps, harmful chemicals, and power-plant waste. We can follow God's example in the care of His creation. Look for ways to recycle some of the waste in your home. Some communities offer recycling pick-up days or recycling centers where you can dispose of plastic, metal, and glass for reprocessing. Some centers offer to help you correctly dispose of harmful chemicals so you don't need to pour them down a drain or onto the street. City drains sometimes flow out to wetlands, streams, and other water sources.

A Biblical Perspective of Conservation

In the study of ecosystems, some people place the importance of animals and plants over the importance of mankind. What does the Bible say?

And God blessed them, and God said unto them, Be fruitful, and multiply, and replenish the earth, and subdue it: and have dominion over the fish of the sea, and over the fowl of the air, and over every living thing that moveth upon the earth.
Genesis 1:28

God does provide homes for many plants and animals through forest ecosystems, but God also provides for people through the plants and animals He has made. Besides giving us food and oxygen, trees also supply us with materials to make paper, tires, and even houses. As trees are cut down for use, more trees can be replanted. We are to use, but care for, all that God has given. Conservation is the management or preservation of a natural resource. Can you think of some other ways to conserve natural resources?

Some people celebrate nature rather than give praise to the One Who created it. How does this kind of thinking compare to God's Word?

Because that, when they knew God, they glorified Him not as God . . . and worshipped and served the creature more than the Creator, who is blessed for ever. Amen.
Romans 1:21, 25

A worshipful attitude toward nature and the idea that people are nature's enemy is unbiblical. What then, is a biblical attitude towards nature? God created the earth for our benefit and His glory. Notice His care for you—in the little ways and in the big ways. Give praise to God for Who He is and what He has done. An important part of our worship of God is in our service to Him. Look for ways to care for the people around you who are created in His image. Be a caretaker of the world you live in, remembering it is God's creation.

Taking wise care of what God has placed in our care is called stewardship. A steward is a caretaker who manages what belongs to someone else. Everything that God made belongs to Him. What could you do to be a good steward of God's creation?

A Wetland Conservation Discussion

A habitat or ecosystem can change or disappear for many reasons. Sometimes, this change happens because of natural causes, such as a forest fire or flood. Sometimes, this change happens because of unnatural reasons, such as an invasive species or land development.

A long time ago, the Everglades was much larger than it is today. Farmers needed more farmland, and people wanted to develop cities and settlements there. People could use the water from the wetlands to irrigate their crops. At that time, there was little discussion of the benefits and disadvantages to draining wetlands. While God wants us to use the resources on Earth, we should always make sure we do so with thought and care. This is part of our job as His stewards. Discuss the positive effects and negative effects of draining a wetland area.

Positive Effects	Negative Effects
____________________	____________________
____________________	____________________
____________________	____________________
____________________	____________________
____________________	____________________

Chapter 6 Concepts Review

Basic Ecology Concepts 6.1–6.3

A. Remember

Fill in the Blank: *Choose the correct word from the box below and write it in the blank.*

1. A(n) habitat________ is the natural home of a plant or animal.
2. A(n) ecosystem________ is a community of organisms interacting with each other and their physical environment.
3. An organism's niche________ is its special job in its environment.
4. A(n) herbivore________ is an animal that eats only plants.
5. A(n) carnivore________ is a meat-eating animal.
6. An animal that eats both plants and meat is a(n) omnivore________.

carnivore	**habitat**	**niche**	**predator**
ecosystem	**herbivore**	**omnivore**	

B. Fun with Terms

Matching: *Write the letter of the correct answer in the blank.*

B____ 1. the weather conditions an area receives over time

D____ 2. saltiness

C____ 3. the transfer of energy from one living thing to another for survival

A____ 4. predator at the top of its food chain with no natural enemies

A. apex predator
B. climate
C. food chain
D. salinity

C. Think Like a Scientist

Identify: *Circle the living parts and underline the nonliving parts of an ecosystem.*

amount of rain	**consumers**	**salinity of water**
amount of sunlight	**decomposers**	**soil type**
atmospheric gases	**producers**	**weather conditions**

Essay: *Write a paragraph that explains why each ecosystem needs an apex predator. Include the terms* sunlight energy, producer, consumer, *and* decomposer *in your essay.*

2 Terrestrial (Land) Ecosystem Concepts 6.4–6.6

A. Remember

Short Answer: *Write the correct answer in the blank.*

1. What causes a tundra to lack trees? permafrost
2. What do we call the altitude at which trees do not grow? tree line
3. What kind of climate does the savanna have? tropical
4. Which savanna consumers feed on grasses? grazers
5. Which savanna consumers feed on trees and shrubs? Browsers
6. Name two ways desert animals manage in extreme temperatures. ____________

Identify: *Label the layers of this forest.*

7. emergent
8. canopy
9. understory
10. forest floor

B. Think Like a Scientist

Short Answer: *List a niche for each living thing in its habitat.*

Forest	**Savanna**
tree ____________________	termite ____________________
____________________	____________________
____________________	____________________
fungi and bacteria ____________	elephant ____________________
____________________	____________________
____________________	____________________
bird of prey ________________	hyena ____________________
____________________	____________________
____________________	____________________

C. Fun with Terms

Matching: *Write the letter of the correct answer in the blank.*

____ 1. variety and amount of life within an ecosystem

____ 2. the movement from one place to another with the change of seasons

____ 3. a deep winter sleep to conserve energy

____ 4. a large, flat, open area of grasses

____ 5. an animal that eats other animals that have already died

____ 6. a dry land with little plant growth

____ 7. fertile place in the desert where there is water

A. biodiversity
B. desert
C. grassland
D. hibernation
E. migration
F. oasis
G. scavenger

3 Aquatic (Water) Ecosystem Concepts 6.7–6.9

A. Remember

Fill in the Blank: *Choose the correct word from the box below and write it in the blank.*

1. A(n) ice cap is a year-round covering of ice and snow over polar land.
2. A(n) ice shelf, or frozen, floating seawater connected to land, forms a giant ring around Antarctica.
3. A(n) glacier is a large, moving body of ice.
4. Sometimes, a(n) iceberg breaks off and floats out into the ocean.
5. A(n) delta is a triangular landform at the mouth of a river.
6. A river's mouth can sometimes open into a body of water called a(n) estuary.

delta	**glacier**	**ice cap**
estuary	**iceberg**	**ice shelf**

Identify: *Draw a picture of the zones of the ocean; label each zone.*

B. Think Like a Scientist

Sort: *Write each living thing from the box in the correct column below.*

clam	**dragonfly**	**orca**	**sawgrass**
cypress tree	**kelp**	**phytoplankton**	**zooplankton**

Producer	Consumer
Kelp	Clam
Sawgrass	orca
Phyo	dFly
cypress tree	ZOO P

Identify: *On a separate piece of paper, choose an ecosystem and draw a food chain with at least three consumers. Be prepared to explain your drawing.*

C. Fun with Terms

Crack the Code: *Find the letter of the correct term and write it on the numbered line below.*

_____ 1. underwater land along the coast of a continent

__T__ 2. period of cold temperatures bringing ice and snow after the Flood

_____ 3. valley-like floor of the ocean between continents

_____ 4. microscopic, floating water plants and animals

_____ 5. land covered with water for at least part of the year

_____ 6. deepest places of the ocean

A. abyssal plain

B. wetland

E. continental shelf

N. plankton

S. trenches

T. Ice Age

__ C O S Y S T E M S   H __ P P E __   I N   H A __ I T A T __

(1) ___ (2) T (3) ___ (4) ___ (5) ___ (6) ___

Unit 3

Earth and Space Science

Understanding the Earth and Its Foundations

Chapter 7

TERMS

geology: the study of the earth and its structure

geologist: a scientist who studies the earth

7.1 The Study of the Earth

Our Home, the Earth

Home. What does that word make you think of? Perhaps you are thinking of the people in your family who live with you. Your family members are the first people you learn to interact with and to love. Are you thinking of the house you live in? Your house is where you are sheltered, protected, and provided for. Now think bigger.

When God created the universe and all that is in it, He planned a special place for people to live. That special place is our planet, Earth. In life science, we learned about the living things God made; now we will learn about the nonliving parts of the home God made for us. This is the study of earth science.

Have you ever traveled away from your own home? What amazing things did you see? If you live in the Midwest region of the United States, your first trip to the ocean filled you with wonder. If you live near the coast, a trip through the prairie landscape helped you see just how big the earth is. If you do not live near mountains, you were in awe upon your first glimpse of a towering peak. If you have ever wished you could understand more about our planet, you will enjoy **geology**, *the study of the earth and its structure.*

Geology

A scientist who studies the earth is called a **geologist**. Geologists study mountains, valleys, caves, rocks, minerals, soil, and anything else that is part of our planet. As you know, not all scientists agree that God created the world. This means there are two basic views of geology. Some scientists believe "the present is the key to the past." They believe that what they can currently observe in nature explains what happened a long time ago. Evolutionists believe that mountains, canyons, riverbeds, and every other part of our planet were formed gradually over millions of years. For example, as the Colorado River flows through the Grand Canyon, the force of its water slowly erodes some of the riverbed. But can scientists know for certain that it was the Colorado River that made the Grand Canyon over millions of years? Is there a scientist who has lived for millions of years who could tell us what he or she observed? Scientists who believe in evolution are basing their ideas on faith in what they can see.

Grand Canyon

Creation scientists base their faith on what God says in His Word. *They believe that current observations help them understand and confirm what has happened in the past.* They believe that God miraculously created the earth and all that is in it in six days. Then many years after Creation, God saw the terrible wickedness of man and flooded the whole earth as a judgment for sin. In His grace, God provided an escape from this judgment—the Ark. By faith, Noah and his family entered into the Ark by its one and only door. In the Ark, God provided the only salvation from the worldwide Flood to come.

When a Creation scientist studies the Grand Canyon, he or she sees evidence of a worldwide Flood as described in the Bible.

Throughout our study of geology, we will learn more about places like the Grand Canyon and the clues on the earth's surface that point to a worldwide Flood. We will also see the evidence of God's grace upon His creation.

The LORD by wisdom hath founded the earth; by understanding hath He established the heavens. By His knowledge the depths are broken up, and the clouds drop down the dew.

Proverbs 3:19–20

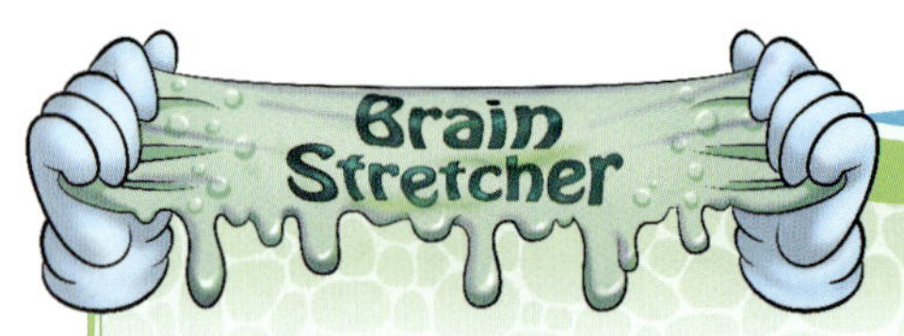

The Grand Canyon: Here are some fun facts about the Grand Canyon.

- ✓ The Grand Canyon is located in the state of Arizona in the United States.
- ✓ It is considered one of the seven natural wonders of the world.
- ✓ At its widest place, it is about eighteen miles across. In its deepest place, it is over a mile deep.
- ✓ It takes about five hours to drive the perimeter of the Grand Canyon.
- ✓ Many Native American tribes have lived in the area for thousands of years. The Paiute tribe named it "Kaibab," which means "mountain turned upside down."
- ✓ A Civil War veteran, John Wesley Powell, mapped the Colorado River and explored the Grand Canyon in the 1800s.
- ✓ The Grand Canyon was made a national park in 1919. Today, many people enjoy exploring, hiking, and river rafting in the area.

Grand Canyon

Landforms

Throughout your study of this chapter, observe landforms in your area. While your family is driving from place to place, notice natural formations of rock or soil; these are landforms. A landform can be a high place, a low place, or a flat place on the surface of the earth. Some landforms are very large while others are smaller.

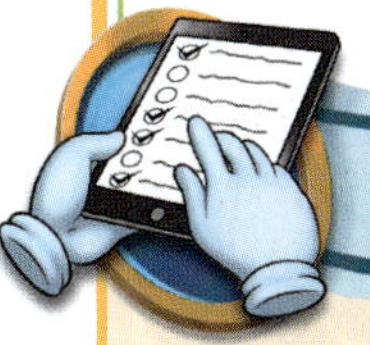

Comprehension Check 7.1

Give the correct answer.

1. What is geology? ______________________________

2. What do we call a scientist who studies the earth? ______________

Think and Conclude.

3. What is the difference between how an evolutionist and how a Creation scientist interpret geology? ______________________________

____

____

____

____

____

____

____

____

____

____

____

____

TERMS

equator: the imaginary line around the earth equally distant from the poles

hemisphere: half of a sphere

crust: a solid layer of rock beneath the soil; made of two kinds—oceanic crust and continental crust

mantle: the layer of the earth beneath the crust

magma: hot, liquid rock within the earth

core: the innermost part of the earth

7.2 The Circle of the Earth

The earth is round like a ball, or sphere. The Bible told us about the earth's shape long before people made this discovery.

> *It is He that sitteth upon the circle of the earth.* *Isaiah 40:22*

The Earth's Shape

Because the earth is shaped like a sphere, special maps called globes can help us to understand what the earth is like. A globe is a sphere-shaped map that looks like a miniature Earth.

We find the North Pole at the top of the globe and the South Pole at the bottom. What do we call *the imaginary line around the earth equally distant from the poles*? The **equator** divides the sphere of the earth in half. Each half is called a **hemisphere**, which means *half of a sphere*. The half of the earth between the North Pole and the equator is the Northern Hemisphere. The half between the South Pole and the equator is the Southern Hemisphere. Which hemisphere do you live in?

To make it easier to find places on the earth, most globes are marked with points and lines. These markings are not actually on the earth, but they are on globes and maps to help us to find places that are north, south, east, or west of other locations. Horizontal lines measure distances north or south of the equator and are called lines of latitude. Vertical lines measure distances east or west of the Prime Meridian in Greenwich, England, and are called lines of longitude.

The diameter of the earth, or the distance from side to side through the center, at the equator is about 8,000 miles. The diameter measured from the North Pole to the South Pole through the center is a little shorter. This means that the earth is not perfectly round. It is slightly flattened at the poles. However, the difference is so small that when the earth is viewed from space, it appears round.

Maps and globes are often updated as new information about the earth becomes available. Satellites orbiting the earth have been very useful to map-makers. The *Landsat* series of satellites began mapping the earth in 1972, revealing information about our planet that had previously been unknown. A photograph taken in 1973 showed an island that no one knew existed off the coast of Labrador, Canada. The island has been named Landsat Island.

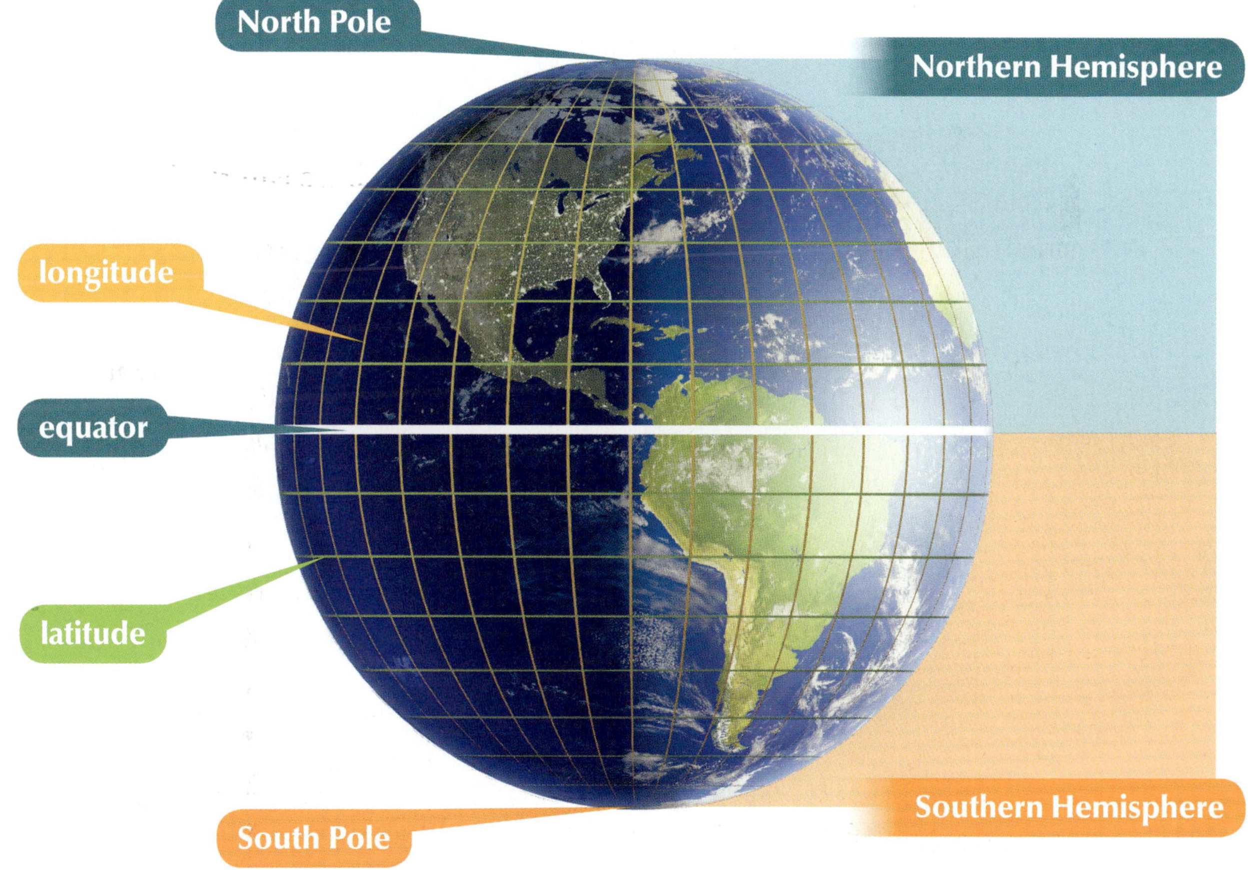

Try This! Measure the circumference of a sphere.

Materials needed:

- ✓ apple
- ✓ long piece of string
- ✓ scissors
- ✓ ruler
- ✓ friend

1. We can measure the circumference of a sphere. The circumference is the distance around the outside of an object. Have your friend hold the apple with one hand at each end while you wrap the string around the center of the apple. Try to place the string around the middle of the apple where the equator would be if the apple were the earth.
2. Cut the string where the ends meet. Measure the string in centimeters; this is the circumference of your apple.

You could not measure the circumference of the earth with a string because you would need a piece of string almost 25,000 miles long! That is the circumference of the earth at the equator.

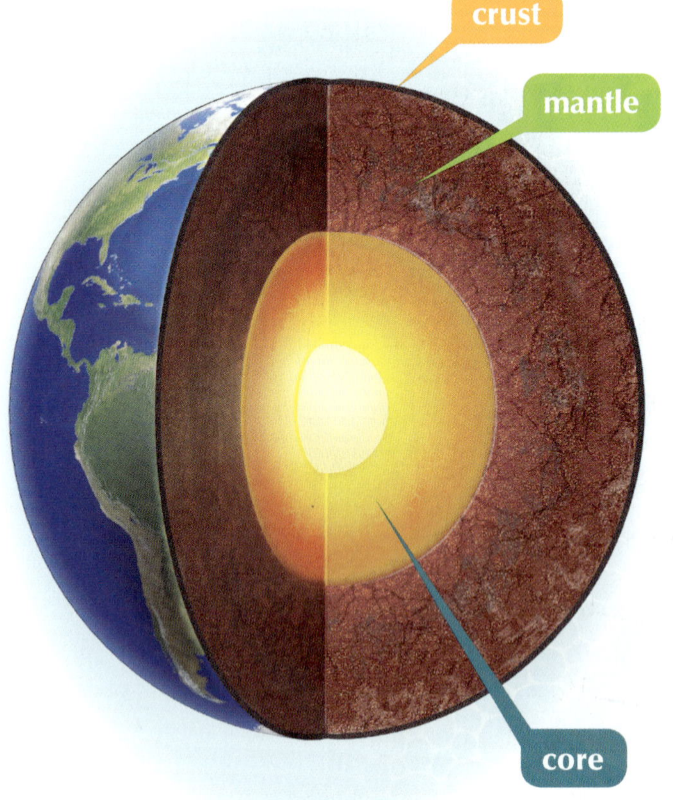

The Earth's Layers

The **crust** of the earth is a *solid layer of rock that begins beneath the soil* and continues about three to forty-five miles below the earth's surface. *There are two kinds of crust—continental* and *oceanic*. The continental crust, which lies beneath the land areas that make up our continents, is very thick. Oceanic crust is much thinner. People have drilled and dug into the earth's crust, but no one has ever gone beneath the crust to the layers below. If the earth were as small as an apple, the crust would be about as thin as the peel of the apple.

The layer of the earth beneath the crust is called the **mantle**. If the earth were apple-sized, the mantle would be about as thick as the white part of the apple. Geologists believe that the mantle is made mostly of *solid, hot, dense rock*. In some places, however, the rock has melted. This *hot, liquid rock* is called **magma**. The mantle is about 1,800 miles deep.

The innermost part of the earth is called the **core**. If you compare an apple's center core to the earth's core, the name will be easy to remember. Geologists think that part of the core is melted iron and nickel. The innermost portion of the core is probably solid because of the great pressure of all the rock above it. *The temperature is higher deeper in the earth. This means the core is the hottest place in or on the planet.*

Comprehension Check 7.2

Matching: *Write the letter of the correct answer in the blank.*

_______ 1. half of a sphere

_______ 2. solid layer of rock that begins beneath the soil

_______ 3. layer of the earth beneath the crust

_______ 4. hot, liquid rock

_______ 5. innermost part of the earth

A. corals
B. core
C. crust
D. hemisphere
E. magma
F. mantle

Give the correct answer.

6. Magma is found in what layer of the earth? ____________________

7. What is the hottest place on the earth? ____________________

Label the layers of the earth.

8. ____________________

9. ____________________

10. ____________________

oceans: the earth's largest bodies of water

groundwater: water found beneath the earth's surface that supplies wells and springs

continents: the largest land masses that rise out of the oceans

crustal plates: large areas of the earth's crust

earthquake: a trembling or shaking in the earth's crust caused by plate activity

fault: a break between two moving plates

volcano: a place in the earth's crust where magma can erupt as lava

landform: a natural formation of rock or soil on the earth's surface

7.3 Water and Land

The Earth's Water

How much water is there on the earth? Just looking at a globe shows you why the earth is sometimes called the Blue Planet. *More than 70% of the earth is covered with water.* Besides oceans, there are lakes, ponds, rivers, streams, and wetlands. The polar ice caps and glaciers make up about 69% of the earth's fresh water. If all the ice in the world melted, there would be enough water to raise the level of the oceans over 250 feet.

There is enough water in the oceans to cover all land on the earth more than a mile deep, if the land were all the same level and the valleys of the ocean were filled.

The **oceans**, which are *the earth's largest bodies of water*, are all connected to each other, forming one vast body of water which we often call the sea.

There are five oceans—the Pacific Ocean, the Atlantic Ocean, the Indian Ocean, the Arctic Ocean, and the Southern Ocean. Locate them on the map below.

Oceans of the World

> *The sea is His, and He made it:*
> *and His hands formed the dry land.*
> *Psalm 95:5*

Water is found not only in oceans, lakes, and rivers, but also in the soil. Have you ever felt water in the moist soil of a garden? The ground soaks up the water from runoff. It seeps into crevices and underground caves, forming a large network of rivers and lakes. The force of gravity causes water to flow downward, and some of it eventually soaks into the ground. We call this groundwater. You could dig a hole anywhere on the earth, and sooner or later you would find water if you dug deep enough. Some groundwater can come to the surface, bubbling up as a spring.

Groundwater *is always found beneath the earth's surface, supplying not only springs but also wells.* Well water is what many people drink.

Mammoth Cave
Kentucky, USA

Mammoth Cave,
Western Australia

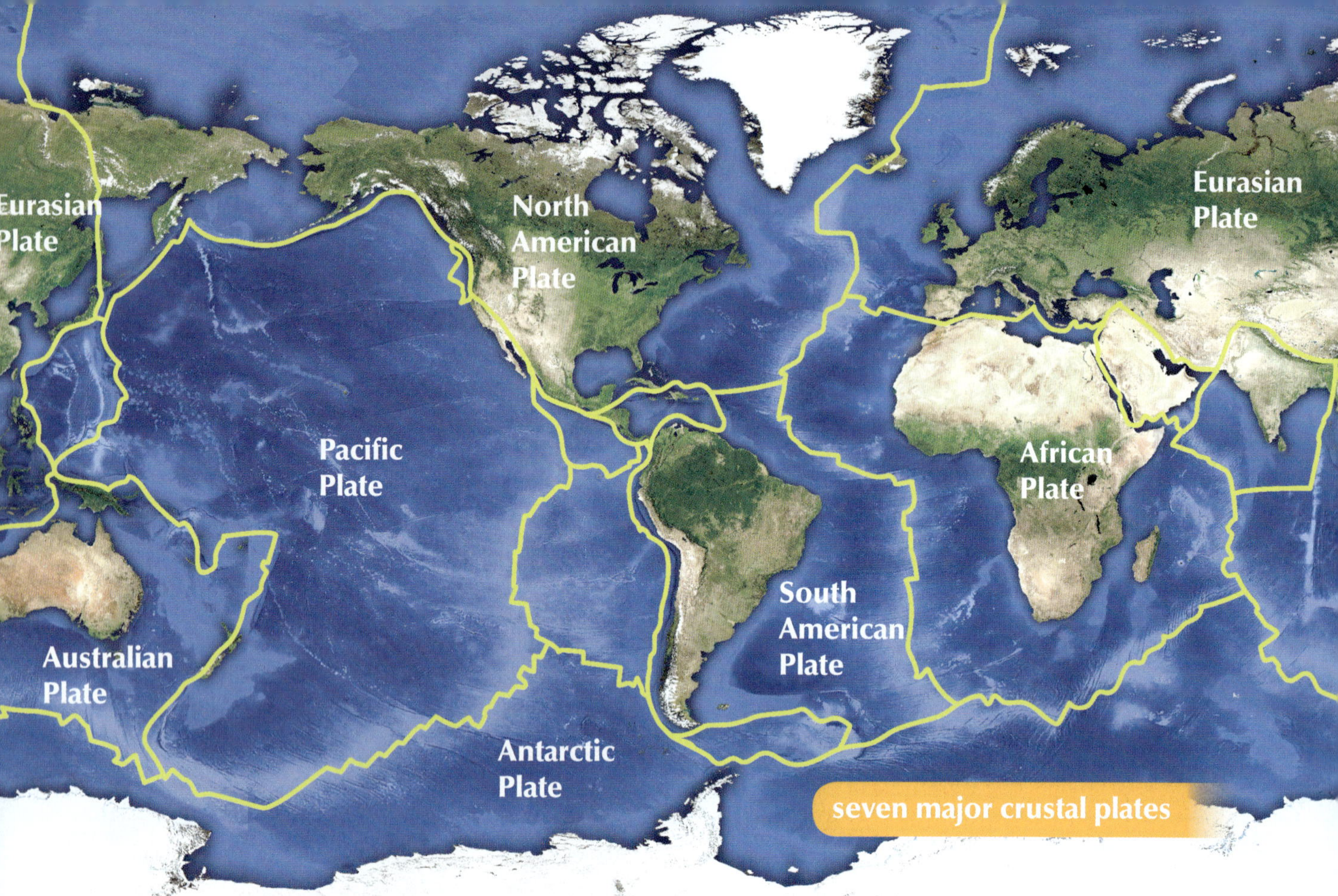

seven major crustal plates

The Earth's Land

The largest land masses that rise out of the oceans are **continents**. *There are seven continents on the earth—Asia, Africa, North America, South America, Antarctica, Europe, and Australia.* Locate the seven continents on a map and on a globe. Did you notice that most of the continental land areas of the world are in the Northern Hemisphere?

Geologists believe that *the earth's outer crust is divided into* seven major *areas called plates* and several minor plates that fit together like a giant puzzle. **Crustal plates** *are large areas of the earth's crust.* These plates float on the molten rock of the earth's mantle. *Where the plates meet with each other, geological events can happen, such as volcanic eruptions and earthquakes.* By studying the plates, some Creation scientists believe that all the continents were connected together at the time of Creation but separated because of Noah's Flood. This could have happened when the "fountains of the great deep" were broken up as described in the Bible. Some scientists say this is a very reasonable hypothesis because of the location of the plates and the way the continents could have fit together. This hypothesis is only a theory, or unproven idea, because there is no way to observe and experiment about something that happened thousands of years ago. Creation scientists believe that plate activity caused major earthquakes and volcanic eruptions to happen during the Flood. *Most geologists believe plate activity still causes earthquakes and volcanic eruptions today*.

> *In the six hundredth year of Noah's life, in the second month, the seventeenth day of the month, the same day were all the fountains of the great deep broken up, and the windows of heaven were opened.*
>
> *Genesis 7:11*

An **earthquake** is *a trembling or shaking in the earth's crust* caused by plate activity. Scientists believe this happens because *plates of the earth's crust push on each other*. The ground can move up or down; sometimes, it even moves sideways. Earthquakes are more likely to happen where there is a **fault**, or *a break between two moving plates*. An earthquake can trigger other natural disasters, such as landslides; avalanches; or tsunamis [tso͞o·nä′mēz], which are sometimes called tidal waves.

Mount St. Helens erupting

A **volcano** *is a place in the earth's crust where magma can erupt as lava*. Most volcanoes are located between the plates. For example, at the edge of the Pacific Plate, there is an area called the Ring of Fire made of hundreds of active volcanoes located along the fault lines. Some volcanoes in the world are not active, but *dormant*. This means they have not erupted in many years. Other volcanoes are considered *extinct*. An extinct volcano has not erupted for thousands of years, and *geologists do not believe that extinct volcanoes can erupt again*.

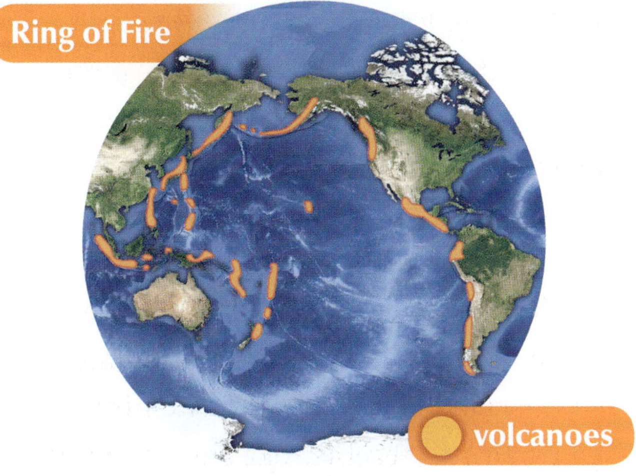

Ring of Fire

Landforms

If you were to travel across the United States or throughout a continent, you would see that the surface of the earth has a variety of landforms. *A* **landform** *is a natural formation of rock or soil on the earth's surface.* A landform can be a high place, such as a mountain, hill, or plateau; or a low place, such as a valley or canyon. A landform can even be a flat place, like the plains. Some landforms hold bodies of water, such as oceans, lakes, and rivers. It is possible that many landforms were made during the Flood and as the earth was recovering after the Flood. As the waters retreated to the places God prepared for them, high and low places were formed.

The waters stood above the mountains. At Thy rebuke they fled; at the voice of Thy thunder they hasted away. They go up by the mountains; they go down by the valleys unto the place which Thou hast founded for them. Thou hast set a bound that they may not pass over; that they turn not again to cover the earth.

Psalm 104:6–9

If you examine a physical map of the world, you will notice that most mountains are part of mountain ranges, or chains. Some Creation scientists believe that mountain chains rose up as the earth's plates were pushing on each other at the time of the Flood.

By studying the rock layers, geologists believe that plate activity caused different kinds of mountains to form. Some mountain chains are called *folded mountains* and may have been made when softer, forming rock folded up together as plates moved. The rock layers of folded mountains curve and bend. The *Rocky Mountains* in the United States are *folded mountains.* Try to locate other folded-mountain chains, such as the Alps, the Andes, and the Himalayas, on a physical map.

Plate activity could have also caused *fault-block mountains* to form. A fault-block mountain might have formed when a part of the earth's crust thrust upward while other parts of the crust fell downward. The Sierra Nevada Mountains in California are fault-block mountains.

Understanding the earth's land and water reminds us that God is powerful. Our God can create and move mountains. Nothing is too difficult for Him.

> *He putteth forth His hand upon the rock; He overturneth the mountains by the roots.* *Job 28:9*

Try This! Demonstrate how folded mountains may have formed.

Materials needed: ✓ slice of bread ✓ cracker ✓ flat surface

1. Place the bread on a flat surface to represent soft, forming rock.
2. Push the sides of the bread in toward its center to represent plate activity. The center should rise upward.
3. Repeat steps 1–2 using a cracker. What happened?

The cracker did not fold because it was hard. Folded mountains must have formed before the rock was hardened.

Comprehension Check 7.3

Fill in the Blank: *Write the correct answer in the blank.*

1. The water that supplies wells and springs is groundwater.
2. Geologists believe the earth's outer crust is divided into plates.
3. Plate activity causes earthquake and volcanic eruptions.
4. A fault is a break between two moving plates.
5. A volcano is a place in the earth's crust where magma can erupt as lava.
6. Geologists believe an cnent volcano will not erupt again.
7. Mountains, valleys, canyons, and rivers are landform.
8. The Rocky Mountains are folded mountains.

Think and Conclude: *Read* Genesis 1:9, 10 *and* Genesis 7.

9. Can you think of a time in the history of the world when the earth was completely covered with water? Where do you suppose all the water came from? the great flood

humus: the soft organic material in the soil made from the decayed remains of living things

mineral: a substance found naturally in the earth that has a crystal-like structure

crystal: a solid made of atoms that are always arranged in a precise, repeating pattern

gem: a rare and valuable mineral

loam: mixture of sand, silt, and clay

soil horizons: the layers contained in most soil—humus, topsoil, subsoil, and bedrock

bedrock: the solid rock beneath the soil; part of the earth's crust

7.4 Soil and Its Horizon Layers

Soil's Ingredients

Humus

Most of the earth that is not covered by water is covered by soil. *Soil contains two main ingredients: humus and minerals*. You will remember that **humus** *is the soft organic material in the soil made from the decayed remains of living things*. Because humus is soft, it can soak up water. Humus also returns nitrogen to the soil. If the soil were not being constantly enriched by humus, it would lose its ability to provide nourishment for plants.

Minerals

Minerals in the soil are tiny pieces of rock. A **mineral** *is a substance found naturally in the earth that has a crystal-like structure*. According to geologists, a substance must have the following characteristics to be called a mineral:

1. A mineral is never manmade.
2. A particular mineral is always made up of exactly the same kinds of atoms, no matter where on Earth that mineral is found.
3. The atoms of a mineral are *always arranged in a precise, repeating pattern* that often forms a **crystal**.

Many substances form crystals. These substances often have smooth, flat surfaces that meet in sharp edges and corners. If you look at salt with a magnifying glass, you will see its little box-like crystals, or cubes. Other minerals form crystals in shapes like pyramids, diamonds, and needles. Crystals are beautiful examples of God's creativity.

There are about 2,000 different minerals in the world, but only about 100 of them are common. Salt, sulfur, and copper are common, useful minerals. The lead in your pencil is made from the mineral graphite. Gold and silver are beautiful and valuable minerals. Some minerals are very rare and costly, such as diamonds, rubies, and emeralds. *These rare and valuable minerals are called* **gems**. Many useful minerals and precious gems can be found in mountain rocks.

Soil Variation

Although you will find soil all over the world, not all soil is the same. Some soils are dark brown. Others are gray, yellow, or red. *The color comes from the different minerals found in the soil.* Do you remember which mineral made the soil of the African savanna red? Which mineral gives desert sand different textures? Iron makes the African soil red, while quartz gives the desert sand its texture.

It is part of God's plan to have a variety of soil. The different soils are designed for different types of plants. Rice needs to grow in soil that can hold a great amount of water while carrots and other root vegetables need well-drained soil that does not hold moisture.

Soil Components

The best topsoil contains **loam**, *a mixture of sand, silt, and clay.* Each element of loam is a different-sized particle that helps the soil in a certain way. *Sand* is made of larger, coarse particles of rock, most commonly quartz. It allows air into the soil. *Silt* is the name for medium-sized particles of soil. Very fine particles of soil are called *clay.* Clay holds water in the soil. Loam and humus work together to make the best topsoil.

Do you remember what makes topsoil different from subsoil? Topsoil is the top layer of soil and is dark because it is rich in humus and nutrients. Topsoil is also softer and more absorbent than subsoil. Subsoil is the layer of soil below the topsoil. It is lighter in color because it does not have as much humus as the topsoil.

Although most land is covered by soil, the depth of the soil is different around the world. In some places, the soil may be very shallow. In other places, you could dig a deep hole before you finally strike rock.

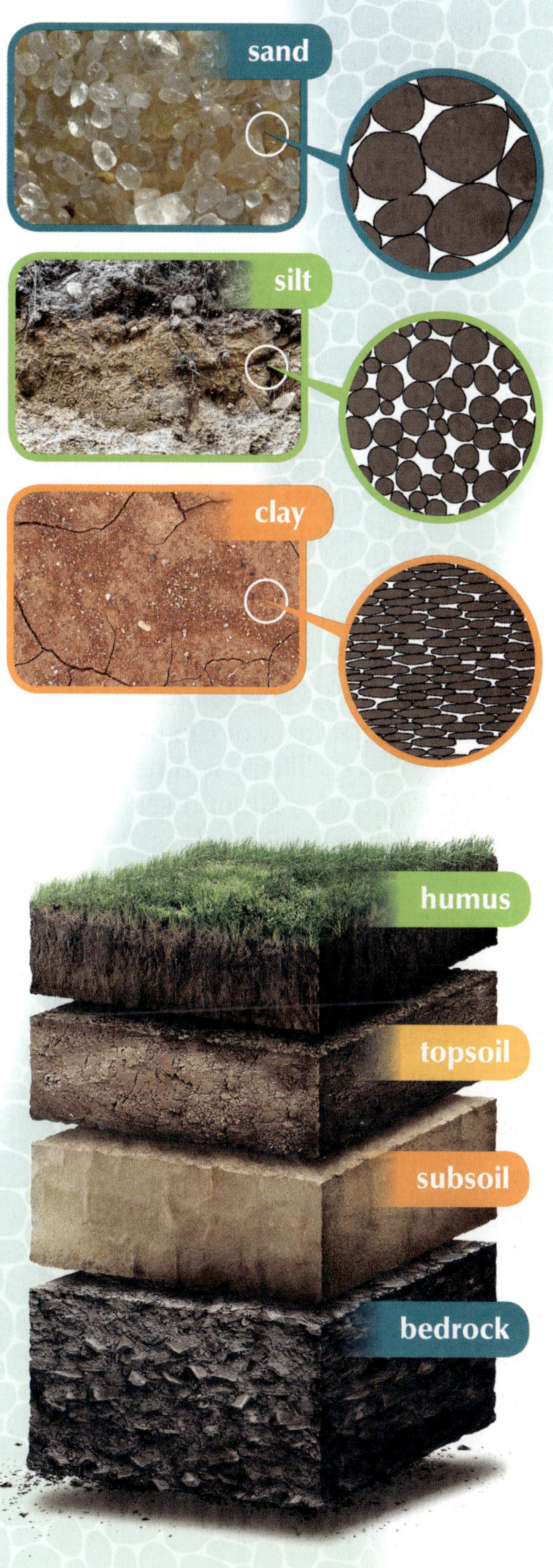

Soil Horizons

Though different places have different kinds of soil, most soil has four layers, or **horizons**: *humus, topsoil, subsoil, and bedrock.* You will remember that *humus* is found in topsoil. The greatest amount of humus is at the very top. Below the humus is what we call *topsoil* itself. Topsoil is rich in nutrients and able to hold water because of its humus. What do we find below the topsoil? *Subsoil* is harder and more packed down than topsoil. It is also lighter in color than topsoil. The last horizon layer is *rock.* Below loose rock we find bedrock. **Bedrock** *is the solid rock beneath the soil and is part of the earth's crust.*

Try This! Make crystals.

Materials needed:

- ✓ drinking glass
- ✓ measuring cup
- ✓ tablespoon
- ✓ boiling water
- ✓ spoon
- ✓ borax
- ✓ pencil
- ✓ piece of cotton yarn
- ✓ goggles
- ✓ mask
- ✓ gloves

1. Wear laboratory safety equipment during this activity. Borax is not harmful, but can cause irritation to some people's skin, eyes, or lungs.
2. Measure 1 cup of boiling water and 3 tablespoons of borax and place in the drinking glass. Stir until most of the borax is dissolved. There will be some borax that will not dissolve in the bottom of the glass.
3. Suspend a piece of yarn into the water from a pencil that is placed across the rim of the glass. You should see some crystal formation within a few hours.

Notice the precise, repeating pattern in the way your crystal's atoms are arranged.

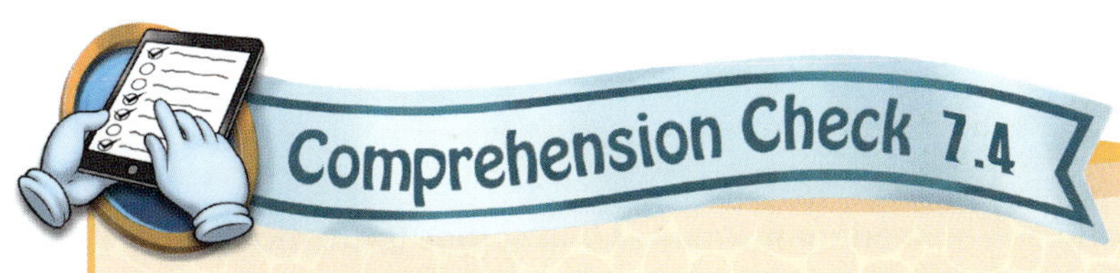

Multiple Choice: *Circle the correct answer.*

1. The organic matter in soil is __?__.

   a. humus   b. loam   c. minerals

2. __?__ are made of atoms which are always arranged in a precise, repeating pattern.

   a. Crystals   b. Energy   c. Humus

3. __?__ are rare and valuable minerals.

   a. Crystals   b. Gems   c. Salts

4. The color in soil comes from its __?__.

   a. gems   b. minerals   c. texture

5. __?__ is a mixture of sand, silt, and clay.

   a. Bedrock   b. Loam   c. Subsoil

Give the correct answer.

6. What are the two main ingredients in soil? ______________________

____

Label the soil horizons.

erosion: the loss of soil by water or wind

weathering: the breaking down of rock by the forces of nature

sediment: matter that settles to the bottom or sides of a body of water

7.5 Water Affects Soil

Soil Erosion and the Weathering of Rock

We know that natural disasters, such as earthquakes, volcanic eruptions, and landslides, can cause the surface of the earth to change very quickly. *Erosion and weathering can change the earth's surface either rapidly or slowly.*

Erosion

Soil **erosion** *is the loss of soil by water or wind.* Water constantly changes the earth's soil and crust. Perhaps you have seen the gullies that have been cut away from the side of a hill as a result of running water. *Water is seldom at rest. The force of gravity pulls it downward.*

Water runoff shows us how gravity works. When water falls to the ground as rain, much of it runs off the surface of the ground to form tiny streams. The streams come together to form bigger streams. The larger streams often end in a pond, lake, or ocean, where the downward flow of water has finally stopped. Gravity gives moving water its energy. As it travels, water easily carries away the topsoil from bare hillsides. As glaciers move slowly over land, they break rocks and soil apart and push them along their paths.

Weathering

Rocks are broken down, or **weathered**, *by the forces of nature,* such as rain, wind, frost, ice, animals, or root systems. Temperature changes are often an important part of the process of weathering. Frozen water seeps into the cracks of rock. Do you remember what water does when it nears its freezing point? As the water freezes, it expands with enough force to crack the rock farther apart.

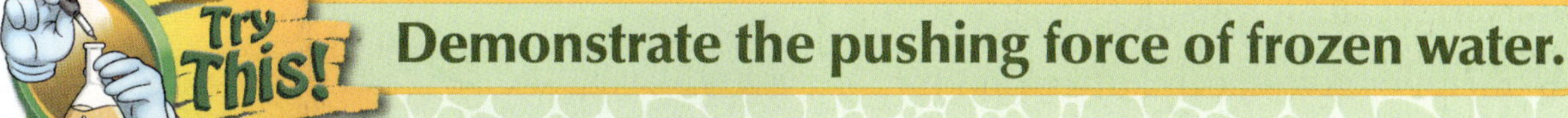

Demonstrate the pushing force of frozen water.

Materials needed:
- ✓ empty plastic water bottle
- ✓ water
- ✓ freezer

1. Completely fill the water bottle before placing its cap back on.
2. Place it in the freezer. After a day, observe the water bottle. What happened?

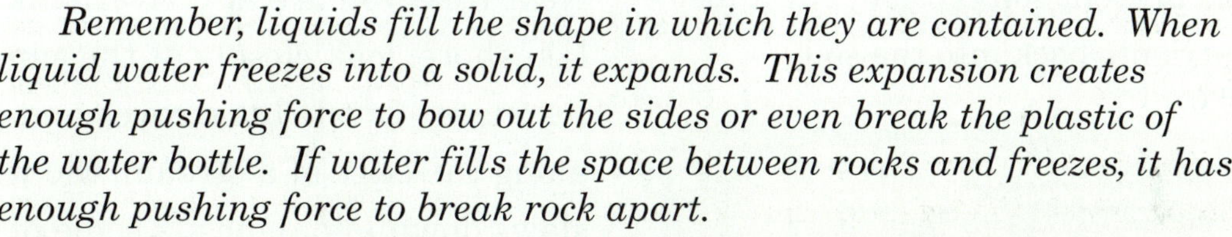

Remember, liquids fill the shape in which they are contained. When liquid water freezes into a solid, it expands. This expansion creates enough pushing force to bow out the sides or even break the plastic of the water bottle. If water fills the space between rocks and freezes, it has enough pushing force to break rock apart.

Soil Conservation

You will remember that conservation is the management or preservation of a natural resource. Soil is one of the many natural resources God has given to us. Like any natural resource, fertile soil is a gift from God and should be used with care. Farmers and gardeners use a variety of methods to conserve soil. Soil erosion is reduced when trees are planted and strips of grass are left between fields. The leaves of trees keep the rain from beating down directly on soil and washing it away. Leaf litter eventually forms humus, which easily soaks up water. Root systems absorb water and hold the topsoil in place. Some farmers also plant crops, such as alfalfa or clover, to cover and protect the soil. These crops return nitrogen to the soil, making it fertile.

Since the soil on hillsides erodes easily, farmers give special attention to the way hillsides are planted, or cultivated. Contour plowing is one way to conserve soil on hillsides. Plowing the slope across instead of up and down forms ridges which slow down the flow of the water. In other places, sloping land is terraced by building the land up into wide, flat areas that make the hillside look like an enormous staircase. Soil built up in this way is able to hold the rain because terraces, or ridges, prevent running water from forming gullies. After the harvest, farmers often use a chisel plow to turn over the plant stubble that is left behind. Not only does this prevent soil erosion, but it also adds nutrients back into the soil.

contour farming

terraced rice fields

Soil Building

Water not only tears things down, but it also works to build things up. Next time you are at a riverbank or lakeshore, look closely at the sand along its edges. In the little coves and pools along the edge of a stream, the water slows down and deposits some of the fine bits of rock, sand, and decaying organic material it has been carrying. Farther downstream, the water's current slows down, and even more of the particles it carried settle in the riverbed. *Matter that settles to the bottom or sides of a body of water is called* **sediment**. A river can carry its sediment all the way to its mouth and deposit it there. Eventually, a delta is built by the collecting sediment. Melting glaciers can also leave sediment behind, making the soil more fertile.

Have you ever thought about how powerful flowing water is? It is always working as it tears down highlands and takes the material from them to build up lowlands. Sometimes, changes happen very quickly because of flooding. Though floods can be destructive, there are some benefits from the soil deposits carried by water. Some people of the world depend upon floods which carry rich soils from higher lands down to lower lands used for farming.

Comprehension Check 7.5

Give the correct answer.

1. What can cause the earth's surface to change rapidly or slowly?
   erosion and weathering
2. What two things cause erosion? wind and water
3. Give an example of something that causes the weathering of rock. __________
4. Give an example of a soil conservation method. __________
5. What do we call the matter that settles to the bottom or sides of a body of water? __________

Think and Predict.

6. What would happen to precipitation falling on a hill or cliff? __________

focus: the place where plates move inside the earth's crust; where the energy of an earthquake comes from

tremor: a minor earthquake

epicenter: where earthquake damage is often greatest; located above the focus

tsunami: a set of giant ocean waves often caused by underwater earthquakes or volcanic eruptions

vent: the opening of a volcano where magma, gases, hot ash, and rock can erupt

lava: melted rock that flows from a volcano

7.6 Geological Events That Change the Earth's Surface

In the South Pacific, a volcano called Krakatoa erupted in 1883. It was a very destructive volcanic eruption. So much volcanic ash was released into the atmosphere that it caused red sunsets around the world. Air temperatures were several degrees cooler for three years. Geological events sometimes cause a chain reaction of other events in our weather resulting in changes to the earth's surface. Do you remember what two geological events are caused by plate activity? *Earthquakes and volcanic eruptions happen because of plate activity and can cause rapid changes to the earth's surface.*

Krakatoa Island

Earthquakes

Earthquakes are most likely to happen where there is a break between two moving plates, or a fault. Two plates meet in western California; their boundary is named the "San Andreas Fault." This fault is about 800 miles long and at least 10 miles deep. In 1906, it caused the powerful San Francisco earthquake and resulting fire. Those who live near the San Andreas Fault cannot prevent an earthquake from happening, but city planners and engineers try to keep people safe while preventing some damage. Wider streets help prevent the spread of fires or many buildings toppling on each other. Earthquake-resistant buildings have strong and flexible foundations. A flexible building frame can sway with the earthquake instead of collapsing. Safety drills also help people to know exactly what to do if there is an earthquake.

An earthquake's energy comes from its **focus**, *or the place where plates move inside the earth's crust.* An earthquake gives off waves of energy that radiate from its focus. Some earthquakes have more energy than others. *A minor earthquake is called a* **tremor**. Sometimes, tremors are a warning that a major earthquake could happen. Major earthquakes are very destructive and can create new landforms very quickly. Large areas of land can sink down or rise up. Parts of the coast, such as cliffs, can fall into the sea.

Above the focus is the **epicenter**, *where the damage caused by the earthquake is often greatest.*

If an *earthquake* happens on the ocean floor, it *can cause a tsunami.* A **tsunami** is a *set of rapidly traveling, gentle waves moving throughout the sea and building up energy.* When a tsnuami reaches shore, it can cause a destructive and deadly wall of water as much as ten stories, or one hundred feet, high.

Earthquake Safety

There are many places in the United States where it is possible to experience an earthquake. Find out if you need to be prepared for one where you live. Major earthquakes are not as common as minor tremors, but it is important to know what to do if one were to occur.

If there is ever an earthquake where you live, the safest place to be is under a table or desk. Remember the words, "Drop, cover, and hold on." Drop down under a table or desk, cover your eyes and face with one arm, and hold on to something stable with the other. Wait until the shaking stops before you move. Remember, there could be *aftershocks*, or smaller earthquakes, and even fires. If you are hurt, get help as soon as you are able.

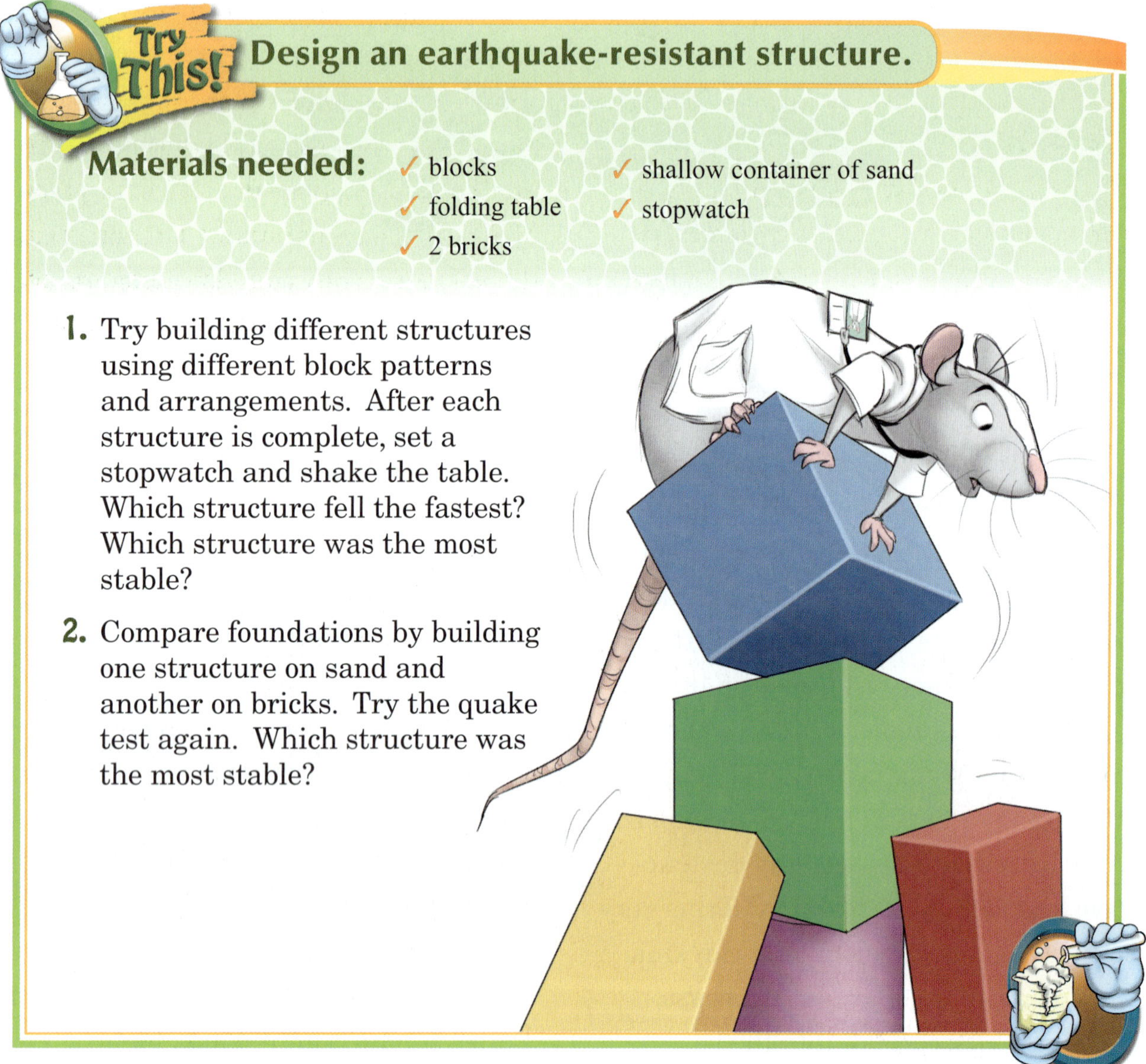

Design an earthquake-resistant structure.

Materials needed:
- ✓ blocks
- ✓ folding table
- ✓ 2 bricks
- ✓ shallow container of sand
- ✓ stopwatch

1. Try building different structures using different block patterns and arrangements. After each structure is complete, set a stopwatch and shake the table. Which structure fell the fastest? Which structure was the most stable?
2. Compare foundations by building one structure on sand and another on bricks. Try the quake test again. Which structure was the most stable?

Volcanoes

Volcanoes can also erupt near places where plates meet. You will remember that geologists believe that beneath the crust, the earth's mantle contains some very hot liquid rock, or magma. Sometimes, *magma erupts out of an opening in the earth's surface*. The place where this eruption can happen is *called a volcano*. Volcanoes usually occur because of a weak spot in the earth's crust. *Magma, gases, hot ash, and rock can all shoot out of a volcano's opening*, or **vent**, during an eruption. When magma erupts out of a volcano, *the liquid rock that flows out is called* **lava**.

A volcanic eruption can cause new landforms to be made very quickly. Islands can rise up out of the sea, and mountains can rise up from the land. The Hawaiian Islands and Iceland were both formed by volcanic eruptions. In Hawaii, the Kilauea [Kē′lou·ā′ə] volcano has been slowly erupting since 1983. Lava keeps spreading out and cooling.

People who live near volcanoes protect themselves by being prepared to evacuate, or leave their homes. Certain roads are marked as an evacuation route, and people make sure they have fuel in their vehicles and emergency supplies ready. Scientists who study volcanoes can often predict that an eruption will take place by monitoring earthquake tremors near the volcano. They can also measure heat and gas coming from the volcano's vent.

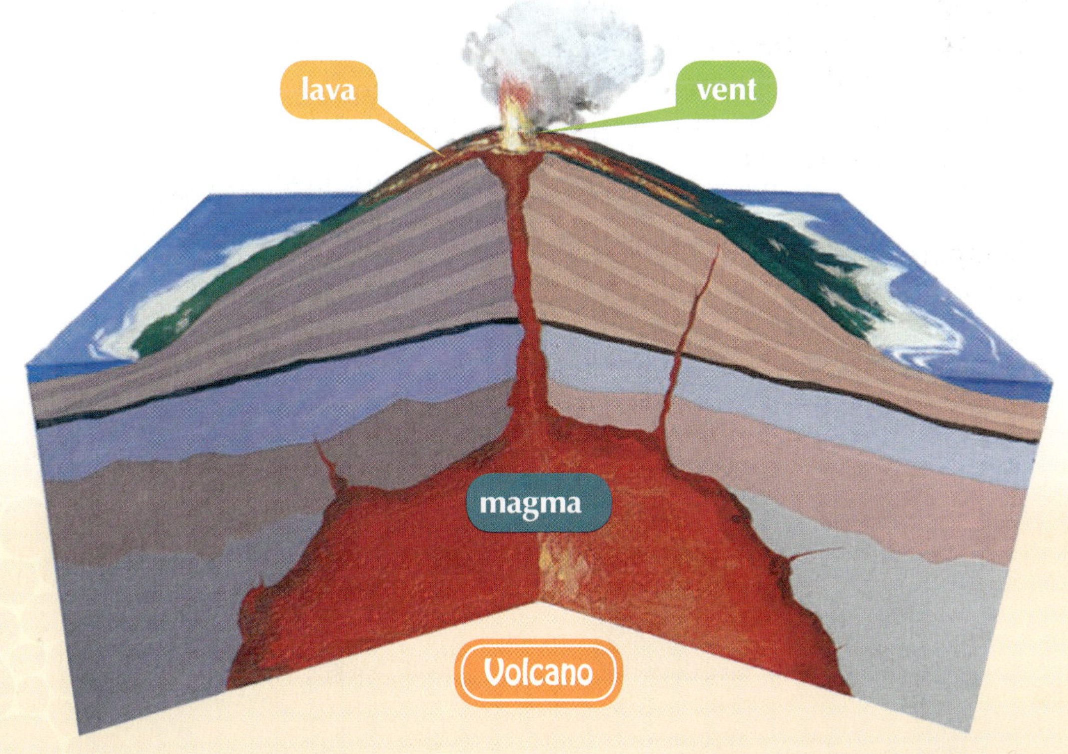

Try This! Construct a volcano.

Materials needed:

- ✓ baking soda
- ✓ tablespoon
- ✓ paper plate
- ✓ funnel
- ✓ play dough
- ✓ beaker
- ✓ vinegar
- ✓ red food coloring
- ✓ safety goggles

1. Place 2 tablespoons of baking soda on the middle of a paper plate. Place the funnel upside down on the plate, covering the baking soda.
2. Using play dough, shape a volcano over the funnel.
3. Mix 50 milliliters of vinegar with a few drops of food coloring in the beaker.
4. Wear goggles and pour vinegar into the funnel. Watch your volcano erupt!

Comprehension Check 7.6

True/False: *If the statement is true, write* true. *If the statement is false, replace the underlined word(s) with a word or phrase that will make the statement true. Do not write* false *in any blank.*

________________ 1. Earthquakes and volcanic eruptions can cause <u>rapid</u> changes to the earth's surface.

________________ 2. An earthquake's energy comes from its <u>epicenter</u>.

________________ 3. The place where the plates moved inside the earth's crust is the <u>focus</u> of an earthquake.

________________ 4. An earthquake's damage is <u>greatest</u> at its epicenter.

________________ 5. <u>Winds</u> cause tsunamis.

________________ 6. The liquid rock flowing out of a volcano is called <u>magma</u>.

________________ 7. A volcano's opening is called a <u>vent</u>.

Think and Conclude: *Discuss your answers.*

8. By looking at the damage, which earthquake had the most energy?

A. 

B. 

7.7 Properties of Rocks

The solid material that forms the earth's crust is rock, making rocks a very important part of God's creation. Can you remember the characteristics of a mineral? Rocks may be made from only one mineral or from a combination of several minerals. Because of the different kinds of minerals found in rocks, they differ from one another in physical properties. *Color, texture, and luster are some of the visible properties geologists use to identify rocks and minerals.*

Visible Properties of Rock

If you have a rock collection, you can easily identify the colors of your rocks by looking at them. Perhaps you have already sorted your rock collection by **color**. It is easy to tell which rocks are shiny or metallic, especially if you place each rock under a light source.

Geologists use light to test a rock for **luster**. Does it sparkle, shine, or reflect light? You can even identify the **texture** of a rock by feeling which ones are smooth and which ones are rough.

Testable Properties of Rock

There are some physical properties of rock that need to be tested to identify the rock, such as hardness, density, cleavage, and streak. You may be thinking, "All rocks are hard. Why do geologists need to identify the hardness of a rock?" Did you know that different rocks have different levels of hardness? Hardness is one of the ways to identify the minerals that make up rocks. *Geologists test the hardness of a rock by the* **scratch test**. Softer rocks can be scratched while harder rocks can scratch other rocks. For example, talc is a mineral that is so soft, parts of it flake off in your hand when you hold it. Talc is used to make talcum powder. A diamond is the hardest mineral of all. It can scratch all other rocks. Some power tools, such as drills and saws, use diamond material to make their blades extra sharp and precise.

Objects with greater density can sink in water. Did you know that some rocks can float? *Pumice* [pŭm′ĭs] *is not dense.* Because it is filled with little air holes, it can float. Geologists can identify the exact **density** of a rock by measuring its volume and mass. If two rocks are the same size and one is heavier than the other, which one is denser? The heavier rock is denser because its molecules are closer together than a lighter rock of the same volume.

Certain rocks always break apart in unique ways. **Cleavage** is when a *rock breaks apart in smooth, flat pieces.* Geologists identify rocks according to their cleavage. Some rocks have cleavage; others do not. If the rock breaks into rough pieces or crumbles, it **fractures**.

The **streak test** *helps geologists identify the true color of the minerals that make up a certain rock.* This is another easy test. To test the streak of a rock, rub it along the rough side of a piece of tile. You will see what true colors are in the minerals of your rock. You may be surprised!

Visible Properties

- ✓ Color—what color is it?
- ✓ Luster—how does it behave under light?
- ✓ Texture—how does it feel?

Testable Properties

- ✓ Hardness—tested with the scratch test
- ✓ Density—tested by measuring volume and mass
- ✓ Cleavage—tested by breaking the rock apart to find fracture or cleavage
- ✓ Streak—tested by the streak test to find the true color of the rock's minerals

Try This! Test a rock for cleavage or fracture.

Materials needed:

- ✓ old pillowcase
- ✓ rocks
- ✓ hammer
- ✓ safety goggles
- ✓ hard surface (such as a driveway or empty parking lot)

1. Protect your eyes by wearing safety goggles during this activity. Place a rock inside a pillowcase and tie it closed; set the pillowcase on a hard surface. Strike the rock several times with a hammer.

2. Remove the rock pieces from the pillowcase and observe how they have broken apart.

If your rock crumbled or came apart in rough pieces, your rock fractured. It lacks the property of cleavage. If your rock separated into flat pieces, your rock has the property of cleavage.

Comprehension Check 7.1

Give the correct answer.

1. What test determines a rock's hardness? scratch test
2. What test determines the true color of the minerals in a rock? streak test

Think and Conclude: *Answer the questions using the choices below.*

3. Which rock has luster? gold and nickel
4. Which rock is not dense? pumice
5. Which rock has cleavage? chalk and nickel and mica

gold **nickel** **chalk** **mica** **granite** **pumice**

igneous rock: rocks that can be formed from hot magma or lava

sedimentary rock: rocks formed from sediment that has been hardened in a cementing process

metamorphic rock: igneous or sedimentary rock that seems to have been changed by heat or pressure

7.8 Three Types of Rocks

Rocks are classified depending upon the way they are formed. Geologists have classified rocks into three main groups: *igneous, sedimentary,* or *metamorphic.* Although rocks are being formed and broken up today by natural processes, we know that God created rock in the beginning when He laid the foundation of the earth. Much of the formation of rock might have taken place on the third day of Creation when God pushed the dry land up out of the seas.

Of old hast Thou laid the foundation of the earth: and the heavens are the work of Thy hands.

Psalm 102:25

obsidian

Igneous Rocks

Rocks that can be formed from hot magma or lava are called **igneous rocks** [ĭg′nē·əs]. The word *igneous* means "fire." This word might remind you of the word *ignite*, which means "to set fire to." Both words come from the same root. *Some common igneous rocks are pumice, obsidian, and granite.*

Pumice and obsidian are igneous rocks that form when lava cools quickly after it comes out of a volcano. The lava actually collects air as it is tossed up before it cools, giving pumice little air holes. Because of its air holes, pumice is not dense and can float on water. Pumice can be ground up and used in soaps and scouring powders. The material dentists use to polish teeth can contain pumice. Obsidian is usually black and shiny and feels smooth with sharp edges. It does not float on water. It forms when lava cools very quickly. Native Americans used it to make arrowheads.

Granite also appears to have formed through the igneous process. It does not come from volcanic eruptions. *Granite forms when magma cools slowly, deep underground.* Most of the earth's continental crust is granite. This abundant and useful rock is extremely hard and can be polished until it is very smooth. It is usually speckled with glittery crystals and can be used for monuments, tombstones, and countertops.

Sedimentary Rocks

Moving water deposits sediment, or fine bits of humus and minerals, on the land over which it flows. **Sedimentary rocks** *are formed from sediments that have been hardened in a cementing process.* Much sedimentary rock is formed from deposits that have settled along the ocean floor. Sediment is pressed together under more sediment until it hardens into rock. If you can imagine how cement hardens into sidewalks and driveways, you can imagine how sediment can harden into rock. Different materials such as sand, mud, or seashells cause different kinds of sedimentary rock to form. *Common forms of sedimentary rock are sandstone, limestone, and shale.*

Sandstone is made of grains of sand cemented together. Its layers are often visible, and if you rub it, you can probably rub off some sand. If you look at sandstone with a magnifying glass, you can distinguish the many different kinds and colors of sand grains. Sandstone is used to make glass, concrete, and sandpaper. Some kinds of sandstone are so hard that they can be used to construct buildings.

Limestone is made of a mineral called calcium carbonate, which is very similar to the baking soda you use to bake cookies. Where did this mineral come from? When some sea creatures die, their shells, which are made of calcium carbonate, settle to the bottom of the sea. Here the shells are pressed together until they form solid rock. Limestone is often crushed and used to make cement.

Shale can be formed from mud or clay. It even smells like mud when it gets wet. Shale is the most common sedimentary rock. It is very soft and can be crushed to make tiles and bricks. It is often found with layers of sandstone or limestone.

shale formation

Try This! Make sedimentary "rock" layers.

Materials needed:

- ✓ jar with lid
- ✓ beaker
- ✓ water
- ✓ Epsom salts
- ✓ scale (or ¼ cup measuring cup)
- ✓ sand
- ✓ pebbles
- ✓ organic material (such as dried grass, leaves, and twigs)

1. Add sand, pebbles, and organic material to the jar. Next, add two ounces, or ¼ cup, of Epsom salts and enough water to fully cover the ingredients plus one or two inches.
2. Tightly close the jar with a lid and shake the water and sediments together. Set the jar on a table and check it every hour. Notice what sediments sink to the bottom of the jar first. When the sedimentary layers are fully settled, remove the lid and carefully pour out the water that is on top. Allow your sedimentary layers to dry into "sedimentary rock."

Metamorphic Rock

What animals undergo metamorphosis? Metamorphosis is the change of form that insects and amphibians undergo as they grow from an egg to an adult. Rocks can also undergo a change of form. **Metamorphic rock** *is igneous or sedimentary rock that seems to have been changed by heat or pressure. Marble and slate are common forms of metamorphic rock.*

Marble is one of the most beautiful kinds of rock. Marble may begin as limestone and be changed by heat and pressure. Like granite, marble can be highly polished. It is often used for statues and for very special buildings. The Washington Monument, the Lincoln Memorial, and the Jefferson Memorial are all made of marble. The most famous deposits of marble are found in Italy, where the great artist Michelangelo carved several magnificent statues from large pieces of marble.

Slate is a useful metamorphic rock. Slate, which may begin as shale, is very hard and smooth. In colonial and pioneer days, children used chalk and a small piece of slate rather than paper for their schoolwork.

Metamorphic rock is found in very large areas all over the world. Where would the heat and pressure needed to form these rocks come from? Most Creation scientists believe that metamorphic rock formation had to do with the drastic changes in the earth's environment during and after the Flood.

***True/False:** If the statement is true, write true. If the statement is false, replace the underlined word(s) with a word or phrase that will make the statement true. Do not write false in any blank.*

_______________ 1. Igneous rocks can be formed from hot magma or lava.

_______________ 2. Granite is a kind of sedimentary rock.

_______________ 3. Sedimentary rocks are formed from sediments which have been hardened in a cementing process.

_______________ 4. Sandstone is a kind of igneous rock.

_______________ 5. Metamorphic rock seems to have been changed by water and wind.

_______________ 6. Marble is a kind of metamorphic rock.

TERMS

fossil: the remains or impression of a living thing found in sedimentary rock that has been hardened

transitional form: a term evolutionists use to describe a "missing link" between two kinds of animals to show how one kind of animal could have turned into another kind of animal

7.9 Fossils in Rocks

If you are ever on a road trip with your family, look for places along the road where rocks have been blasted out and carved away to make room for the road. You will see layers of sedimentary rock in these areas. Much *sedimentary rock is formed in the sea,* and yet we find sedimentary rock *on almost 75% of continental land.* Sediments are formed in liquid. How do you suppose the continental land got its sedimentary rock? Creation scientists believe that this is easy to explain when we consider that Noah's Flood covered the entire planet.

Fossils

mammal and fish

fish eating fish

The destructive events triggered by the Flood would have caused mountains to split apart and volcanoes to erupt, spouting lava, ash, and rock everywhere. There would also be a massive amount of organic matter from dead plants and animals. For months, the flood waters churned all this matter together until it settled into layers of sediment. This sediment turned into sedimentary rock.

There are billions of fossils buried in sedimentary rock layers. *A* **fossil** *is the remains or impression of a living thing most often found in sedimentary rock that has hardened.* Fossilization is rare today because unique conditions have to be present for a fossil to form.

What normally happens to an animal's body when it dies in the wild? Scavengers break up the carcass, eating what meat they can find. Decomposers break down what remains, and nutrients return to the soil. *A plant, animal, or footprint can be fossilized only if it has been buried at the same time sedimentary layers were forming.* It would have to happen very quickly. The fossil record shows a sudden, catastrophic event. Some animals have even been fossilized in the middle of a meal. All over the world, scientists find clam fossils with shells clamped down tightly. Clams only do this when they are in danger and want to protect themselves. Looking at fossils shows scientists a "live-action picture" of an event in an animal's life.

Scientists are always finding fascinating things when they study fossils. Animals from different habitats are often buried together, such as when sea creatures are buried with land creatures. Creation scientists are not surprised by this. During the Flood, all kinds of animals would have been buried together. Did you know that fossils are even found on mountaintops? The highest mountain in the world, *Mt. Everest, has ocean-animal fossils in sedimentary rock at its peak.* The rocks that make Mt. Everest would have had to be underwater at one time. Can you think of a time when mountains were covered with water? Mt. Everest is part of the Himalayas, which are folded mountains. Just like every other place on Earth, the Himalayas were once under the waters of the Flood. Creation scientists believe the Himalayas were lifted up as the softer, saturated land and sediments folded up together when crustal plates moved. The fossils in the sediments moved upward along with the rapidly forming rock.

Another fascinating fossil find is the dinosaur tracks and eggs buried in sedimentary rock. The tracks are unusual because they are always found in straight lines, often as if in search of higher ground. Animals run in straight lines only when they are afraid and running away from something.

Egg-laying land animals, even reptiles, shelter their eggs in some way after laying them. Some make nests; others bury them. Yet the fossil record shows piles of dinosaur eggs laid without any shelter. Could it be that these animals had no time to shelter their eggs because they were running away from a catastrophic event?

There is one kind of fossil that scientists have not been able to find. *There has not been one true transitional form in the fossil evidence.* Evolutionists use the term **transitional form** *to describe a "missing link" between two kinds of animals to show how one kind of animal could have turned into another kind of animal*. Scientists have never found a fossil of the missing link between a whale and a dog or a reptile and a bird. Why do you suppose transitional forms have not been found? God fully designed and created each kind of animal. There may be different variations within a kind of animal, such as different breeds of dogs, but a dog cannot evolve into another kind of animal.

dragonfly

coelacanth

pterodactyl

sauropod

Fossils are fascinating because they help us study animals that were alive in Noah's time but are now extinct. Even after surviving a global flood, animals that left the Ark would have entered a world very different from the one they lived in before. The climate would have changed, habitats would have been lost or changed, and some animals might have experienced a food shortage. Whatever the reason, some animals, such as dinosaurs, became extinct.

Imagine what it was like for Noah and the people of his time who lived in a world with dinosaurs. Some were very small, but some were very large. Sauropod [sôr′ə·pŏd′] fossils show that these dinosaurs were the largest land animals that have ever lived. Some sauropods were taller than a five-story building, which is about fifty feet. There are many species of sauropods, but they have a common design—a very long neck and tail and column-like legs.

herd of sauropods

Try This! Make a "fossil."

Materials needed:

- cellulose sponge
- *scissors
- bowl
- hot water
- Epsom salts
- spoon
- plate

1. You may want to cut your sponge into the shape of an animal or a bone.
2. Stir Epsom salts and hot water together in a bowl. Place the sponge in the bowl so that it can soak up the salt water. Wait one day.
3. Take the sponge out of the water and place it on a plate. Wait three days. What happened?

The sponge hardened in a similar way to the way that fossils harden. Fossils can quickly harden if they are buried in sediments with the right chemicals that can fill up the fossil's spaces or pores. Scientists call them petrifying chemicals. The Epsom salts acted as a petrifying chemical to harden your sponge as it filled its air holes.

*optional

"Leonardo": duck-billed dinosaur mummy

brachylophosaurus: duck-billed dinosaur

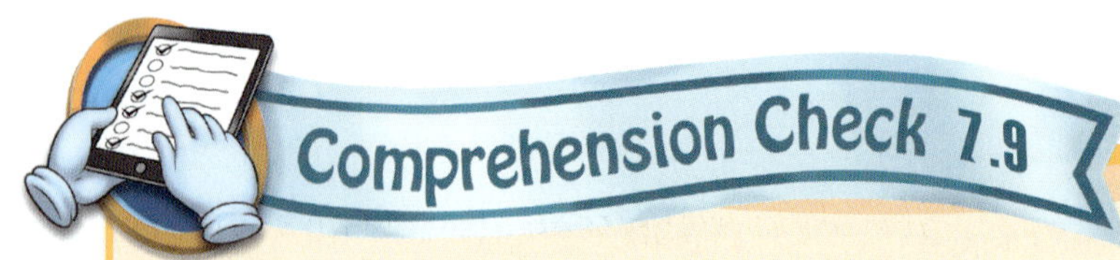

Give the correct answer.

1. Where is sedimentary rock found? in the sea and on 75% of land
2. Where are fossils found? ____________
3. What conditions must be present for fossilization to occur? ____________
4. How do we know Mount Everest was once underwater? ____________
5. Has a true transitional form ever been found in the fossil record? ____
6. Why do you suppose God engineered sauropods with column-like legs and such a large tail? ____________

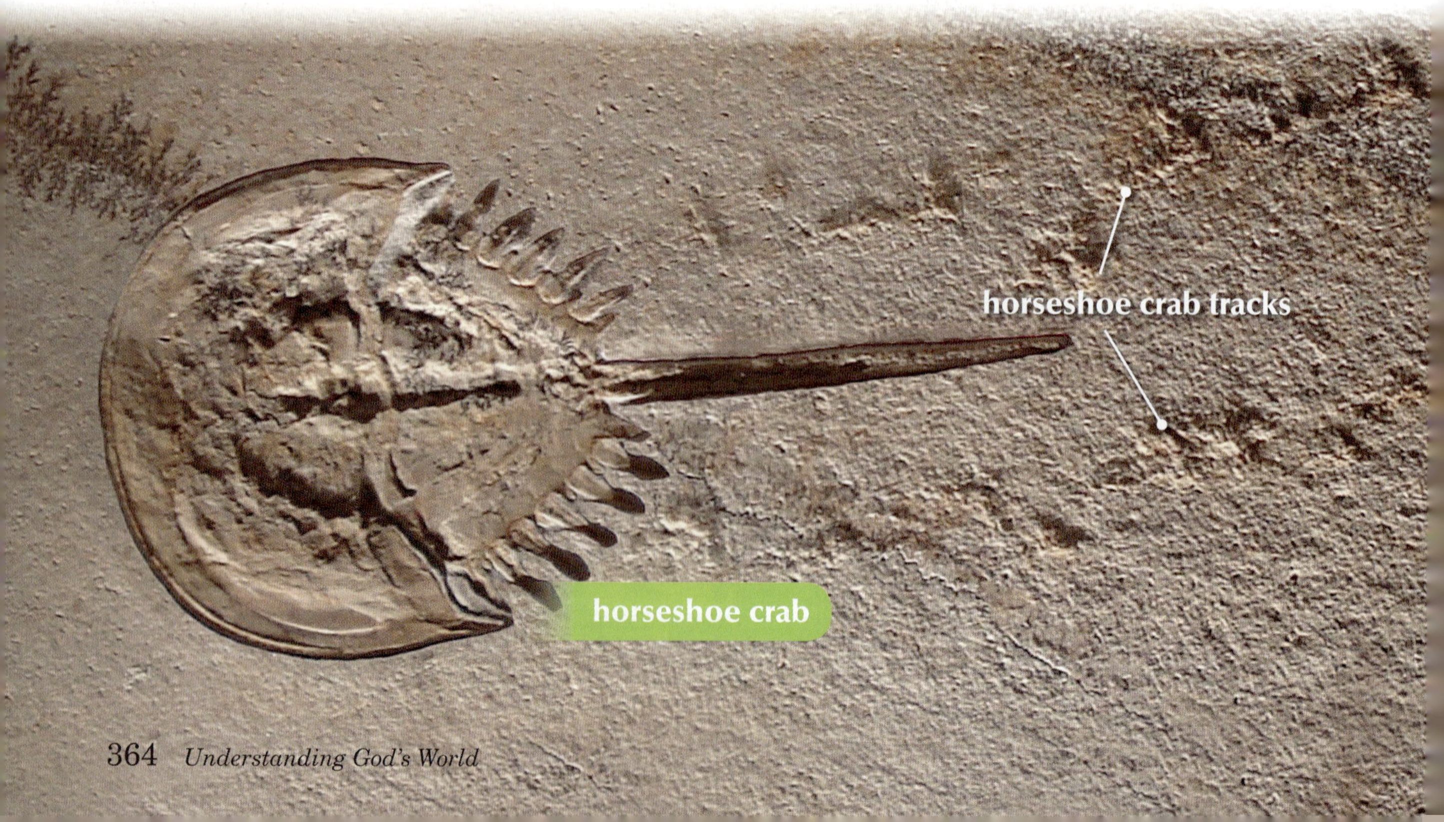

horseshoe crab tracks

horseshoe crab

fossil fuel: a natural fuel, such as coal, oil, or natural gas; made from remains of living organisms

nonrenewable energy source: an energy source which cannot be used again, such as fossil fuels

renewable energy source: an energy source which cannot be used up, such as wind, water, and solar energy

natural resource: a material found in or on the earth which is helpful to people

7.10 Earth's Energy Sources

Because of the Flood, a new kind of energy source was stored away. For some time, no one knew about this kind of energy or had a use for it. Eventually, people began using fuel found deep underground. By the 1800s, coal was used for many things, such as powering steam locomotives. Oil and natural gas helped to light homes and streets. *Coal, oil, and natural gas are sources of energy called* **fossil fuels** *because of the way they were formed.* Fossil fuels are *buried in bedrock.* The pressure of weight and heat turned *the remains of living organisms*, such as plants and animals, into the fossil fuels we use today.

The Formation of Coal

Coal is a type of sedimentary rock that is formed from living things instead of minerals. Its sedimentary layering and hardening is very similar to what we see in sedimentary rock formation. Scientists who believe in evolution claim that coal was formed in swamps, like the Florida Everglades, over millions of years. Dead plant matter settles to the bottom of the swamp, and without oxygen it cannot decompose like other plants do. After some time, the plant matter turns into a material called peat. Those same scientists believe that the peat hardens and turns into coal after millions of years.

Organic material, such as dead plants, simply needs the right conditions to be made into coal. Those conditions include the removal of oxygen and the addition of enough heat and pressure. Did you notice which condition is *not* listed, but is part of the evolutionist's theory? It does not take millions of years or even a long time to make coal. A scientist can make coal in a laboratory in a very short time. Though we do observe that peat is forming in swampy areas, there is not enough heat or pressure for it to become coal. Most Creation scientists believe that coal formed because of the Flood. The rapidly changing environment would have provided the right conditions for coal to form.

People have been using fossil fuels in small amounts for thousands of years and in large amounts for over two hundred years. You might be wondering how there could be this much fossil fuel in the world for people to use. Before the Flood, the land on Earth was filled with vegetation. Fossil records indicate that the earth's climate was much more tropical than it is today. There were types of giant trees and plants which do not grow today. Remember that the Flood was global. A massive amount of plant and animal remains would have made the fossil fuels we use today.

Nonrenewable Energy Sources

Fuel sources need to be burned to release energy. As fuel energy converts to heat energy, it can give other things the power to work. Many vehicles depend on fuel to burn for energy. *As fossil fuels are used, they are burned up and cannot be used again.* There is such a great amount of fossil fuels in the world that it would take a long time to burn them all up. However, once they are gone they will not be replenished. This is one of the reasons scientists try to find ways to use other energy sources. Another concern that some scientists have is the release of carbon into the air when fossil fuels are burned. Some believe this carbon is harmful to the earth's atmosphere; others do not.

Fossil fuels are **nonrenewable energy sources** *because they cannot be used again.* God has so wisely given us **renewable energy sources** as well, *which cannot be depleted, or used up.* Energy sources like *wind, water, and solar energy will not run out as long as the earth remains.*

coal digger

Renewable Energy Sources

Wind

The wind's energy can be used to do work. Long ago, people used the wind's energy to sail large ships across the ocean. Many people still enjoy sailing for pleasure.

Windmills use the wind's energy to provide power for machines. Windmills help collect wind energy. A windmill works when a wheel driven by the wind turns a shaft. The shaft is connected to a machine that does the work. Since the wind does not always blow at a constant speed, windmills cannot be used if a steady supply of power is needed. Windmills are still used today in some places to grind grain and to pump water into storage tanks, but they are mostly used to generate electrical energy.

Water

Water is filled with energy. When it is moving, it has the power to rotate a water wheel. We call water energy *hydropower*. Water dams are a way of collecting water energy that can be changed or converted into electrical energy. Areas in California, Nevada, and Arizona all use the Hoover Dam's hydropower to make electricity to power cities.

hydroplant

windmills

solar panels

Solar

The sun is the earth's main source of light and heat energy. In fact, all energy sources can be traced back to the sun's energy in some way. Wind currents happen because of the sun. The water cycle depends on the sun's energy. Even fuel energy can be traced back to the sun's energy. Fossil fuels would not be buried deep inside the earth if it were not for the living things they once were before the Flood.

Solar panels can collect the sun's energy and convert it into electrical energy. Some people have solar panels on their homes. There are communities that have large solar fields of panels to collect enough energy for many people.

Caring for the Earth's Resources

God has provided us with many natural resources. *A* **natural resource** *is a material found in or on the earth which is helpful to people.* Can you think of some other natural resources we have already studied? God has given us water sources, soil, and minerals which make metals, rocks, and gems. We are blessed with an atmosphere of life-giving gases. Our planet is filled with a variety of food. God has given us many energy sources to power our homes, vehicles, and electronic devices.

Our biblical view of stewardship and conservation encourages us to care for and manage the resources God has entrusted to us. As we use natural resources, we are thankful to our good God Who provides for our every need.

Comprehension Check 7.10

Give the correct answer.

1. Where are fossil fuels found? ______________________________
2. What are fossil fuels made of? ______________________________

   ____
3. Name a nonrenewable energy source. ______________________________
4. Name a renewable energy source. ______________________________
5. What is the earth's main source of light and heat energy? ____________
6. Name a natural resource. ______________________________

Brain Stretcher

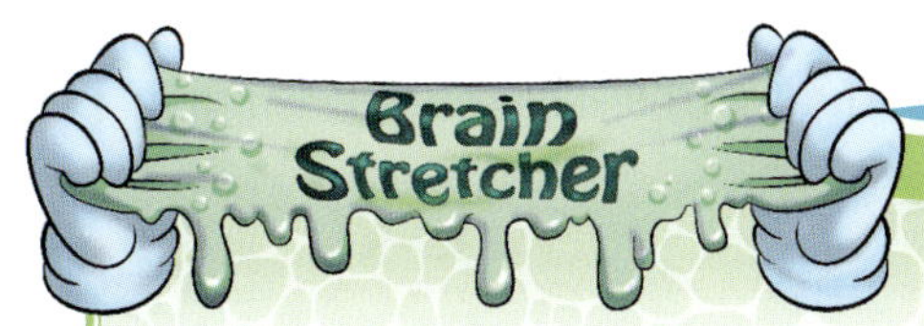

Stewardship: As stewards of God's creation, there are many ways to care for it. We can conserve water or electricity by using only what we need. Turning off faucets and lights have probably become habits to you. Here are some activities which may help you think through how stewardship works in your community.

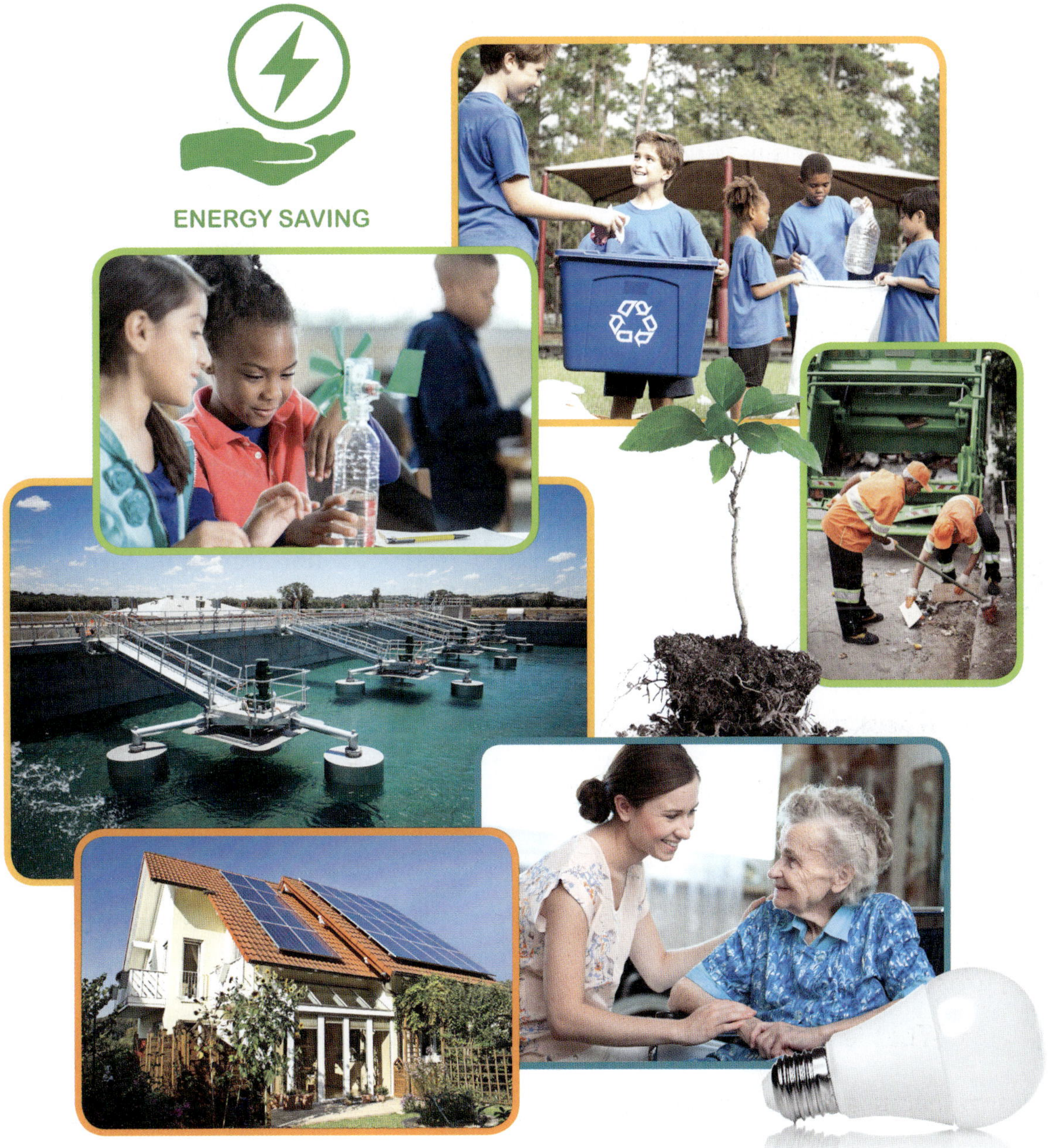

ENERGY SAVING

Renewable Energy Discussion

Each energy source comes with both positive and negative effects on the world around you. Fossil fuels are nonrenewable, and some people are concerned about the gases they give off when burned. Scientists are trying to find more ways to use renewable energy sources.

While God wants us to use the resources on Earth, we should always make sure we do so with thought and care. This is part of our job as His stewards.

Study the chart. Consider the resources, communities, and ecosystems in your area. Discuss the positive and negative effects of each renewable energy source. What could be the positive effects in your area? What could be the negative effects in your area?

Energy Source	Positive Effects	Negative Effects
Solar power	✓ little or no pollution ✓ low-cost maintenance	✓ high construction/start-up costs ✓ expensive to produce electricity ✓ cannot store energy ✓ not accessible for all
Wind power	✓ low operating cost ✓ no pollution ✓ no lack of wind	✓ high construction/start-up costs ✓ cannot store energy ✓ not accessible for all
Water power	✓ can be stored ✓ low operating cost ✓ no pollution ✓ accessible for many	✓ high construction/start-up costs ✓ can destroy homes and habitats

A Geological Case Study

Mt. Saint Helens

In 1980, Mt. Saint Helens erupted in the state of Washington. At the top of this coned mountain was an ice cap. When the volcano erupted, all this ice quickly melted. A blast of steam and a flood of water caused landslide material to rush down the mountain. Some of this material went to the nearby Toutle River, and some flowed into Spirit Lake, which was on the north side of the mountain.

In a matter of hours, layers of sediment were deposited. Within a very short time, layers of ash, lava, and mud formed. It took only a few years for these layers to harden into almost 600 feet of sedimentary rock. The rock layers that were created look very similar to the rock layers evolutionary scientists believe took millions of years to form. This observable case study helped Creation scientists confirm something they already believed—that sedimentary layering can happen in a very short time. It does not take millions of years to make layers of rock. Creation scientists continued to observe Mt. Saint Helens and the area around it as a living laboratory to study the effects of geological, catastrophic events.

continued

mudflows from Mt. St. Helens

Mt. Saint Helens even taught Creation geologists how the Grand Canyon could have formed. In 1982, only two years after the eruption, a mudflow burst forth from a debris dam. The mudflow cut through the newly formed layers of sedimentary rock carving out a canyon. Some people called it the "Little Grand Canyon." A new stream began to flow through the new canyon.

He cutteth out rivers among the rocks; and His eye seeth every precious thing. Job 28:10

Mt. Saint Helens also helped Creation geologists explain how coal could have formed because of the Flood. A large landslide caused a 200-foot layer of rocks to sink to the bottom of Spirit Lake. Thousands of fallen trees from the mountainside landed in the lake as well, creating a giant, floating log mat. As the logs moved about and rubbed against each other, the tree bark and other organic material sank to the bottom of Spirit Lake. This organic material turned into peat very quickly. If another eruption were to happen, more sedimentary layers would be dumped on top of the peat. Enough heat and pressure might cause coal to form.

Spirit Lake

Some of the trees in Spirit Lake also show us how petrified forests could have formed. A petrified forest looks like a fossilized forest. Some trees that landed in the lake are turned vertically as if they had been planted there the whole time. We know from observation, however, that the trees actually grew on the mountainside. They did not grow in the sediment they now sit in; it just looks that way. If the lake were drained and the sediments hardened, it would look just like the petrified forests that are found in other places on Earth.

Scientists were surprised by the way ecosystems have recovered so quickly after the eruption. Trees and plants began to grow back. Gophers helped many of these plants by making the soil ready again. Many rodents, which are tunneling animals, protected themselves from the eruption by burrowing. As they dug new tunnels after the eruption, their digging brought good soil back to the surface. Seeds grew in the new soil. Besides gophers, there are many elk now living in the area, and birds are flourishing. New wetland areas have attracted birds that had not been spotted in that area before. It took some time, but Spirit Lake is recovering.

Spirit Lake

God's Promise

The Great Flood of Noah's day brought much death and destruction. Yet today, we look upon the earth God made and see that it is filled with life. After the Flood, God caused the earth to go through a recovery process. Creation scientists have studied the continental shelves and slopes in the ocean to get clues about how some of this might have happened. They believe that ocean basins sank down and new landforms, such as mountains, rose up. This caused the water to drain off the continents and into the oceans. New land boundaries were established. Noah and his family stepped out of the ark with the faith that God would take care of them. Dormant plant seeds began to germinate and grow. Animals spread throughout the world to the new habitats they were to live in. Noah and his family were commanded to fill the earth with people as well. God made a new covenant, or promise, with Noah that He would never again destroy the earth with water.

> *And I will remember My covenant, which is between Me and you and every living creature of all flesh; and the waters shall no more become a flood to destroy all flesh. And the bow shall be in the cloud; and I will look upon it, that I may remember the everlasting covenant between God and every living creature of all flesh that is upon the earth.* *Genesis 9:15–16*

The earth would never be exactly the same after such a catastrophic event. Habitats and ecosystems may have changed, but God equipped many animals to survive in harsh environments. Many kinds of animals even survived the Ice Age. The water cycle that we observe today probably worked differently before the Flood, but it is exactly what we need. New energy sources were stored away, too. Though imperfect because of sin, the world we live in today is still beautiful because of God's mercy. At the very places where catastrophic plate activity may have happened, we see breathtaking mountain ranges and peaceful coasts.

What does understanding the earth and its foundations teach you about its Creator? We can trust that God is faithful to us and true to His promises.

Chapter 7 Concepts Review

Basic Geology Concepts 7.1–7.3

A. Remember

***Short Answer:** Write the correct answer in the blank.*

1. What does a geologist study? ______________________

2. Name two kinds of geological events that are believed to be caused by plate activity. ______________________

***Identify:** Label the diagram of the globe and the earth's layers below; then circle the name of the earth's layer that contains magma.*

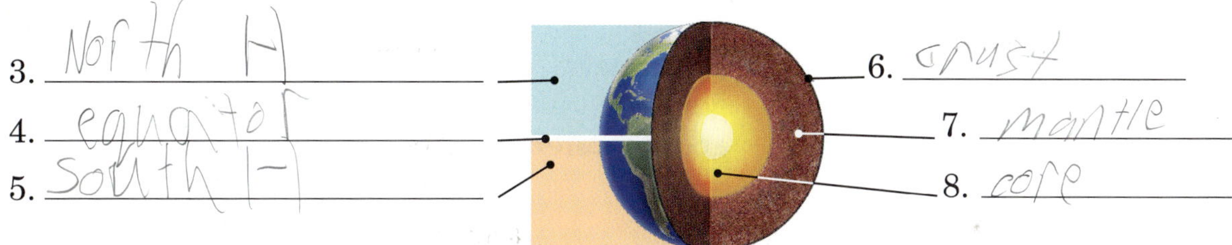

3. ______________________
4. ______________________
5. ______________________
6. ______________________
7. ______________________
8. ______________________

B. Think Like a Scientist

***Short Answer:** Write the correct answer in the blank.*

9. Which layer of the earth has the highest temperature? ______________________

10. Is the earth covered with a greater percentage of water or land? ______________________

11. Where could you predict an earthquake to happen? ______________________

12. Name a type of landform. ______________________

13. Which type of mountain chain was probably made when softer, forming rock folded up together as crustal plates moved? ______________________

C. Fun with Terms

Puzzle: *Fill in the answers to find the word in the starred column.*

★

1.
2.
3.
4.
5.
6.
7.
8.
9.

continents	**earthquake**	**mantle**
core	**geology**	**oceans**
crust	**groundwater**	**volcano**

1. the study of the earth and its structure
2. a trembling or shaking in the earth's crust
3. the earth's largest bodies of water
4. a place in the earth's crust where magma can erupt as lava
5. the innermost part of the earth
6. water found beneath the earth's surface which supplies wells and springs
7. the largest land masses that rise out of the oceans
8. a solid layer of rock beneath the soil; made of two kinds—oceanic and continental
9. the layer of the earth beneath the crust

 a scientist who studies the earth

2 Soil Concepts 7.4–7.5

A. Remember

Identify: *Label the soil horizons.*

B. Think Like a Scientist

Short Answer: *Write the correct answer in the blank.*

5. What is the difference between humus and minerals? ______________________

____

____

6. Name a gem. Ruby

Matching: *Write the letter of the correct answer in the blank. Be prepared to explain your answers.*

A 7. Which picture shows the effects of erosion?

C 8. Which picture shows the effects of weathering?

C. Fun with Terms

Look It Up: *What do you know about these words? If you are not sure, look up the definitions and study them. Be ready to tell what each word means.*

bedrock	**erosion**	**sediment**
crystal	**loam**	**weathering**

3 Geological Events, Rocks, and Resource Concepts 7.6–7.10

A. Remember

Fill in the Blank: *Write the correct word or phrase in the blank.*

1. Earthquakes and volcanic eruptions happen because of plate __activity__ and can cause __rapid__ changes to the earth's surface.
2. An earthquake is most likely to happen at a __fault__.
3. An earthquake's energy comes from its __focus__.

B. Think Like a Scientist

Label: *Write* I *for igneous,* S *for sedimentary, and* M *for metamorphic. Be prepared to explain your answers.*

__I__ 1. granite	__M__ 3. marble	__M__ 5. sandstone
__S__ 2. limestone	__I__ 4. pumice	__S__ 6. slate

Identify: *Circle the properties of rock that are visible. Be prepared to explain how the other properties can be tested.*

cleavage	**density**	**hardness**	**streak**
color	**fracture**	**luster**	**texture**

Sort: *Write each energy source from the box in the correct column below.*

coal	**natural gas**	**solar power**
hydropower	**oil**	**wind energy**

renewable	**nonrenewable**
______________	______________
______________	______________
______________	______________

Short Answer: *Write the correct answer in the blank.*

7. What test would you use to find a mineral's true color? Streaks test

8. What test would you use to find a mineral's hardness? Scracth test

C. Write

Essay: *Write a paragraph to answer the following questions:*

In what kind of rock would you most likely find a fossil? Why?

D. Fun with Terms

Crossword: *Fill in the answers to complete the crossword puzzle. Leave shaded boxes blank.*

Across

1. a type of rock formed from sediments that have been hardened in a cementing process
2. the remains or impression of a living thing found in sedimentary rock that has been hardened
3. a minor earthquake
4. a material found in or on the earth which is helpful to people

Down

5. a set of giant ocean waves caused by underwater earthquakes or volcanic eruptions
6. a natural fuel, such as coal, oil, or natural gas; made from the remains of living organisms
7. either igneous or sedimentary rock that seems to have been changed by heat or pressure
8. rocks which can be formed from hot magma or lava
9. a term evolutionists use to describe a "missing link" between two kinds of animals to show how one kind of animal could have turned into another kind of animal
10. where earthquake damage is often greatest; located above the focus

epicenter	**natural resource**
fossil	**sedimentary**
fossil fuel	**transitional**
igneous	**tremor**
metamorphic	**tsunami**

Understanding Weather

TERMS

climate: the weather conditions an area receives over time

weather: the condition of the air

atmosphere: the air surrounding the earth

8.1 The Atmosphere and Weather

Have you ever wished you could control the weather? When ball games, picnics, and parades are canceled because of rain, there is nothing you can do to clear the sky. Only God can control the weather. Jesus showed His disciples that He is God by calming the stormy sea. God governs all matter and the energy that makes weather happen. Weather gives us a glimpse of our all-powerful God.

But the men marvelled, saying, What manner of man is this, that even the winds and the sea obey Him!
Matthew 8:27

Have you ever wished you understood how weather works? As Junior Scientists, we can learn how God's laws of nature make weather. We can even predict some weather because of God's orderly design.

Weather Happens in the Atmosphere

Weather is different from climate. **Climate** *describes weather conditions a certain area receives over time.* **Weather** *describes the condition of the air* closest to the earth, which is constantly changing. To learn about weather, we must first learn about the invisible blanket of air that completely surrounds us. **Air** *is a mixture of colorless, odorless, and tasteless gases.* You cannot see, smell, taste, or pick up a piece of air; but air is there. You can feel air when it moves. You can see and hear some of the effects of moving air, such as leaves rustling in the trees or seeds being scattered by the wind. There is no place on the surface of the earth that is without air.

The air surrounding the earth is called the **atmosphere**. No other planet has an atmosphere that can support life the way the earth's atmosphere can. People, plants, and animals depend on air to live. You can live for weeks without food and for days without water, but you cannot live for more than a few minutes without air.

The Gases in the Atmosphere

Air is made up of several different gases. The atoms of these gases do not combine to form a molecule of air in the same way that atoms of hydrogen and oxygen combine to form a molecule of water. Instead, they remain separate substances. This is why we call the atmosphere a mixture of gases.

About 21% of air is *oxygen*. What living things depend on oxygen for life? People and animals need oxygen to breathe. But, did you know that plants also need oxygen? While plants need carbon dioxide to make food, they also need oxygen to use the food they have stored.

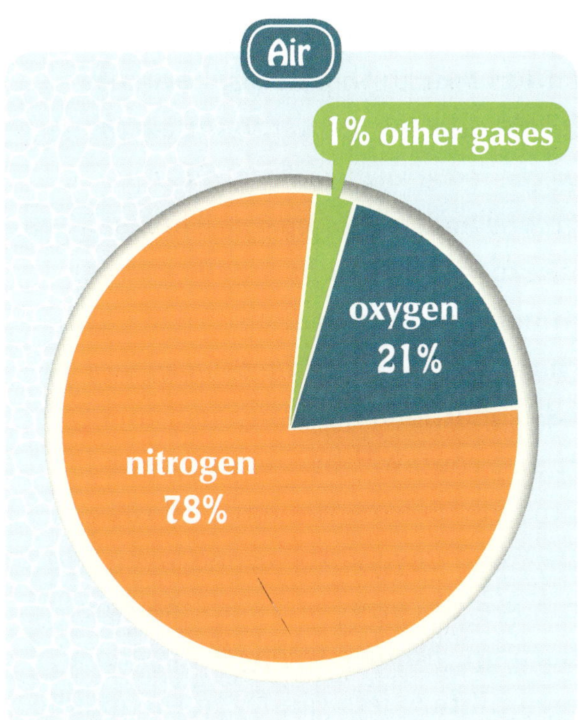

Another gas, nitrogen, makes up most of the air. Air is about 78% nitrogen. Nitrogen from the air also goes into the soil, where the bacteria on the roots of certain kinds of plants turns it into important nutrients that help plants grow.

Very small amounts of *other gases* make up the remaining 1% of the air. One of those gases is *carbon dioxide*, which plants use. Do you remember the process that plants go through to make sugars, or food, by using carbon dioxide, sunlight, water, and chlorophyll? Without photosynthesis, we would not have the food and oxygen we need for life.

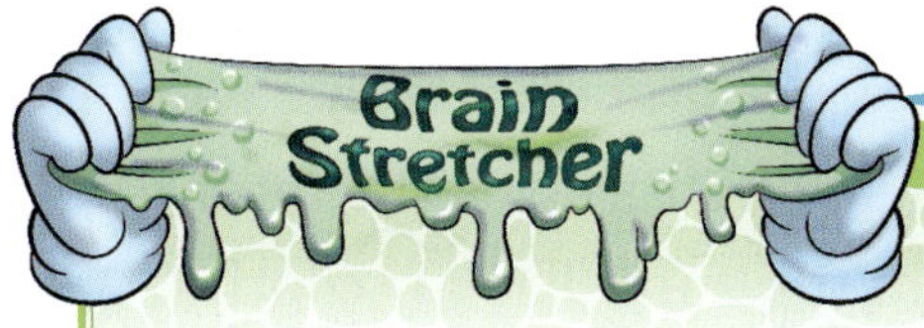

Earth's Atmosphere: Earth's atmosphere is made of nitrogen, oxygen, and carbon dioxide. If it had too little oxygen, we could not survive. If it had too much oxygen, Earth would be in danger of uncontrollable fires across the planet. People would suffer physical problems from the high oxygen levels.

When God created our universe, He created it in the perfect order to meet the needs of His creation. On day two, God created the atmosphere with the right combination of nitrogen, carbon dioxide, and oxygen to allow plants to conduct photosynthesis on day three. Earth's air was perfect for the birds and animals God would create in the next few days. It was perfect for human life. God's design for Creation reminds us that we can trust His perfect plan for us every day.

The Lord by wisdom hath founded the earth; by understanding hath He established the heavens. Proverbs 3:19

Levels of the Atmosphere

The atmosphere extends many miles above the earth. What force keeps the gases of the atmosphere from escaping into outer space? Gravity keeps the earth's atmosphere in place. Because of the force of gravity, the air closest to the earth is the densest air. This means that the molecules of air close to the earth's surface are packed together more tightly than the molecules of air high above the earth. *This difference in density is why the air gets thinner the higher up in the atmosphere, or altitude, you go; the air molecules are less dense.*

It is part of God's wise plan that the densest air is closest to the earth. If you traveled even six miles above sea level, you would suffocate in a few minutes because the higher air is less dense. The molecules of oxygen are too far apart, and you would not get enough oxygen to breathe.

We can divide the different densities of air in the atmosphere into levels. *The air lowest to the earth, or the lower atmosphere, has the greatest density of gases*. This is where weather *happens*. The lower atmosphere's conditions are always changing. The *middle atmosphere* has conditions that stay the same. It also contains an ozone layer that protects us from the harmful rays of the sun. Beyond the ozone layer, you can see "shooting stars" as meteors burn up. Some passenger airplanes and military jets can fly in the middle atmosphere. The *upper atmosphere* has the lowest density of gases. Its temperature is high because this part of the atmosphere receives the most energy from the sun. The air continues to thin until outer space begins. Satellites can orbit the earth in the upper atmosphere.

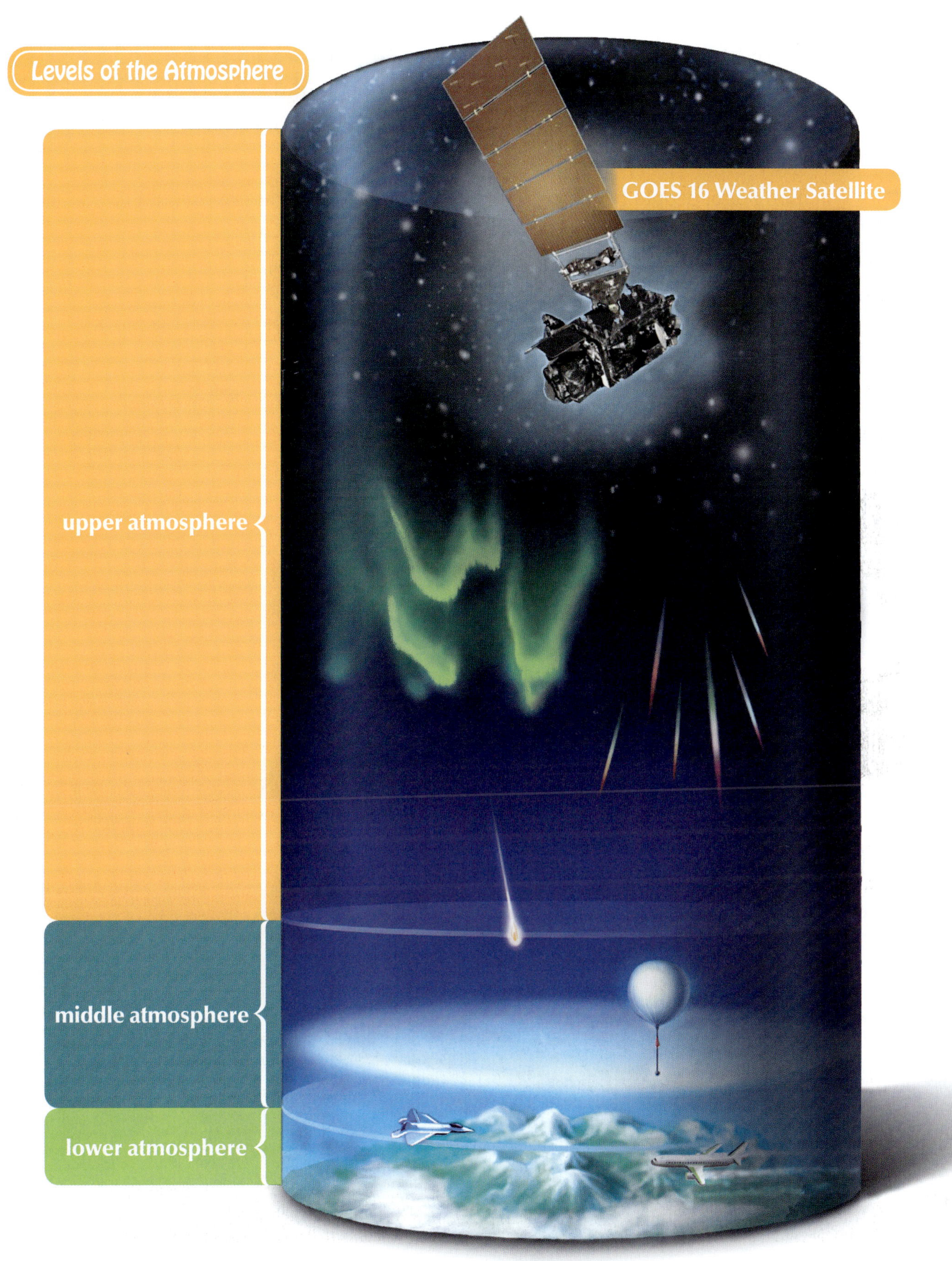
Levels of the Atmosphere
GOES 16 Weather Satellite
upper atmosphere
middle atmosphere
lower atmosphere

The Sky

Long ago, people began watching the sky to predict weather. They watched its color and the kinds of clouds they saw. Weather scientists, or meteorologists, use the terms cloudy, sunny, partly cloudy, and partly sunny to describe basic weather conditions. Meteorologists also predict rainy and stormy weather. A weather prediction is called a *forecast*. Throughout your study of this chapter, use your senses to observe the weather conditions in the sky. Use these words to describe what you see:

- ✓ Sunny—the sky is mostly clear with few clouds.
- ✓ Partly sunny—there are clouds in the sky, but the day is mostly sunny.
- ✓ Partly cloudy—there is some sunshine but more clouds than sunshine.
- ✓ Cloudy—the sky is almost all cloudy.
- ✓ Rainy—rain is falling.
- ✓ Stormy—lightning can be seen and thunder is heard. You can also feel more wind.

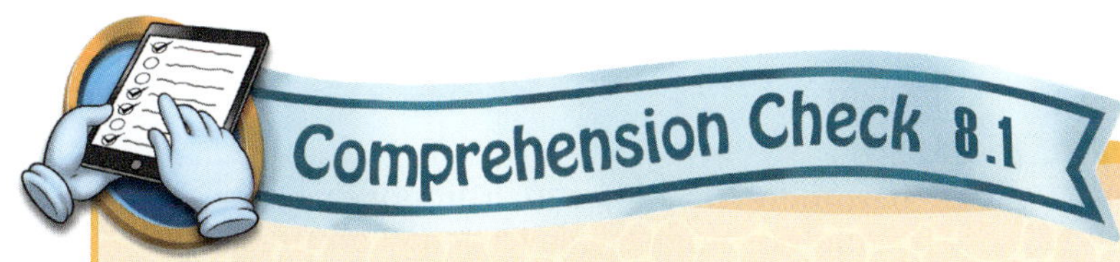

Comprehension Check 8.1

Give the correct answers.

1. What is the difference between weather and climate?

   ____

   ____

2. What is air made of? mixture of gases

3. What is the air surrounding the earth called? atmosphere

4. Name two gases in our air. etc. nitrogen

   ____

5. Which level of the atmosphere has the highest density of gases?

   lower

6. In which level of the atmosphere does weather happen?

   lower

7. Which level of the atmosphere has the lowest density of gases?

   upper

Label the levels of the atmosphere.

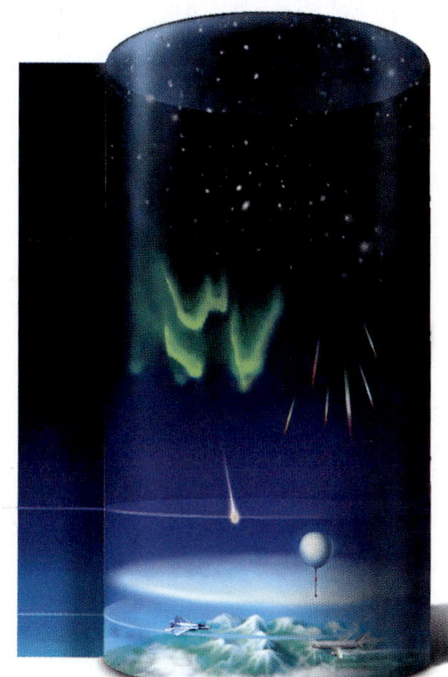

8. 

9. 

10. lower

pressure: the measurement of the force exerted against an object

meteorologist: a scientist who studies the weather

barometer: weather instrument meteorologists use to detect air pressure changes

forecast: a weather prediction

8.2 Air's Weight and Pressure

Air Has Weight

You have probably heard the expression "as light as air." You may have thought that air does not weigh anything because air does not feel heavy. Just like anything made of molecules, however, air does have weight. A cubic foot of air weighs just about one ounce. The Bible mentioned "the weight for the winds" long before scientists learned that air has weight. Since *air has weight and takes up space, air is matter.*

For He looketh to the ends of the earth, and seeth under the whole heaven; to make the weight for the winds; and He weigheth the waters by measure. *Job 28:24–25*

Air Has Pressure

Because air has weight, it creates a pushing force called pressure. **Pressure** *is the measurement of the force exerted against an object*. Matter can exert pressure because of its weight. If you have ever pressed flowers, you have probably used the pressure of solid books to flatten the flowers. Liquids exert pressure, too. You will remember that the different zones in the ocean have different water pressures. The midnight zone has a much greater amount of water pressure than the sunlight zone because of the amount of matter, in the form of water, pushing downward. You can feel the force of water pressure if you put your hand over the end of a garden hose when it is turned on.

Because gases have weight, they can also exert pressure. Right now, about one ton of air is above you. How is it that you do not feel that ton of air? Since the air outside your body is exerting the same amount of force as the pressure inside your body, you do not feel the air's pressure. There is a balance of the forces of pressure. If you have ever flown in an airplane, you may have experienced what it feels like when air pressure becomes unequal. When the airplane gets to a certain height, or altitude, you can feel the air inside your ear pressing against your eardrums. This is because the air inside your ear has a higher pressure than the air outside your ear. When your ears "pop," the pressure has suddenly become equal on both sides of the eardrum.

Try This! Observe air pressure.

Materials needed: ✓ balloon

Inflate the balloon and tie it. Place it on a surface and push down on it with your finger. Did you feel the balloon pushing back on your finger? What happens when you release your finger?

A higher density of air gives the balloon a greater amount of pressure. By pushing air inside a balloon, inflating it, you are increasing the air pressure. As you pushed on it with your finger, the balloon felt "tight" because of the air pressure inside pushing against your finger. As you released your finger, the air pressure pushed back, allowing the balloon to regain its shape.

Air Pressure Changes

Do you remember how high the atmosphere extends? Every bit of that air, which goes up for miles and miles, has weight. The weight of all the air in the atmosphere creates pressure as it presses down. Air pressure is greatest close to the surface of the earth. *As you travel higher into the atmosphere, air pressure gradually lowers, or decreases.* This happens because there is less air above to press down. Air pressure near the earth varies. At high altitudes, such as on a mountaintop, air pressure is less than at sea level.

Air pressure changes in the lower atmosphere cause weather events to happen, such as stormy or rainy weather. Meteorologists watch for air pressure changes. *Cold air is heavier than warm air.* As cold air sinks down, it pushes the warm air away, creating the right conditions for rain to fall. *A* **meteorologist** *is a scientist who studies weather.* Meteorologists use a *weather instrument*, called a **barometer**, *to detect air pressure changes. If the air pressure decreases, meteorologists can predict precipitation*, such as rain. If the pressure increases, they can *predict*, or **forecast**, fair *weather*.

Try This! Make a barometer.

Materials needed:

- ✓ jar
- ✓ balloon
- ✓ scissors
- ✓ rubber band
- ✓ toothpick
- ✓ index card
- ✓ marker
- ✓ tape
- ✓ glue

1. Cut the neck from the balloon, stretching the remaining part over the top of the jar. Use a rubber band to keep the balloon in place.
2. Glue the toothpick on top of the balloon, half of it protruding over the edge of the jar.
3. Tape the index card to the side of the jar so that the toothpick is positioned in the middle. Mark this place. At the top of the card, write "high pressure"; at the bottom, write "low pressure."
4. Place your barometer in a room away from heat or moisture. Notice if the pressure is higher (the toothpick rises) or lower (the toothpick lowers) and write down the resulting weather.

The air outside the jar changes pressure. On low-pressure days, the balloon will inflate a bit and you will notice more rain. On high-pressure days, the balloon will deflate slightly and you will notice more fair weather.

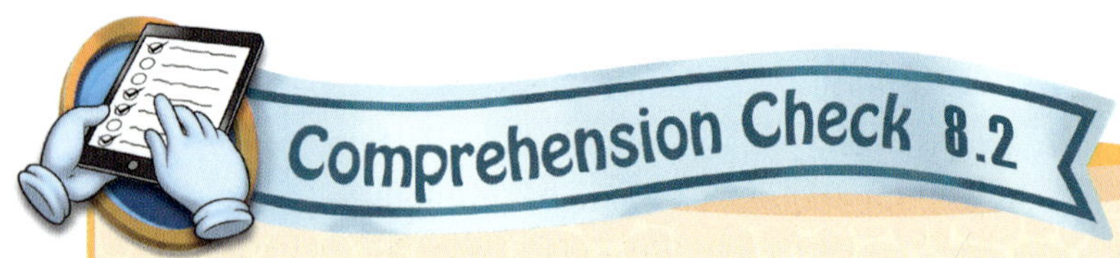

Comprehension Check 8.2

True/False: *If the statement is true, write* true. *If the statement is false, replace the underlined word(s) with a word or phrase that will make the statement true. Do not write false in any blank.*

___T___ 1. Air has weight and takes up space because it is made of <u>matter</u>.

___can___ 2. Gases <u>cannot</u> exert pressure.

___decrease___ 3. Air pressure <u>increases</u> the higher in the atmosphere you go.

___T___ 4. Cold air is <u>heavier</u> than warm air.

___barometer___ 5. A <u>thermometer</u> is an instrument used to measure air pressure changes.

___decrease___ 6. <u>Increased</u> air pressure helps meteorologists predict precipitation.

Think and Predict.

7. The air pressure has decreased. What kind of weather can you expect? ___rain___

8. The air pressure has increased. Make a prediction. ___sunny___

greenhouse effect: the atmosphere's ability to keep heat from easily escaping into space

global winds: bands of wind that usually flow in the same predictable direction at certain latitudes

jet stream: a band of wind high above the earth that flows from west to east around the globe

8.3 Moving Air

Have you ever noticed that the earth and water become warmer during the day than the air surrounding us? Since the atmosphere is closer to the sun, it seems as though air should receive heat from the sun first. However, this does not happen. The sun heats the earth and water directly, but it warms the air very little. Why is this so?

Air Has Temperature

If you remember how light behaves with different types of matter, you will see how this is possible. Air is transparent. It allows sunlight to pass straight through it, meaning little sunlight energy can be absorbed by air. Sunlight passes straight through the air to the earth and the water. Sunlight energy is converted to heat energy as the earth and water absorb rays from the sun.

We know that the air has warmth, but how is air heated? The molecules of matter are always moving. Molecules of gases move faster than molecules of solids or liquids. Molecules in the air move very rapidly. They bump into the earth, and some of the heat passes from the molecules of the earth to the molecules in the air.

In the daytime, the earth absorbs heat energy from the sun; but at night, the earth radiates, or gives off, heat, keeping the air around the earth warm. The atmosphere keeps the heat from escaping back into space. If it did not, all life on the earth would freeze at nighttime and become unbearably hot during the daytime. The atmosphere helps to keep the earth's temperature fairly stable. People have used this principle to design greenhouses for plants to grow in the right temperatures. *The atmosphere's ability to keep heat from easily escaping into space is called* the **greenhouse effect**.

Temperature and Pressure Cause Wind

The earth is tilted as it spins, causing its surface to be unevenly heated. The part of the earth tilted toward the sun receives the most heat. The air above this part of the earth becomes warmer than the air in other places. This uneven heating is the reason we have winds.

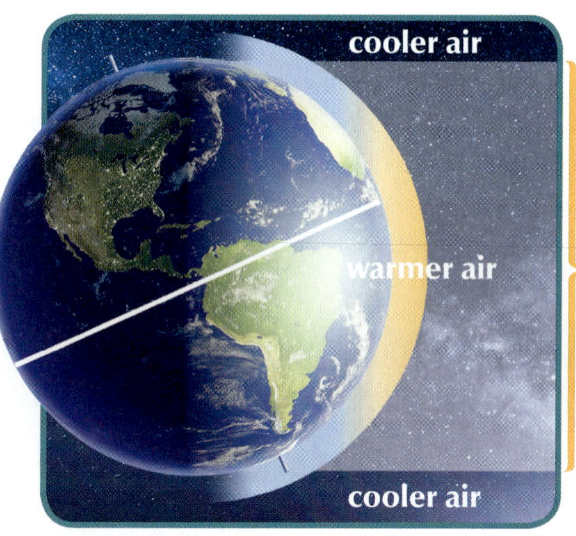

When air is heated, its molecules spread out. As warm air expands, it becomes less dense than cool air. Warm air rises because it is less dense. *When warm air rises, it leaves an area of low pressure near the earth's surface*. Cooler, denser air flows into the low-pressure spot to replace the warm air that rose. As air continually heats and rises, cooler air continually flows in to take the place left by warmed air. *This constant motion of the air is what causes wind.*

Wind Has Direction and Speed

We name the wind by the direction it is blowing from; a northwest wind is coming from the northwest. We can also measure the power of wind by its speed. For example, a twenty-mile-per-hour northeast wind is stronger than a ten-mile-per-hour northeast wind.

Meteorologists measure wind speed and direction to help them forecast weather. Greater wind speed usually means a change of weather. If you live in the northeastern United States, a northwest wind brings cold arctic air; but a southwest wind brings warmer air. Meteorologists around the world make different predictions based on wind.

Because the air is always colder at the earth's poles and warmer at the equator, there are *bands of wind that usually flow in the same predictable direction*. These bands are called **global winds**, and they help regulate the earth's air temperatures. *Global winds blow in different directions at certain latitudes.* Sailors throughout history have depended on the wind to carry their sailing ships across the ocean as they formed trade routes.

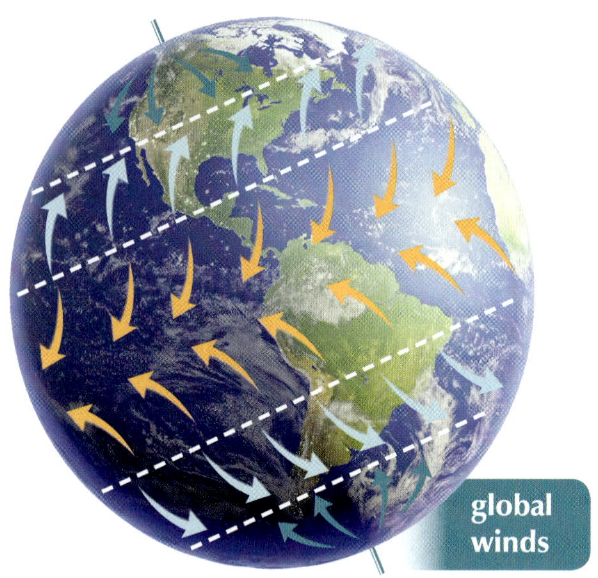

High above the earth, warm air and cold air clash with each other. This collision *creates a band of wind that flows from west to east around the globe.* If you think of a river's continual current, you will remember what this upper wind is called—the **jet stream**. The jet stream is powerful enough to *cause weather systems to move in a predictable pattern—from west to east, dipping up and down.* This predictability helps meteorologists forecast weather.

Most meteorologists did not know about the jet stream until the 1940s. World War II pilots flew at higher altitudes to avoid enemy aircraft but found that their rate of speed became dramatically unpredictable. If they were flying with the jet stream, they arrived at their destination too quickly; if they were flying against the jet stream, they arrived too late or ran out of fuel.

> *He causeth the vapors to ascend from the ends of the earth; He maketh lightnings for the rain; He bringeth the wind out of His treasuries.*
>
> *Psalm 135:7*

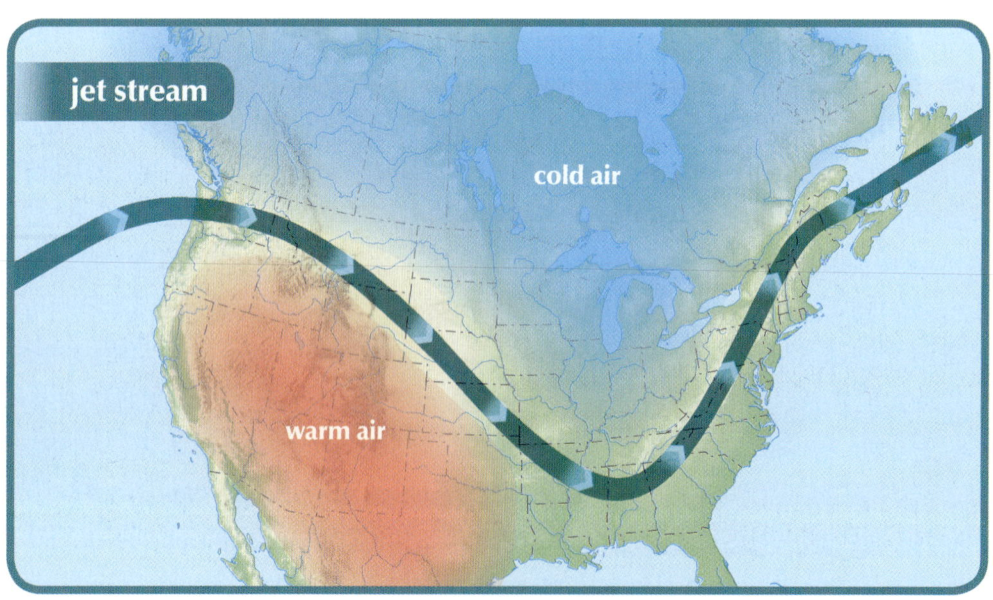

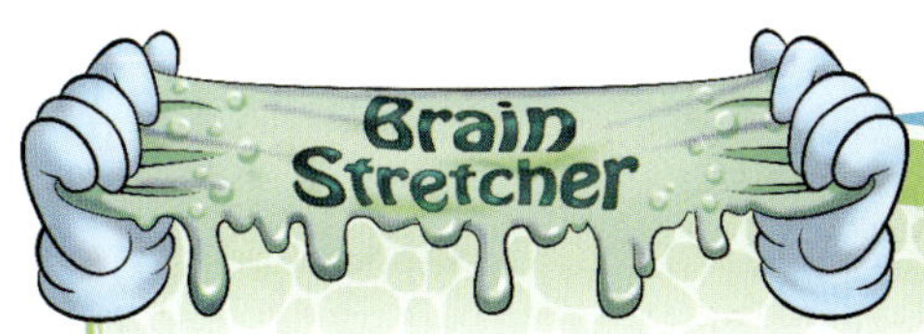

Other Types of Winds: Winds can be seasonal, local, or come from a storm. *Seasonal winds* change when the season changes. *Local winds* are predictable in certain areas, such as a sea breeze that always comes from the direction of the ocean toward land during the day. A land breeze happens at night when the land air blows toward the ocean. Local winds can even be the wind that blows down the slope of a mountainside or up from a valley. *Storm winds* occur during thunderstorms, hurricanes, and tornadoes.

Make a weathervane to observe wind direction.

Materials needed:

- ✓ plastic cup
- ✓ permanent marker
- ✓ nail
- ✓ sharpened pencil with an eraser
- ✓ modeling clay
- ✓ plate
- ✓ marbles (or rocks)
- ✓ paper triangles
- ✓ tape
- ✓ drinking straw
- ✓ pin
- ✓ compass

1. Turn the cup upside down and write *N*, *E*, *S*, *W* around the bottom of the cup. Insert the pencil into the center of the bottom of the cup. (You may choose to poke a hole using a nail first.)
2. Use clay to stabilize the cup on the plate. Surround the plate with marbles or rocks for added stability.
3. Make an arrow by taping paper triangles to each end of the straw. Use a pin to attach the straw to the eraser end of the pencil.
4. Place your weathervane outside on a stable surface. Turn the plate until it matches the directions on your compass.

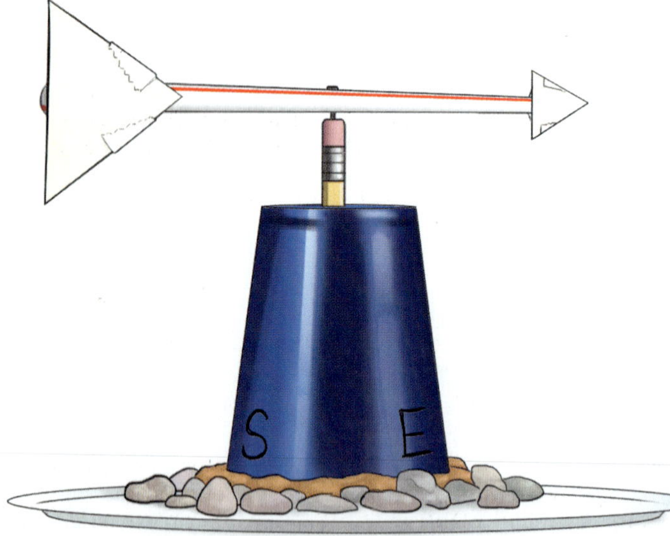

Observe your weathervane for the next several days, writing down the wind direction. Remember, wind is named for the direction it is coming from. Note any changes in direction. Do you notice any rain or storms after the wind changes?

Comprehension Check 8.3

Matching: *Write the letter of the correct answer in the blank.*

C 1. the atmosphere's ability to keep heat from escaping back into space

A 2. constant motion of warm air rising, cool air taking its place

B 3. bands of wind that usually flow in the same predictable direction at certain latitudes on the earth

D 4. a band of wind high above the earth which flows from west to east around the globe

A. cause of wind
B. global winds
C. greenhouse effect
D. jet stream
E. seasonal winds

Think and Predict.

5. The wind has just changed its direction. What could you predict?

comimy strom

relative humidity: a measure of the amount of water vapor in the air

hygrometer: weather instrument that measures the relative humidity of air

fog: a cloud at ground level

8.4 Water in the Air

Do you remember which state of water cannot be seen? *Gaseous water, or water vapor, is always in our air.* We cannot see it, but sometimes we can feel it. Some of this water is given off by plants which release it into the air through the thousands of pores in each of their leaves. This all happens as a part of photosynthesis. An acre of corn can give off three to four thousand gallons of water each day.

People and animals also release water into the air. Every day, each man, woman, and child gives off almost 3½ cups of water. Not only do we give off this water as perspiration from our skin, but we also breathe out moisture from our lungs.

Most of the water vapor in our air comes from the water sources of Earth, such as oceans, lakes, rivers, and streams. Later, we will learn how water from Earth can get into our air. For now, we will learn to recognize water in the air and observe it using our senses.

Humidity Is Water

The amount of water vapor, or humidity, in the air is changeable. If the air has high humidity, there is a lot of water vapor in the air. You might recognize a humid day by feeling sticky, damp air. The temperature of the air affects humidity. Warm air is able to hold more water vapor than cool air. When we say "hold," we are saying different temperatures allow for different amounts of water vapor to be present in air. *The measure of the amount of water vapor in the air is called* **relative humidity**. If the air has a relative humidity of 100%, then the air is "holding" all of the water vapor that it can at that temperature. When the air is very humid, rain can fall. Meteorologists *measure humidity levels using an instrument called a* **hygrometer** so that they can forecast weather. *If the humidity is low, they can predict fair weather; if the humidity is high, they can predict rain.*

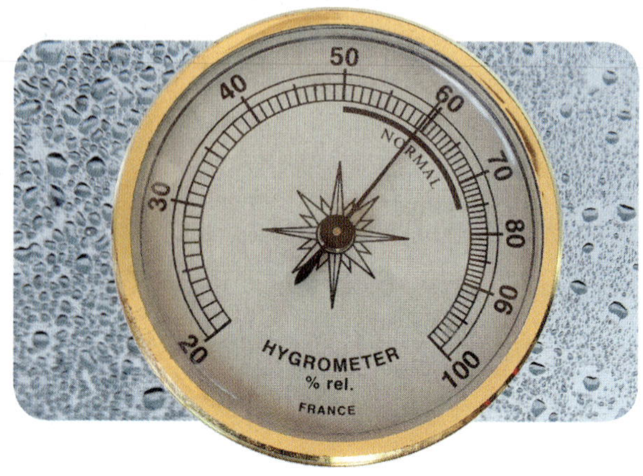

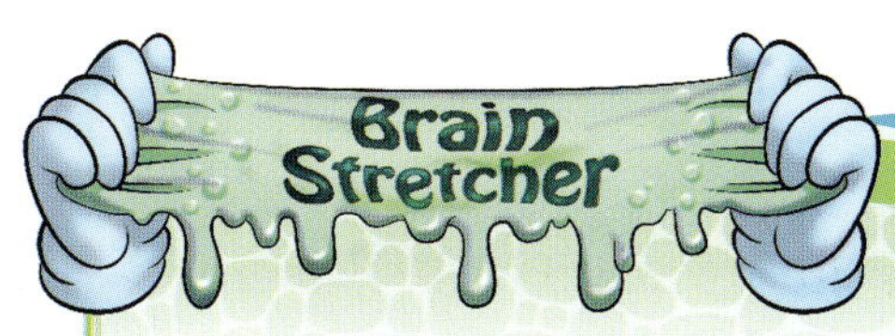

Cloud Types: Clouds are always changing shape because whenever warm air touches a cloud, some of the cloud evaporates. Clouds have been divided into three main groups, according to their shapes. You will learn to recognize if a cloud is *cirrus*, *cumulus*, or *stratus* by looking at its shape.

The word *cirrus* means "curl." *Cirrus clouds* are very high, thin, feather-like clouds. Because they are very high in the sky, they are made entirely of ice crystals and air. Cirrus clouds look hazy and delicate. They usually appear in dry weather, but they often mean that a storm is on its way.

The word *cumulus* means "heap." A *cumulus cloud* is large and puffy with a flat base. You see these clouds most often on hot summer afternoons. Usually cumulus clouds are bright white and appear during fair weather, but sometimes they become tall, thick, and dark. They are then called *cumulonimbus clouds*, or *thunderheads* because of the heavy rain, thunder, and lightning that they bring. You may even notice *the flat top and base of these clouds in the shape of an anvil.*

While cirrus clouds are high and thin, *stratus clouds* are low, thick layers of clouds that look like a blanket. They are flat and often gray, and they frequently bring drizzling rain or fine snow. Fog is a stratus cloud that is very close to the earth's surface.

Clouds Are Made of Water

Sometimes, you can see clouds move across the sky. Have you ever lain down on the ground to watch them? If you use your imagination, you can see the shapes of things. If you look carefully, you may be able to see places in the distance where a raincloud has burst and water is coming from the cloud to the earth. This clue tells you what clouds are made of. Just like anything else in the physical world, *clouds are made of matter; they are mostly water and air.* Although most clouds look fluffy and light, a single cloud may contain more than one million tons of water.

How high up are the clouds? Each kind of cloud has its own range of height. The height of the clouds also depends upon the condition of the atmosphere. The more humid the air is, the lower the clouds will be. *When a cloud is so low that it is at ground level*, we call it **fog**.

Clouds that are high in the sky are made of tiny ice crystals. Clouds that are low are made up of very tiny droplets of water. Each of these droplets is so small that it would take about one million of them to make a raindrop.

Try This! A Week of Weather

Fill out the weather chart each day.

The high temperature, low temperature, and relative humidity can be found on your local news.

To describe the sky conditions, look at the sky at the same time each day and choose from the following: sunny, partly sunny, partly cloudy, cloudy, rainy, stormy.

To determine the cloud type, choose from the following: cirrus, cumulus, stratus.

Day	High temp.	Low temp.	% of humidity in the air	Precipitation amount and form	Sky conditions	Cloud type	Wind speed
Monday	63°	34°	58%	none	sun		0
Tuesday	66°	37	58%	none	sun		13
Wednesday	64°	59°	81%	some	cloldy		12
Thursday	45	57°	51%	some	clouldy		14
Friday	64%	46°	27%	none	sun		15

At the end of the five days, answer the following questions.
To find the answers to questions 1–4, add the five numbers together; then divide the total by five.

1. What was the average daily high temperature? 66 °F
2. What was the average daily low temperature? 47 °F
3. What was the average daily relative humidity? ________ %
4. What was the average wind speed? 13 mph
5. With the information that you have gathered Monday–Friday, predict what the weather will probably be on Saturday. ________

Observe to Understand: Clouds

Throughout your study of this chapter, observe the sky for clouds. Try to name each cloud correctly—cirrus, cumulus, or stratus. Can you find any thunderheads? You may even try to make a weather forecast based on the clouds you see.

Comprehension Check 8.4

Multiple Choice: *Circle the correct answer.*

1. Water vapor is a __?__ which is always in our air.
   a. dust b. gas c. liquid

2. __?__ is a measure of the amount of water vapor in the air.
   a. Atmosphere b. Fog c. Relative humidity

3. A __?__ is an instrument used to measure humidity.
   a. barometer b. hygrometer c. thermometer

4. Clouds are mostly made of __?__.
   a. fog b. minerals c. water

5. Clouds at ground level are called __?__.
   a. fog b. humidity c. thunderheads

Think and Predict.

6. The humidity increased. What could you forecast? ________________

the water cycle: the movement of water from the earth to the air and back again

evaporation: the process by which liquid water becomes water vapor

condensation: the process by which water vapor changes back into very tiny droplets of liquid water

precipitation: any form of water which falls from the sky to the earth

water runoff: precipitation that eventually trickles into streams

8.5 The Water Cycle

The earth and its atmosphere are constantly exchanging water. As *water moves from the earth to the air and back again*, it changes from a liquid to a solid or a gas. This continuous process is called the **water cycle**. The water cycle is part of God's plan for keeping the earth supplied with water.

Because of the water cycle, we are able to continually use the water that God made when He created the world. The raindrops that fell on you the last time it rained were made up of water that has existed for thousands of years. That same water has probably fallen as raindrops many times before. The rain falls into the rivers, the rivers flow into the sea, and yet the sea never overflows because of the way the water cycle works.

There are four main processes of the water cycle, and they always happen in this order: *evaporation, condensation, precipitation, and water runoff*.

Water Cycle

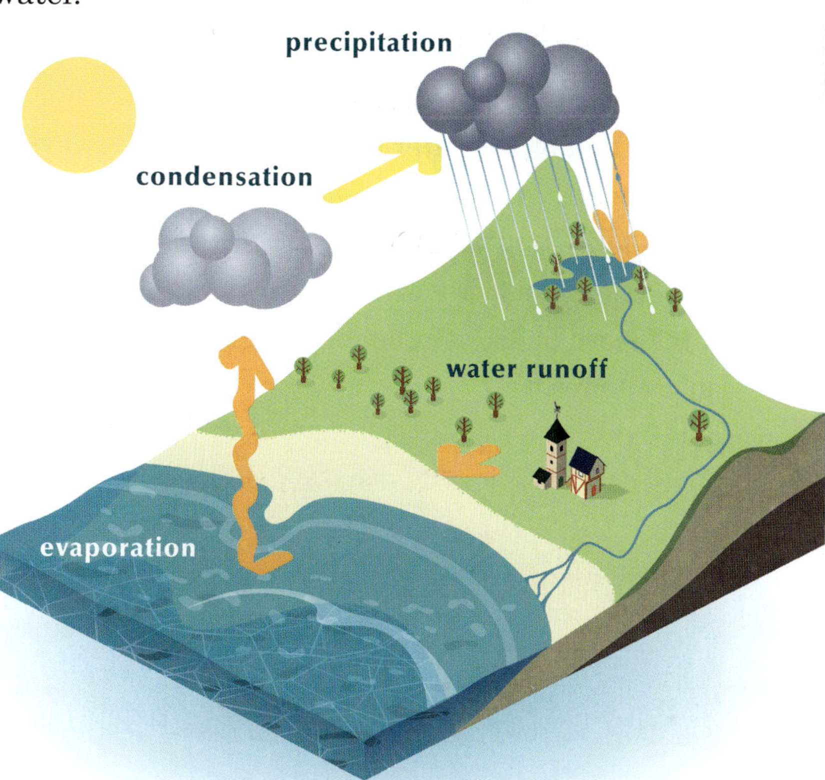

Evaporation

How does water vapor get into the air? Molecules of water are always moving. When heat energy is added to water, its molecules move faster. The molecules bump into each other and bounce. Sometimes, when molecules at the surface of the water are bumped by other molecules below, they are thrown out of the water and escape into the air. We say they evaporate, or become vapor. They are no longer a part of the liquid. The escaped molecules of water move into the spaces between the molecules of gases in the air. The water has turned into a gas called water vapor. *The process by which liquid water becomes water vapor is called* **evaporation**.

What heat-energy source causes oceans and other water sources to evaporate? *The sun's powerful heat energy controls the water-cycle process*. This is why water evaporates more quickly on hot days than on cool days.

Condensation

What do you like most about a rainy day? It is fascinating to watch the sky change colors. Sometimes, a rainy day is very windy. Watching tree branches as they sway back and forth during a storm may remind you to be thankful for God's protection. Perhaps your favorite thing about a rainy day is listening to the rain as it falls from the sky. What causes water vapor in the air to change back into liquid water and fall to the earth as rain?

Remember, warm air is able to "hold" more water vapor than cool air. Because water vapor is less dense than the other gases in air, it makes the warm, moist air very light. The warm, moist air rises because it is less dense than both warm, dry air and cool air. As it moves farther away from the earth's surface, the warm, moist air begins to cool.

As the air cools, something happens to the water vapor in it. Because cool air cannot "hold" as much water vapor as warm air can, the *water vapor begins to turn back into very tiny droplets of liquid water*. This process is called **condensation**.

Water does not condense easily unless it has something on which to condense. *Tiny droplets of water condense on the particles of dust that are in the air, causing clouds to form*.

Precipitation

As the clouds become colder, more and more water condenses. The droplets of water bump into each other and stick together to form larger drops. When the drops of water become too large and heavy to stay in the cloud, the force of gravity causes them to fall to the earth as rain. If the temperature of the air near the earth's surface is lower than the freezing point of water, sleet or snow is formed instead of rain. Under certain stormy conditions, the water droplets become hailstones. *Any form of water that falls from the sky to the earth is called* **precipitation**. What forms of precipitation can you think of? Rain, snow, sleet, and hail are all precipitation that can fall back to the earth.

Rain is the only form of liquid precipitation. Hail, sleet, and snow are all solid forms of precipitation. Hail comes down in frozen pellets, sleet falls as beads of ice, and snow is crystalized water vapor.

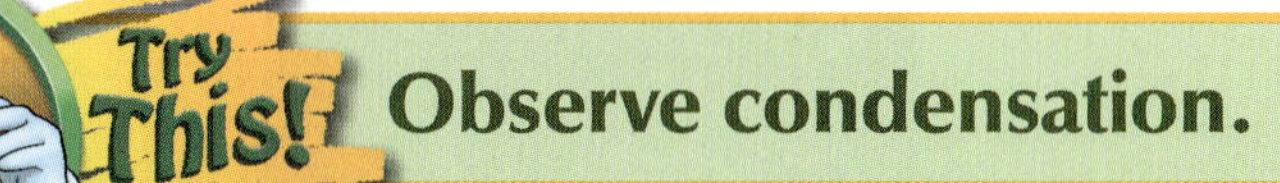

Observe condensation.

Materials needed:
- ✓ a glass
- ✓ ice cubes
- ✓ water

1. Fill a dry glass with ice cubes and water. Let the glass of ice water stand in a warm room for a while.
2. Notice what forms on the outside of the glass. Where did the droplets of water come from?

Because the surface of the glass is colder than the air around it, water vapor from the air condensed on the outside of the glass.

Water Runoff

Water runoff *is precipitation that eventually trickles into streams*. Where do streams travel? Streams flow into bigger rivers, returning water to the ocean where the water cycle begins again. We can be thankful for God's continual provision of water. This renewable resource will continue as long as the earth exists. Are you doing your part to steward it well?

> *Thou visitest the earth, and waterest it: Thou greatly enrichest it with the river of God, which is full of water.* Psalm 65:9

Try This! Make a rain gauge.

Materials needed: ✓ wide-mouthed jar ✓ masking tape ✓ marker ✓ ruler

1. Place a piece of tape vertically on the side of the jar.
2. Place the ruler vertically against the side of the jar. Mark each inch measurement on the tape.
3. Place your rain gauge outside away from a building or a tree from which extra rainwater could fall.

The next few times it rains, check your rain gauge after the sky clears. Write down the closest measure; then empty the jar so that it is ready to measure the next rainfall.

Try This! Measure a snowfall.

Materials needed: ✓ ruler

You may live in a region where the precipitation at this time of year is snow. The next time it snows, use a ruler to measure how many inches of snow your area received. Be sure to measure where the area is level and snow is not drifting.

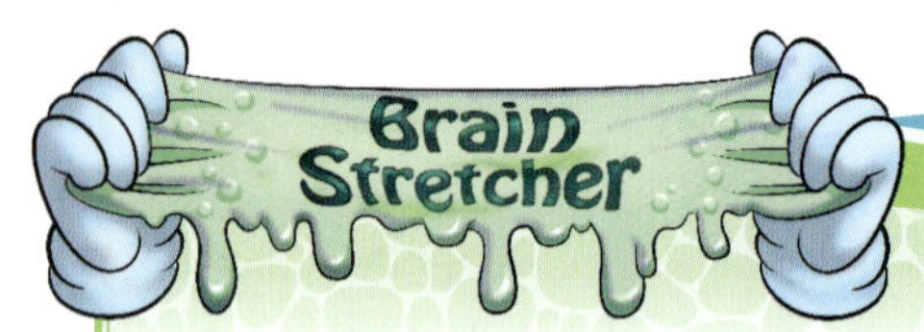

Are Dew and Frost Precipitation?: Have you ever walked outside early in the morning and found that the grass was wet, yet you knew that it had not rained the night before? The moisture you found on the grass was dew. Since dew does not fall from the sky, dew is *not* a form of precipitation. It is, however, a product of condensation.

When the sun sets, the earth and everything on the earth becomes cooler. The air close to the earth also begins to cool, but it does not become quite as cool as the earth itself. The air touches the cooler surfaces of things on the earth, such as the grass and other plants. *Dew forms when the moisture in the air condenses on cool surfaces during the night.*

Sometimes, the temperature drops below *32°F, the freezing point of water.* If the air comes in contact with a freezing cold surface, the water vapor in the air does not condense into dew. It changes directly from a gas to a solid. *The tiny crystals of ice that are formed by frozen water vapor are known as frost.* Have you ever seen frost on your windowpane after a cold night? Even if you live in a place where it is not cold enough for frost to form outside, you may have seen frost in your freezer. The water vapor inside your freezer can change to solid crystals of ice.

Because dew and frost do not fall from the sky to the earth, they are not considered forms of precipitation.

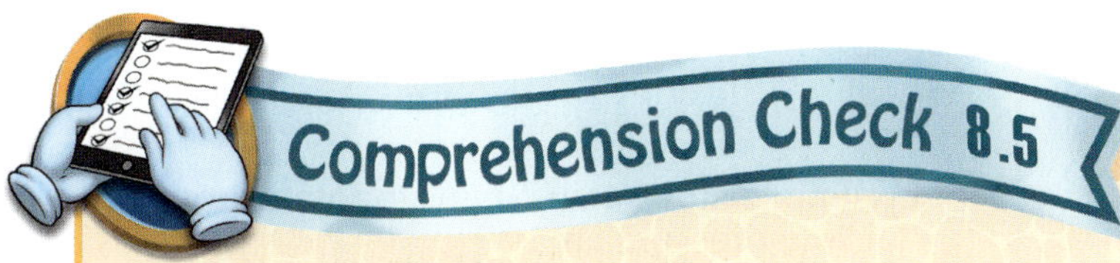

Comprehension Check 8.5

Give the correct answer.

1. List the four main processes of the water cycle in order. ____________

____

2. Which process turns liquid water into water vapor? ____________

3. Name the energy source that controls the water-cycle process. ______

____

4. Which process causes water vapor to change back into tiny liquid droplets of water? ______________________

5. Name a form of precipitation. ______________________

6. What do we call the temperature at which water freezes? ________

____

Think and Predict: *Circle the correct answer.*

7. Which rainfall amount could cause more erosion to take place?

   1 inch of rain   10 inches of rain

Think and Conclude.

8. What forms of precipitation are water in its solid form? ________

____

Label the water cycle.

9. ______________________

10. ______________________

11. ______________________

12. ______________________

weather phenomenon: a weather event caused by specific conditions

thunderstorm: a storm that brings heavy rain, strong winds, and lightning

thunderhead: a thick, dark storm cloud

lightning: electricity that travels from one cloud to another or between a cloud and the ground

cyclone: a storm that begins to rotate around a low-pressure area

hurricane: the most severe type of tropical cyclone; begins over a warm ocean

storm surge: the rise of ocean water surrounding a hurricane

8.6 Severe Weather Phenomena

The conditions of the lower atmosphere, where weather happens, are always changing. This means that no matter where you live on the earth, you will experience some type of weather phenomenon. *A* **weather phenomenon** *is a weather event caused by specific conditions*. This means that cloud formation, rain, fog, and wind are all examples of weather phenomena.

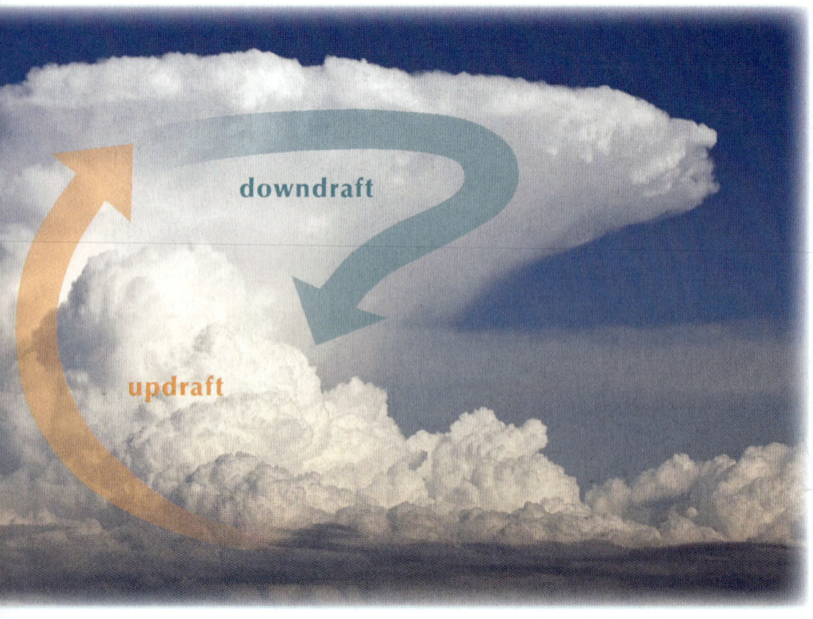

Thunderstorms

The most common type of all storms is thunderstorms. **Thunderstorms** *bring heavy rain, strong winds, and lightning*. If you observe the clouds, you can see some signs of a thunderstorm on its way. We often call a *thick, dark storm cloud a* **thunderhead**.

Thunderheads are very tall, thick clouds with dark undersides and towering tops. They bring quick, heavy downpours of rain. Meteorologists look for the vertical formation of rain clouds to predict thunderstorms. Some thunderheads are so large that they are eight miles high from top to bottom. They often have anvil-shaped tops made of ice crystals. Within a thunderstorm cloud, warm air rises to create an *updraft* current of air while cool air drops with precipitation. This is called a *downdraft*. *Summer afternoons* are most likely to *provide the right conditions for downdrafts to happen.*

You will remember that lightning is a form of static electricity. The sun heats the moist air near the earth's surface. This warm air rises and cools, filling thunderhead clouds with plenty of water droplets. Powerful winds cause the water droplets to rub against each other. What can happen when atoms rub against each other, creating friction? Atoms can lose or gain electrons, giving them a positive or negative charge. Scientists believe that this is how static electrical charges are created in thunderheads. The static electrical charges grow bigger and bigger until a **lightning bolt** *of electricity travels from one cloud to another or between a cloud and the ground*. This quickly heats the air where the lightning happens. The hot air around the lightning gets compressed and becomes too dense, causing the hot air to "explode." What sound do you suppose is caused by this "explosion of hot air"? *Thunder is the sound caused by lightning when it heats the air.*

In the United States, *thunderstorms* are common and *usually move from west to east because of the jet stream*. Thunderstorms are the most common in tropical climates. Some tropical locations have a thunderstorm almost every day. In polar climates, it is usually too cold and dry for thunderstorms to be common, although they can still occur.

Demonstrate that thunder is the sound of air caused by lightning's heat.

Materials needed:
- ✓ paper lunch bag
- ✓ string
- ✓ scissors

1. Blow air into the lunch bag; then twist it to keep the air inside. Tie it with a string to ensure air does not escape.
2. Pop the bag by clapping your hands around it; be sure to use a quick force. Write down a description of the sound you hear.

Sound is caused by vibrations. By popping the bag, you made air move very quickly, causing a loud sound. As lightning strikes, it heats up the air causing it to expand quickly. As the air moves quickly, it causes a loud sound.

Tropical Cyclones

A **cyclone** *is a storm that begins to rotate around a low-pressure area*, or calm center, sometimes called an eye. A tropical cyclone *begins over warm, tropical seas* and often develops into dangerous storms called hurricanes or typhoons, depending on the part of the world in which they occur. Pacific tropical cyclones move toward Asia and are called typhoons. Tropical cyclones in the Atlantic move toward North America and are called hurricanes. **Hurricanes** *are the most severe type of tropical cyclone.* They usually form during or

hurricane

after a hot summer. Warm, moist ocean air rises while cool air rushes in to take its place. This creates the perfect conditions for a tropical cyclone.

When a cyclone first forms, meteorologists call it a *tropical disturbance*. As wind speed picks up, the cyclone is then referred to as a *tropical depression*. Sometimes, the cyclone system weakens and dies out, but other times, it strengthens into a *tropical storm*. This is usually when the National Hurricane Center in Miami, Florida, gives the storm a name and watches to see if it will develop into a hurricane. Tropical storms are always given a person's name, beginning with the first letter of the alphabet. This is why the first tropical storm of the season always begins with the letter *A*. The storms continue to be named alphabetically throughout hurricane season. The National Hurricane Center sends out hurricane hunters, or specially equipped airplanes, that actually fly through hurricanes to take measurements and send data back to Miami. They also check with other weather stations, ships at sea, satellite readings, and radar. This helps the meteorologists to predict the path of each tropical storm.

A tropical cyclone is classified as a hurricane when it has a wind speed of over seventy-three miles per hour, though some hurricanes can have wind speeds up to two hundred miles per hour. *Sometimes, a hurricane will weaken over colder water or over land.* Why do you think this happens? *A hurricane gets its energy from a warm ocean.* Without its energy source, the storm loses its power.

Hurricanes often weaken over cooler waters before reaching land, but other times they gather strength over warm waters before making landfall. As the hurricane comes closer and closer to making landfall, meteorologists are constantly gathering wind-speed data. This helps them predict how catastrophic a hurricane could be.

If a hurricane approaches an area, the people living there may be given warning to secure things easily moved by strong winds or to evacuate their homes and seek shelter in a safe place. Advance warnings of severe storms have saved many lives.

The most damage caused by a hurricane is usually from its storm surge. **Storm surge** *is the rise of ocean water surrounding the hurricane.* As the hurricane pushes down on the ocean, it creates a wall of water, pushing the water toward land like a giant snowplow or bulldozer. Sometimes, a storm surge can create forty foot waves. Water sources many miles inland can also become flooded with the extra water.

In places where hurricanes and tropical storms are common, people have found ways to change how they design buildings and plan cities to minimize damage. Structures are built along bodies of water to hold back flooding.

In the United States, it is rare for the West Coast to experience a hurricane. *This is because tropical cyclones travel over the ocean from east to west.* Global winds push the tropical cyclones along in this direction. Sometimes, the jet stream will cause a hurricane to suddenly turn north or south, but it still moves toward the west rather than the east.

Because the conditions of the lower atmosphere are always changing, weather events are happening all the time. Some events are a usual part of normal weather, such as cloud formation and winds. Other events are serious, such as thunderstorms and hurricanes.

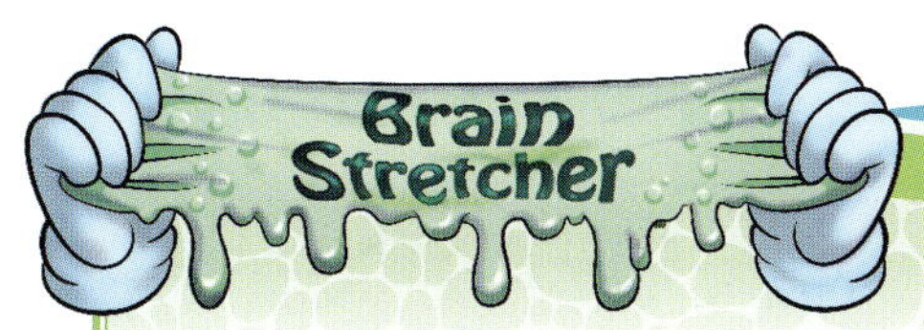

Hurricanes: A hurricane is given a category number on the Saffir-Simpson Scale from 1 to 5. Study the pictures below to help you understand how the Saffir-Simpson Scale works.

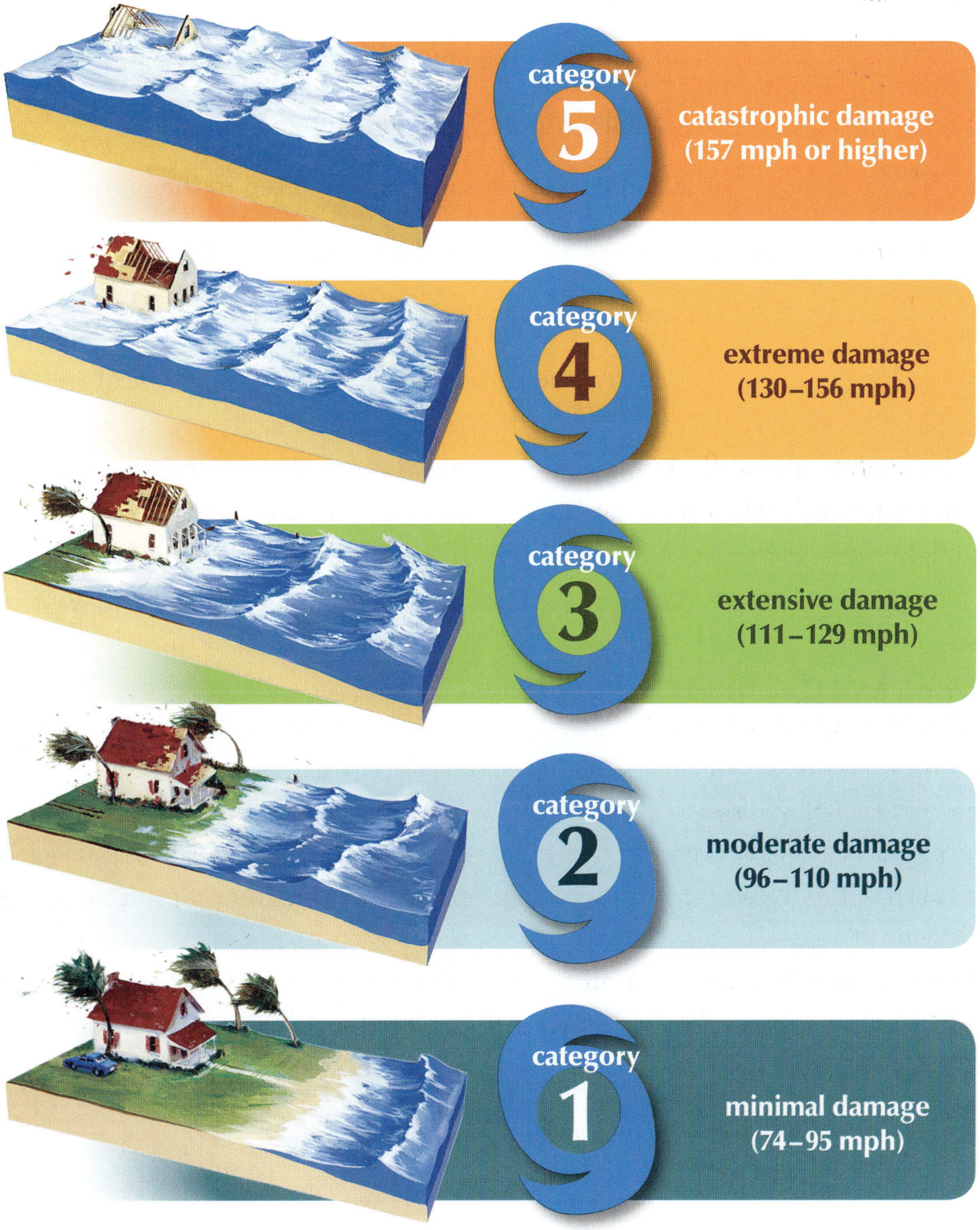

category 5
catastrophic damage
(157 mph or higher)

category 4
extreme damage
(130–156 mph)

category 3
extensive damage
(111–129 mph)

category 2
moderate damage
(96–110 mph)

category 1
minimal damage
(74–95 mph)

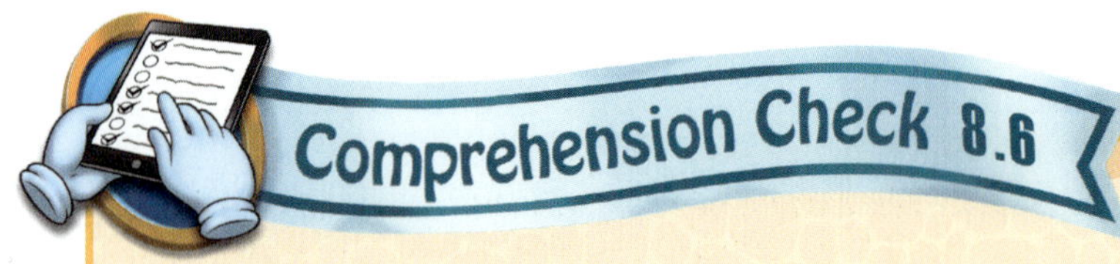

True/False: *If the statement is true, write* true. *If the statement is false, replace the underlined word(s) with a word or phrase that will make the statement true. Do not write* false *in any blank.*

________ 1. Thunderstorms, hurricanes, tornadoes, and fog are all weather phenomena.

________ 2. Thunderstorms bring lightning.

________ 3. Thunderstorms happen most often in the morning.

________ 4. Thunder is the sound caused by lightning.

________ 5. Thunderstorms usually move from east to west.

________ 6. A hurricane is the most severe type of tropical thunderstorm.

________ 7. A hurricane gets its energy from a cold ocean.

________ 8. Most hurricane damage usually comes from its eye.

________ 9. Tropical cyclones travel over land from east to west.

Think and Conclude: *Circle the letter of the correct answer.*

10. Which hurricane is more likely to cause catastrophic damage?

A. category 1

B. category 3

C. category 5

Think and Predict.

11. A thunderstorm is west of you. Will it probably travel toward you or away from you? ________

12. Will a hurricane get weaker or stronger over land? ________

____

tornado: a cyclone that develops over hot land

tornado watch: notification that conditions are right for tornado formation

tornado warning: notification that a tornado has been spotted

blizzard: a severe snowstorm causing colder temperatures, strong winds, and blowing snow

monsoon: a seasonal wind that can bring heavy rain to some places in southern Asia

flooding: an overflow of water

drought: a prolonged period of dryness

8.7 Other Weather Events

Tornadoes

Tornadoes, sometimes called twisters, bring the fastest wind speeds of any storm. A **tornado** is *a cyclone that develops over hot land*, creating a funnel cloud from a thunderstorm to the ground. Tornadoes usually happen in late afternoon, during the hottest part of a summer day, when a very warm air mass meets a very cool air mass. They begin like thunderstorms as the hot air rises and cold air rushes in to take its place. But the rising air forms a swirling funnel cloud that can reach down to the ground. When the tornado touches down, it becomes very dark because of all the debris its vacuum force is pulling up. Tornadoes can uproot trees and lift houses, railroad cars, vehicles, animals, and people.

Long ago, there was no accurate way to predict tornadoes. What might look like a severe thunderstorm one moment could become a deadly tornado the next. Modern meteorology helps many people stay safe. *When conditions are right for tornado formation*, the National Weather Service will issue a **tornado watch**. During a tornado watch, people check their news stations often in case an actual tornado develops. Meteorologists study the radar looking for hook-shaped storms that could develop into a tornado.

If a tornado has been spotted, the National Weather Service will issue a **tornado warning**. At this time, people should take shelter immediately. Some cities have tornado sirens that sound the alarm. Tornado alerts are also sent to cellphones. Many lives have been saved because of these warnings.

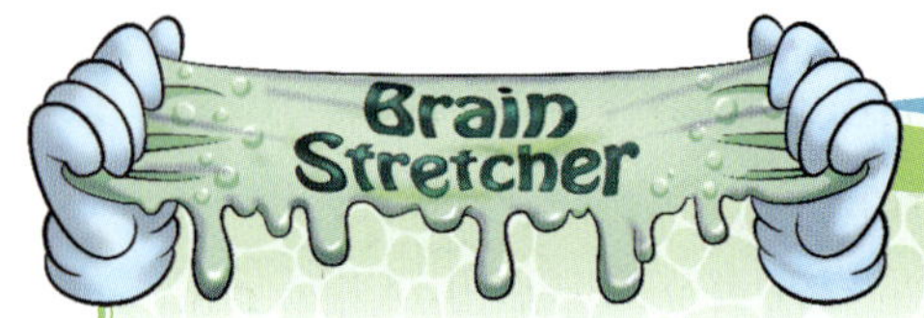

Tornadoes: Tornadoes are more common in the United States than anywhere else, especially in *Tornado Alley*, the Great Plains region. From April to June, weather conditions are more likely to bring tornadoes. Meteorologists work hard during "tornado season" predicting tornadoes in time for people to take shelter.

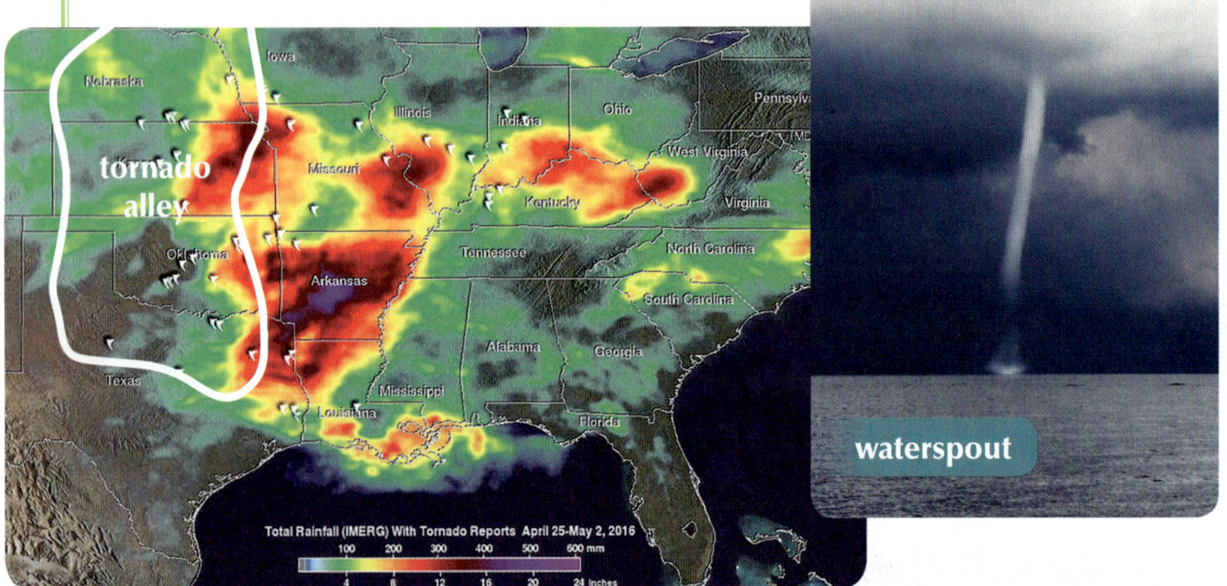

Sometimes, a tornado can hover over a lake or ocean, pulling up a column of water called a *waterspout*.

Tornadoes are measured by the damage they could cause, using the EF-Scale. An EF-0 represents the weakest tornado; an EF-5 represents the strongest. An EF-5 tornado would be catastrophic. To survive an EF-5, a person would need to be in a shelter specifically designed for tornadoes.

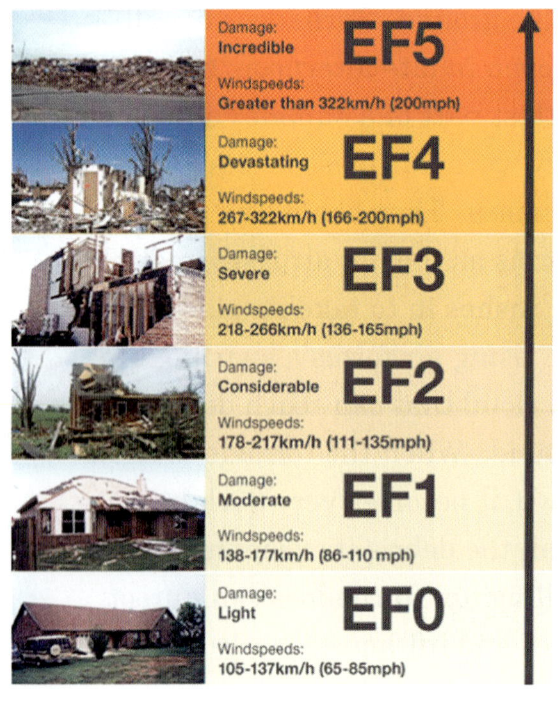

EF-Scale

Blizzards

Do you live where winter weather conditions can bring blizzards? *A* **blizzard** *is a severe snowstorm. It causes colder temperatures, strong winds, and blowing snow.* Sometimes, a blizzard blows snow that has already fallen; other times a blizzard brings its own snow. Blizzards can cause *whiteout conditions*; people cannot see what is in front of them because of whirling snow. During early America, pioneers would tie a rope from their house to their barn. By holding the rope, they could take care of livestock without the risk of getting lost and freezing to death on the prairie.

Because of the jet stream, *blizzards usually move from west to east* as thunderstorms do. After a period of mild winter weather, the jet stream causes a mass of cold, arctic air to bump into a mass of warmer, milder air. As these two air masses meet, the temperature drops, and the winds begin to swirl. A blizzard has begun.

blizzard

Monsoons

Do you remember what two seasons occur in tropical climates? Places near the equator have a rainy summer and a dry winter. *A* **monsoon** *is a seasonal wind that can bring heavy rain to some places in southern Asia.* During the dry winter, seasonal winds blow from the land to the sea. Monsoon rains happen when the wind abruptly changes its direction, coming from the sea to the land instead.

How does the wind shift? Seasons change from winter to summer in tropical climates when the area is positioned directly under the sun. In tropical summers, the land absorbs more heat from the sun than the sea does. Hot air over land rises, and the cool sea air rushes in to take its place. The cool sea air brings heavy rain toward land.

Flooding and Drought

People in tropical countries, such as India and areas of Southeast Asia, welcome the first monsoon rains with celebrations. This marks the beginning of the tropical growing season. It is important that the monsoons bring plenty of rain, but not too much. Not enough rain brings **drought**, *a prolonged period of dryness*, and **famine**, a shortage of food, but too much rain

brings destructive **flooding**, *the overflow from rivers and other bodies of water.*

Flooding causes water to cover areas that are usually dry land. In severe floods, homes, vehicles, animals, and people can be swept away.

Monsoons are strongest and heaviest in Southeast Asia. A monsoon can cause rivers and other bodies of water to overflow. Many people of Southeast Asia prepare for flooding conditions by building their homes on stilts.

What happens if the monsoons do not bring enough rain? Without the rainy season, a drought can follow. In a drought, crops cannot grow without enough water. Loose topsoil blows away in the wind without plant roots holding it in place. A drought can last months or even years, causing people to suffer famine or starvation. Though you may not imagine days and days of monsoon rain as your favorite weather, you can imagine why people of tropical countries celebrate its arrival.

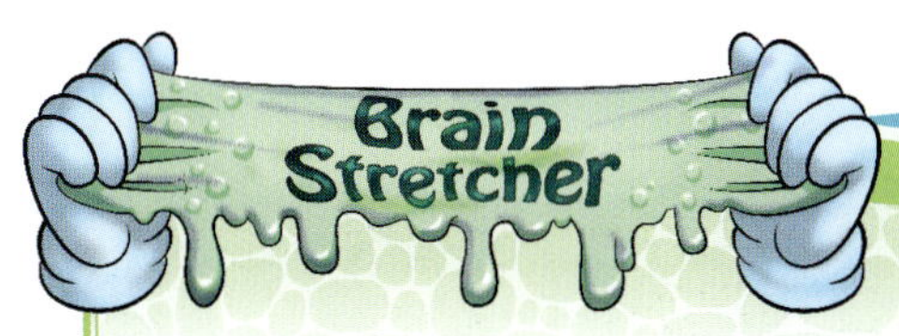

Ways to Control Flooding: Floods are caused when a great deal of water flows into an area, causing a river to overflow its banks. Floods are often caused when large amounts of snow melt quickly or there is a sudden heavy rainfall. Flooding waters have changed the courses of rivers, left people homeless, destroyed crops, and caused death. To prevent damage, people have found ways to control flooding. Sometimes, *canals* are dug to allow excess water to flow quickly out to sea. In some places, the riverbanks are built up higher to protect the land on either side of the river. These high embankments are *levees*. *Dams* are also built to hold back floodwaters. When danger is over, the water behind the dam can be slowly released. Do you remember what natural body of coastal water can prevent flooding inland? Estuaries and other wetlands absorb the extra floodwater from storms and slowly release it back into the sea.

dam

levee

canal

Comprehension Check 8.7

Fill in the Blank: *Write the correct answer in the blank.*

1. A tornado is a cyclone that develops over hot land.
2. A tornado warning means a tornado has been spotted.
3. During a blizzard, whiteout conditions make it difficult for people to see.
4. Blizzards usually move from west to east because of the jet stream.
5. A monsoon is a seasonal wind that brings heavy rain.

Think and Predict.

6. A blizzard is happening to the east of you. Will it most likely move away from you or toward you? away

air mass: a large body of air that has the same temperature and humidity

front: the boundary between two different air masses

8.8 Weather Forecasting

Today's weather where you live began in some other part of the world several days ago. By tracking the weather and observing the changes which occur as it travels, meteorologists are able to forecast weather a few days ahead.

Weather forecasting is a matter of knowing and understanding God's laws of nature about weather. Because weather is changing all the time, we cannot always forecast weather perfectly, even when all the facts are available; but when the proper information is available, weather can be forecasted with some degree of accuracy. Meteorologists are careful to base their predictions on facts, but only God can know exactly how and when the weather will change. How does a meteorologist work? Like all scientists, meteorologists begin with observation.

thermometer— measures temperature

Gathering Data

First, data must be gathered from weather observation stations over a widespread area. In the United States, the Storm Prediction Center uses radar to study all kinds of storms. They also send out planes and weather balloons to gather data about the condition of the air.

There are weather observation stations all over the world recording information all the time. Each hour, meteorologists measure these conditions: *temperature*, *air pressure*, *humidity*, *precipitation*, *wind direction*, and *wind speed*.

hygrometer—measures humidity

anemometer—measures wind speed

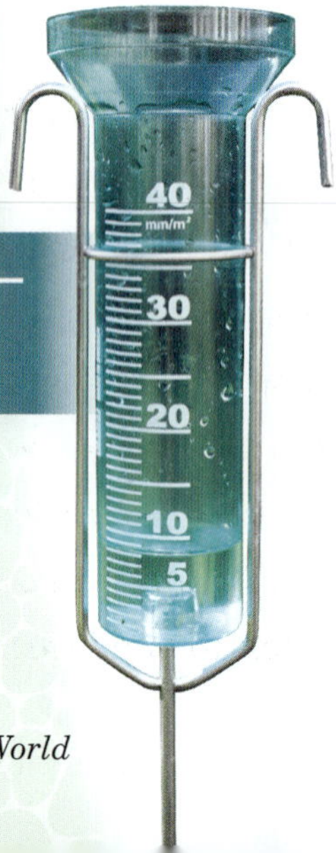

rain gauge—measures precipitation

Some weather stations send balloons into the upper air twice a day to measure weather conditions. After the data is automatically transmitted back to the observation station, the balloon rises higher into the atmosphere where it pops. The equipment on board the balloon parachutes back to Earth. Airplane pilots report weather by radio, and ships at sea send weather reports regularly. Weather satellites orbiting in the upper atmosphere have cameras and other instruments that record cloud patterns and air mass activity. This data is especially helpful in tracking severe storms that begin over the sea. Weather satellites can even determine temperatures, humidity levels, and wind speed. All of these methods—balloons, ships, planes, and satellites—are used by meteorologists to make accurate observations of the condition of the air.

Meteorologists from all over the world exchange the weather data they have collected. This sharing of information helps them do their job of forecasting weather.

Predicting the Weather

After weather observations are gathered from many places, the meteorologists think carefully about the new data. They use computer programs and other forms of technology to process the information. Using computers saves time in organizing and studying this data. This organized data is used to make weather maps that forecast the results of many observations. Weather forecasts are made by comparing the data that is gathered each day with data on past weather. The meteorologist will then make a prediction, or sensible guess, based on data and mathematical formulas. What do we call a sensible guess made by scientists? A meteorologist's forecast is a type of hypothesis. Like a hypothesis, a forecast may be correct sometimes, but other times it is incorrect.

There are several kinds of forecasts, depending on the information available. Short-range forecasts predict the weather only twelve to forty-eight hours in advance. *Short-range forecasts are usually accurate because they are based on current information*. Meteorologists continually update short-range forecasts. Long-range, or extended, forecasts predict the weather six days to a month ahead. *Extended forecasts are not as detailed nor as accurate as short-range forecasts*. Information given in the forecast is based on current

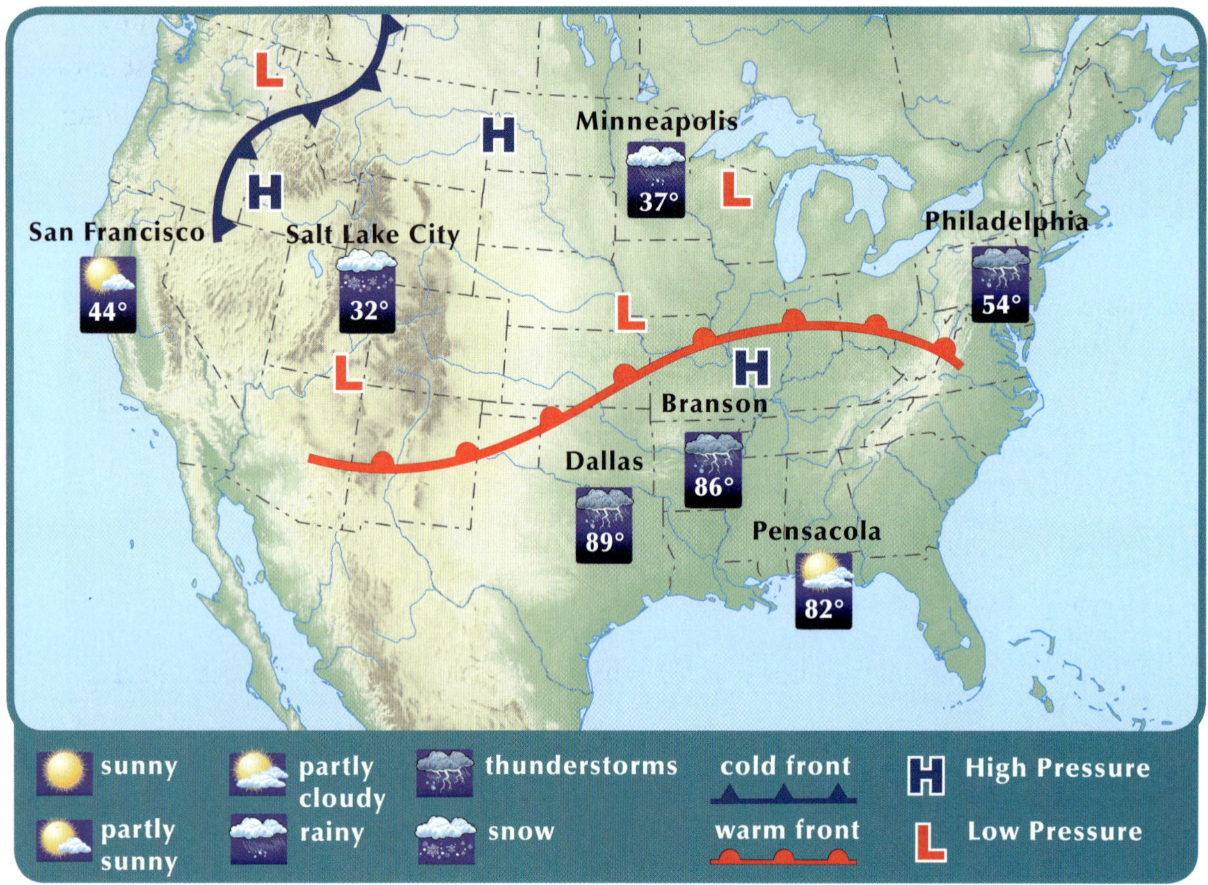

weather observations and averages based on normal patterns, or *trends*, of weather. Computer databases keep records of weather trends for many years. When meteorologists can study data about past weather, they are able to make more accurate forecasts.

A very important part of weather forecasting is observing and measuring the weather conditions of air masses. *An* **air mass** *is a large body of air that has the same temperature and humidity*. Because of the jet stream, most air masses in the United States and in southern Canada move from west to east.

Sometimes, a warm-air mass meets a cold-air mass, or a cold-air mass meets a warm-air mass. When this happens, a front develops. *A* **front** *is the boundary between two different air masses*. A front is usually associated with *a low-pressure system* and precipitation. When a cold-air mass replaces a warm-air mass, we call this a *cold front*. A cold front *can bring heavy precipitation or severe storms* at the boundary between both air masses. A *warm front* happens when a warm-air mass replaces a cold-air mass. A warm front *can bring light precipitation*, such as drizzling rain.

Meteorologists provide weather maps in their forecasts, allowing you to see what weather events are happening. If you understand a weather map's symbols, or key, you can better understand a weather forecast.

Now that you understand how weather works, you can forecast some weather. Will all of your predictions be correct? Of course not, but it is exciting to try.

Comprehension Check 8.8

Multiple Choice: *Circle the correct answer.*

1. _?_ forecasts are the most reliable.
   a. Long-range   b. Short-range   c. Trend

2. A(n) _?_ is a large body of air which has the same temperature and humidity.
   a. air mass   b. cloud   c. front

3. A _?_ is the boundary between two different air masses.
   a. cloud   b. front   c. jet stream

4. A front usually causes _?_ pressure and precipitation.
   a. high   b. low   c. water

5. A _?_ front can bring heavy precipitation or storms.
   a. cold   b. hot   c. warm

6. A _?_ front can bring light precipitation.
   a. cold   b. hot   c. warm

Think and Conclude.

7. If you were a meteorologist, what weather conditions would you need to measure to make a forecast? ______________________________

____

____

____

Robert Boyle

The Father of Chemistry

Our understanding of meteorology begins with our understanding of *chemistry*, or the study of matter—its composition and the way it can change or interact with other matter. We can thank Robert Boyle for opening this field of science.

Robert Boyle was born in Ireland in 1627, the fourteenth child of the wealthy and powerful Earl of Cork. When Robert was eight years old, he was sent to school at Eton in England. During his years at Eton, Robert developed a love for reading that lasted throughout his life.

When Robert was twelve years old, he and his tutor set off for mainland Europe. Robert was an eager scholar. He studied Latin, French, Italian, and mathematics with his tutor and studied science on his own. While he was in Switzerland, thirteen-year-old Robert accepted Christ as his Savior and decided that God was calling him to be a scientist.

Robert did not return home until 1644. He settled on the beautiful estate his father had left to him and spent much of the next ten years reading, studying, observing nature, and writing.

While visiting his sister Katherine in London, Robert Boyle met a number of brilliant scholars who encouraged him to make the study of science his life work. Although he was so wealthy that he did not need to work, Boyle chose to work as a scientist because he believed that the proper study of creation demonstrates the power and wisdom of God to the world.

In 1654, Boyle went to live in Oxford, England. There he met Robert Hooke, a scientist and inventor who became his assistant and friend. Boyle and Hooke began a series of experiments that dealt with the nature and characteristics of air. Together they built an air pump, a machine that can create a vacuum.

Unlike many scientists of his day, Robert Boyle believed that a scientist should do more than think of hypotheses; a scientist must conduct experiments to determine if his ideas are correct. During his experiments, Boyle found that the space that a certain amount of gas takes up is directly opposite to the amount of pressure placed upon the gas as long as the temperature stays the same.

For example, if the pressure upon a gas is doubled, the gas will take up half its original space. This is known as Boyle's law.

Boyle was also the first scientist to understand that the air contained an element necessary for all living things. This same element is necessary for burning. That special element is oxygen.

Boyle experimented with many substances other than air. In one experiment, he proved that water expands when it freezes.

As a result of his many experiments, Robert Boyle concluded that matter consists of particles too small to be seen. Boyle called the particles "corpuscles"; today we call them "atoms."

Boyle published his observations and conclusions in a number of books. His experiments and writings brought him great honor and fame. *Today he is considered "the Father of Chemistry."*

In 1668, Boyle returned to London, where he spent the rest of his life. Although his health was poor, he continued to supervise experiments. Boyle believed that God is glorified through the study of His creation, but he also knew that people must have the Scriptures to rightly know God. He spent large sums of money to have the Bible translated into a number of languages. Boyle was the chief supporter of John Eliot, a missionary to the Native Americans. Boyle was also concerned for the people in his own country who were turning away from God toward atheism. Atheism is the belief that there is no God. Boyle set money aside to pay for a series of lectures to be given in his own country to preach the truth about God and the Good News of the Gospel.

It is a tribute to Robert Boyle that all who knew him considered him to be an honest, upright man whose chief goal in life was to influence as many people as he could to accept the truth of Christianity. When he died in 1691, he was greatly mourned. Today, we can celebrate God's work in his life not only as a scientist but also as a proclaimer of the Gospel.

Lismore Castle, Boyle's birthplace

air pump

Chapter 8 Concepts Review

Basic Weather Concepts 8.1–8.5

A. Remember

Matching: *Write the letter of the correct answer in the blank.*

_____ 1. the condition of the air

_____ 2. weather conditions of a certain area over time

_____ 3. air surrounding the earth

_____ 4. weather prediction

_____ 5. scientist who studies weather

_____ 6. instrument used to detect air pressure changes

_____ 7. atmosphere's ability to keep heat from easily escaping into space

A. atmosphere
B. atmosphere effect
C. barometer
D. climate
E. forecast
F. greenhouse effect
G. meteorologist
H. weather

Identify: *Label the gas that makes up 78% of air.*

8. ______________

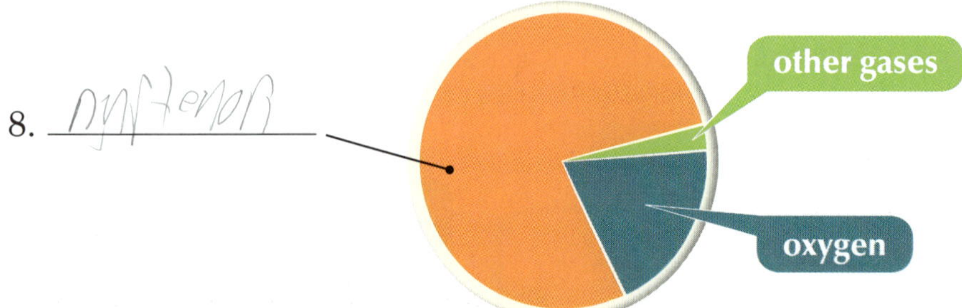

B. Think Like a Scientist

Label: *Write* J *for jet stream; write* G *for global winds.*

_____ 1. causes weather systems to travel from west to east

_____ 2. helps regulate the earth's temperature

_____ 3. band of wind high above the earth

_____ 4. helped sailors form trading routes

_____ 5. discovered during World War II

_____ 6. bands of winds which change at certain latitudes

C. Write

Essay: *Write a paragraph describing the water cycle, beginning and ending at a mountain stream.*

____

____

____

____

____

____

____

D. Think Like a Scientist

Look It Up: *What do you know about these words? If you are not sure, look up the definitions and study them. Be ready to tell what each word means.*

barometer	**evaporation**	**hygrometer**	**relative humidity**
condensation	**fog**	**precipitation**	**water runoff**

2 Severe Weather Phenomena and Forecasting Concepts 8.6–8.8

A. Remember

Fill in the Blank: *Write the correct answer in the blank.*

1. A thick, dark storm cloud is a ________________________.
2. A ________________ is a storm that begins to rotate around a low-pressure area.
3. A ________________ is a rise of ocean water surrounding a hurricane.
4. A ________________ is a severe snowstorm.
5. The seasonal wind, bringing heavy rain to some places in southern Asia, is called a ________________________.

B. Think Like a Scientist

Label: *Write* TS *for thunderstorms,* H *for hurricanes, and* T *for tornadoes.*

_____ 1. gets its energy from a warm ocean

_____ 2. most common of all storms

_____ 3. given a person's name

_____ 4. moves west to east

_____ 5. cyclone that develops over hot land

_____ 6. can weaken over colder water or land

_____ 7. begins over a warm ocean

_____ 8. creates a funnel cloud from a thunderstorm to the ground

_____ 9. has a storm surge

_____ 10. travels east to west

Short Answer: *Write the correct answer in the blank.*

11. List the weather conditions meteorologists measure to make forecasts.

____

____

12. Why is a short-range forecast the most accurate? ____________________

____

C. Fun with Terms

Crack the Code: *Find the letter of the correct term and write it on the numbered line below.*

_____ 1. boundary between two different air masses

_____ 2. large body of air having the same temperature and humidity

_____ 3. notification that a tornado has been spotted

_____ 4. notification that conditions are right for tornado formation

_____ 5. overflow of water

_____ 6. prolonged period of dryness

_____ 7. weather event caused by specific conditions

G. drought

J. front

M. tornado watch

N. air mass

R. tornado warning

S. weather phenomenon

T. flooding

__ U __ I O __   __ E __ E O R O L O __ I __ T

1 2 3 4 5 6 7

D. Think Like a Scientist

Short Answer: *Use the map to answer the questions.*

1. What is the temperature in Salt Lake City? ____________________

2. Which kind of front brings heavy precipitation? ____________________

3. Why do you think there is snow in Salt Lake City? ____________________
____

4. Which kind of front is stretching across most of the United States? ______
____

5. Using your knowledge of the jet stream, predict the weather for Pensacola. ____________________
____
____
____

6. Draw a sun next to the warm front and a snowflake next to the cold front on the map.

Chapter 9 Understanding the Great Expanse of Outer Space

TERMS

constellation: a group of stars that forms a picture

galaxy: a huge star system

Milky Way: the galaxy our sun and solar system belong to

9.1 The Wonders of the Night Sky

How radiantly the stars shine on a clear night! At first glance, it seems as if there is no pattern or order to them. However, God, Who created the universe, does nothing by chance. He placed each star in its proper place in the sky. In ancient times, philosophers who studied the stars saw patterns in their placement. They gave the **constellations**, or *groups of stars that form pictures*, names of objects they resembled. Throughout time, people have found the constellations mysterious and fascinating. They have thought that the constellations told stories.

The Bible tells us that the stars do have a message to tell the world. They speak of the power and glory of God.

> *When I consider Thy heavens, the work of Thy fingers, the moon and the stars, which Thou hast ordained; what is man, that Thou art mindful of him? and the son of man, that Thou visitest him?*
>
> *Psalm 8:3–4*

The Immensity of the Heavens

Most of the stars you can see without a telescope are in a *huge star system*, or **galaxy**, called the Milky Way. *The* **Milky Way** *is the galaxy our sun and solar system belong to.* From far into space, *the Milky Way looks like a giant swirling spiral.* Just how big is the Milky Way? To give you an idea, think of how the sun, like all the stars of the Milky Way, orbits its center. Yet its path is so huge that it would take about 230 million years for the sun to complete this orbit. Besides our sun, there may be as many as 400 billion other stars that belong to the Milky Way. If it were possible to travel at the speed of light across the Milky Way from one side to the other in a spaceship, it would take about 100,000 years. What an enormous galaxy our solar system is a part of!

What do we find beyond the Milky Way? Looking into a telescope from the Southern Hemisphere, we could see other galaxies, such as the *Large and Small Magellanic Clouds.* On a dark and moonless night, we could even see the *Andromeda Galaxy* [ăn·drom′ĭ·də] without a telescope. There are billions of other galaxies that make up the universe. Each galaxy contains an innumerable amount of stars.

It is impossible for us to understand the size of the universe. As we think about the immensity of God's creation, our admiration for God's handiwork grows. God not only created the heavenly bodies, but the Bible says that He also calls them all by name.

> *He telleth the number of the stars;*
> *He calleth them all by their names.*
> *Great is our Lord, and of great power:*
> *His understanding is infinite.*
>
> *Psalm 147:4–5*

The Night Sky

Throughout your study of this chapter, observe the heavenly bodies in the night sky. A *heavenly body* is a star, planet, moon, or other object in space. Notice the moon's changing appearance throughout the month. When the sky is clear, you can observe more stars. Later, you will learn how to identify some *constellations*, or star pictures. You may even spot a shooting star. A shooting star is not really a star at all but is actually a *meteor*. A meteor is a piece of dust or rock that burns up in the earth's atmosphere as it falls to the ground. Most of the time, a meteor burns up before it hits the earth's surface. If it does not completely burn up, we call the rock that is left a *meteorite*.

Comprehension Check 9.1

Give the correct answer.

1. What do we call groups of stars that form a picture? ________________

____

2. What is a huge star system called? ________________

3. Our solar system belongs to which galaxy? ________________

____

4. What is the shape of the Milky Way? ________________

solar system: a star and all the heavenly objects that orbit it

planet: a heavenly body that orbits a star

inner planets: the four planets closest to the sun, having solid surfaces—Mercury, Venus, Earth, and Mars

outer planets: the four planets farthest from the sun, made mostly of gases—Jupiter, Saturn, Uranus, and Neptune

9.2 The Beauty of the Solar System

The glory of the night sky is only a small part of the beauty we would see if we could take a trip through space. *A* **solar system** *is a star and all the heavenly objects that orbit it.* If we were to soar through our own solar system, we would view many magnificent sights, beginning with the planets. **Planets** *are heavenly bodies that orbit a star. The eight planets of our solar system orbit our star, the sun.*

Mercury is the first of the inner planets. *The* **inner planets** *are closest to the sun and have solid surfaces.* Because Mercury is closest to the sun, its orbit is the shortest. There are only 88 days in Mercury's year.

Mercury

On *Venus*, we might see lightning storms crashing over its mountains. Its rotation is different from other planets because Venus spins backward. After *Earth*, we would see *Mars*. Mars is called the Red Planet because its iron-rich surface is rusty in color. Its landforms include an extinct volcano that is taller than Mt. Everest, Earth's tallest mountain.

The **outer planets** *are the four planets farthest from the sun.* They do not have solid surfaces as the inner planets do because *they are made mostly of gases*. If we were to pass *Jupiter*, its white, pink, and orange clouds would draw our attention to the Great Red Spot—an enormous hurricane-like storm larger than two Earths. Next, we would see *Saturn's* majestic, banded rings. After Saturn, we would see a surprising sight—a planet that seems to be tipped on its side. *Uranus* rotates in the most unusual way. One of its poles is always pointing toward the sun while the other is pointing away. Even its rings are tilted as they orbit Uranus's equator. Far in the distance, we would see one more planet. *Neptune* is a beautiful blue color because of one of the gases in its atmosphere. Farthest from the sun, Neptune takes 165 Earth years to make a complete orbit.

If we were to make a large loop back to the sun, we would find its brightness too great to look at without special eye protection. We would watch explosions of gases spray thousands of miles upward from the sun's burning, fiery surface.

But none of these sights would be as welcoming as the most beautiful planet of all—the only planet that supports life. As we approach Earth, we would see the green and brown continents surrounded by deep blue oceans. Nowhere else did we see water as on Earth, which nourishes and sustains life. As our spacecraft readies for landing, we see rivers, forests, and rolling hills. We have returned to the beautiful planet specially designed by God to be our home.

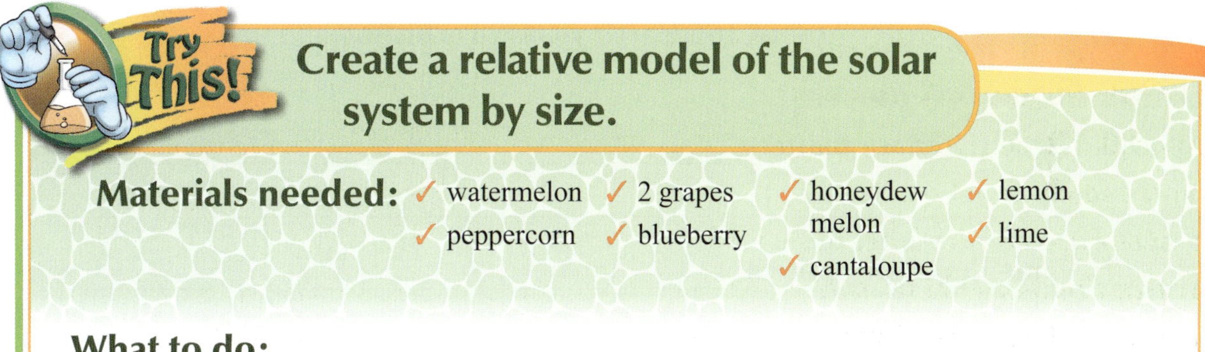

Try This!

Create a relative model of the solar system by size.

Materials needed:
- ✓ watermelon
- ✓ peppercorn
- ✓ 2 grapes
- ✓ blueberry
- ✓ honeydew melon
- ✓ cantaloupe
- ✓ lemon
- ✓ lime

What to do:

Use each object to create a model of the solar system. Arrange each object around the watermelon, which serves as the sun.

Planet	Object
Mercury	peppercorn
Venus	grape
Earth	grape
Mars	blueberry
Jupiter	honeydew melon
Saturn	cantaloupe
Uranus	lemon
Neptune	lime

Comprehension Check 9.2

***Fill in the Blank:** Write the correct answer in the blank.*

1. A planet is a heavenly body that orbits a ____________________.
2. The __________ planets have solid surfaces.
3. ____________ orbits the sun in 88 days.
4. The __________ planets are made mostly of gas.
5. ____________ takes 165 Earth years to orbit the sun.

Give the correct answer.

6. Name the inner planets. ________________________________

____

7. Name the outer planets. ________________________________

____

Label the planets of the solar system.

A. Earth
B. Jupiter
C. Mars
D. Mercury
E. Neptune
F. Saturn
G. Uranus
H. Venus

Think and Predict.

16. Which planet's surface temperature is hotter—Earth or Mercury? Why? ________________________________

____

____

____

rotate: to spin

axis: an imaginary line through the center of the earth from the North Pole to the South Pole

sundial: an instrument using the earth's position to tell time

revolve: to orbit a central point

9.3 How Movement Determines Time

Have you ever wondered what life would be like if there were no calendars? How would you know when it was your birthday? How would anyone know when it was time for school to be over and summer vacation to begin? What if there were no clocks? Would you be able to tell the time? Although our modern time-measuring devices are useful, we would not be lost without them. We would simply do what people did a long time ago—watch the sky and its heavenly bodies. Because of the earth's predictable motions, we have a sense of time.

And God said, Let there be lights in the firmament of the heaven to divide the day from the night; and let them be for signs, and for seasons, and for days, and years.

Genesis 1:14

Day, Night, and Earth's Rotation

Although you cannot feel it, the earth is moving right now. Have you ever watched a spinning top? The earth spins in the same way that a top does. We call this *spinning motion* of the earth its **rotation**.

Notice how your globe rotates. A rod goes through the middle of the globe to hold it in place. One end of the rod is at the North Pole, and the other end is at the South Pole. The globe rotates on this rod. The rod represents *an imaginary line that runs through the earth*. This imaginary line is called the earth's **axis**. The axis does not point straight up and down; it is tilted in the same way your globe is tilted.

The earth rotates on its axis every 24 hours, making one day. The surface of the earth near the equator rotates at a speed of about one thousand miles per hour.

The earth rotates on its axis from west to east. This is why the sun seems to rise in the east each morning and set in the west each evening. However, the sun is not really moving across the sky each day. Because the earth rotates, people on one side of the earth face the sun during daylight hours while people on the opposite side of the earth are facing away from the sun during nighttime hours. As the earth spins, the parts of the earth in darkness gradually rotate until they are facing the sun. When it is daytime in your part of the world, it is nighttime on the opposite side of the world.

People in ancient times often used an instrument called a **sundial** to know the time of day. Remember, a shadow is a dark area or shape cast by an opaque object. In the middle of the sundial was a pin, or gnomon [nō′mŏn′]. *Depending on the earth's position during the day, the gnomon's shadow told the people what time it was by pointing to a number on the dial plate.*

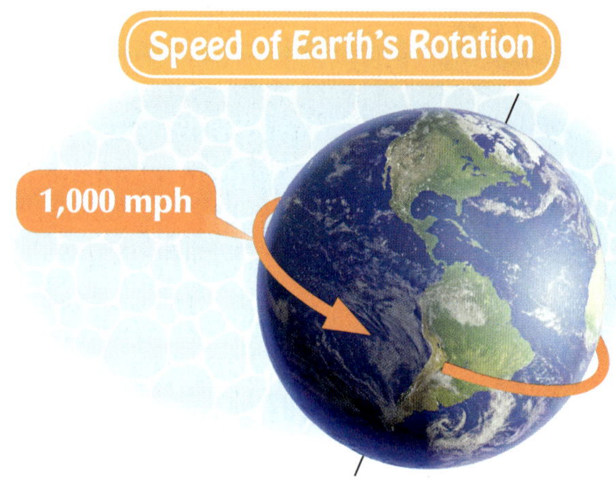

The Year and Earth's Revolution

As the earth rotates on its axis, it also travels in a path around the sun. We say that the earth revolves around the sun. *To* **revolve** *is to orbit a central point.* The pathway, or orbit, that the earth follows is 584 million miles around. The earth travels at a speed of 67,000 miles per hour as it revolves around the sun.

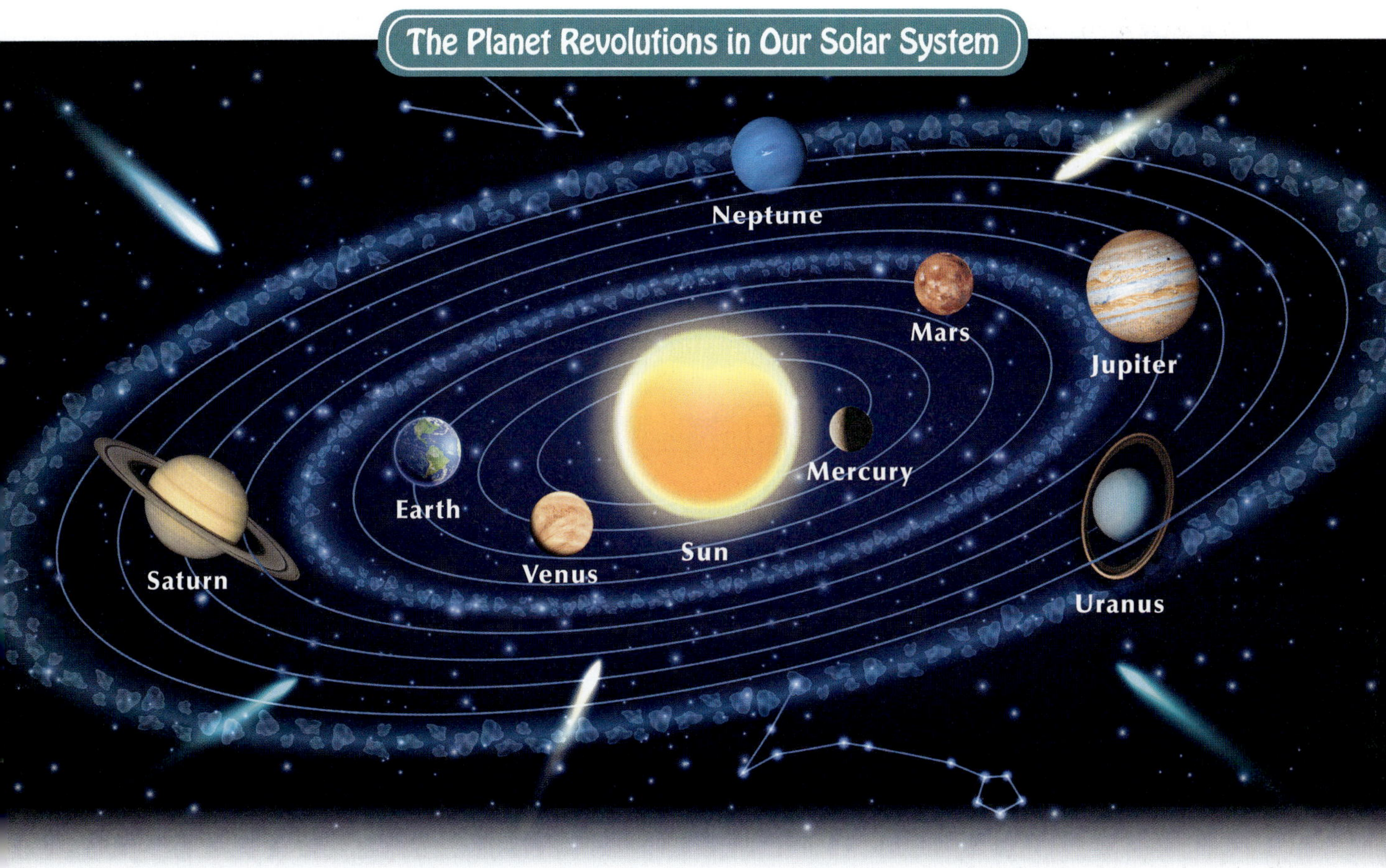

The earth revolves around the sun in 365¼ days, the period which we call a year. Every four years, we add a day to February to make up the extra quarter of a day left over each year.

God has designed the earth to have exactly the right length of each day and each year. If our day were as short as Jupiter's day, which is less than ten hours, we would not have time for all the things we wanted to do. If it were as long as Mercury's day, 59 Earth days, we would have to sleep several times before the day was over. Instead, we have a 24-hour day, which gives us just enough time to work, play, and sleep. If our year were as short as Mercury's year, 88 days, we would not have the right growing seasons for plants to grow and reproduce. What would happen if our year were 687 Earth days like planet Mars? Some countries would suffer from very severe winters. Everywhere we look, we see evidence of a loving God Who created the world. He knows what is best for us.

Discover the cause of day and night.

Materials needed: ✓ globe ✓ flashlight

1. Darken your room and shine the flashlight at the globe. Keep the flashlight still.
2. Turn the globe, observing how the dark and light areas change as the globe rotates.

What movement of the earth is the cause of day and night? The earth's rotation gives us our day and night.

The Earth's Tilt and Seasons

As the earth revolves around the sun, *its axis remains tilted at an angle that does not change.* This tilt causes both light and heat energy from the sun to strike the earth in different ways at different times of the year, *giving our planet its seasons.* Certain constellations are visible in different parts of the night sky during certain months or seasons. Farmers in ancient times could tell if it was planting or harvest time by watching the movements of certain constellations.

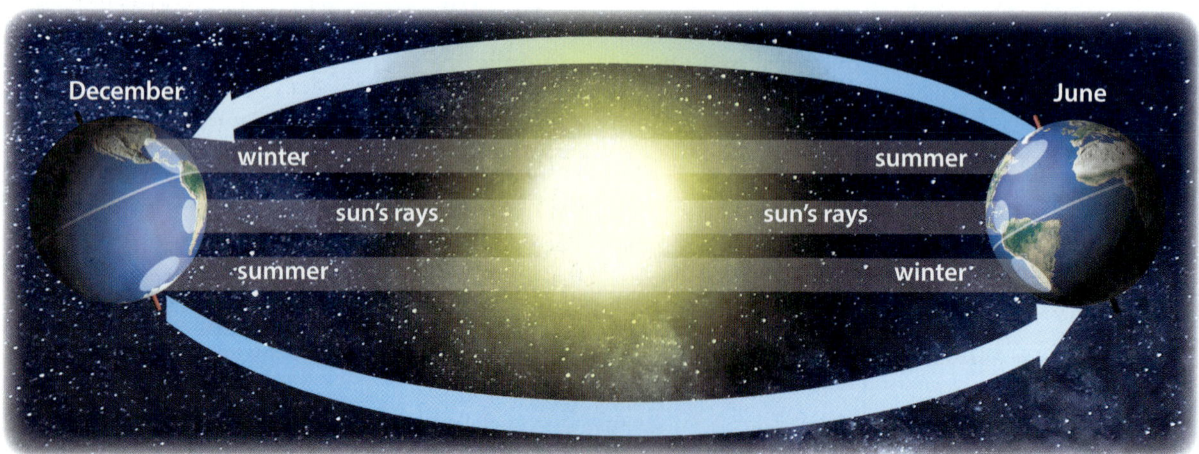

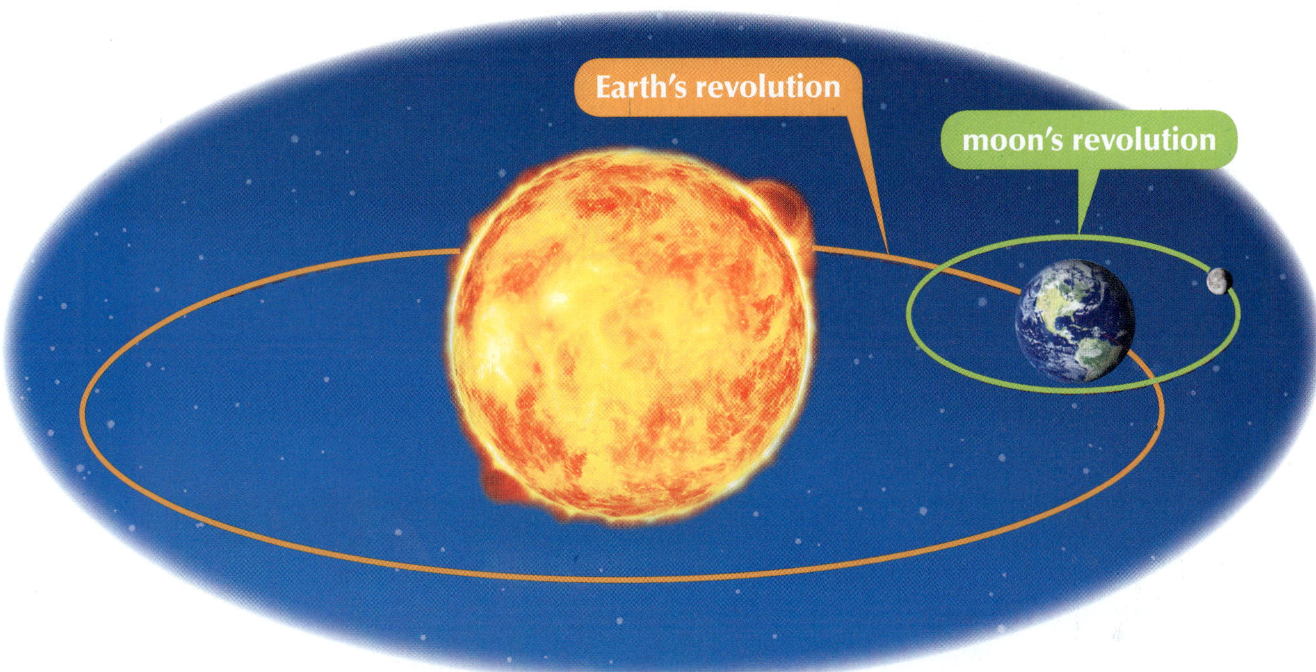

Months and the Moon's Revolution

As the earth is revolving around the sun, the moon is revolving around the earth. *A long time ago people could determine the day of the month by watching the moon.* The moon completes a full revolution around the earth about every thirty days, twelve times a year. Therefore, we have twelve months in a year. Some months or years have an extra day.

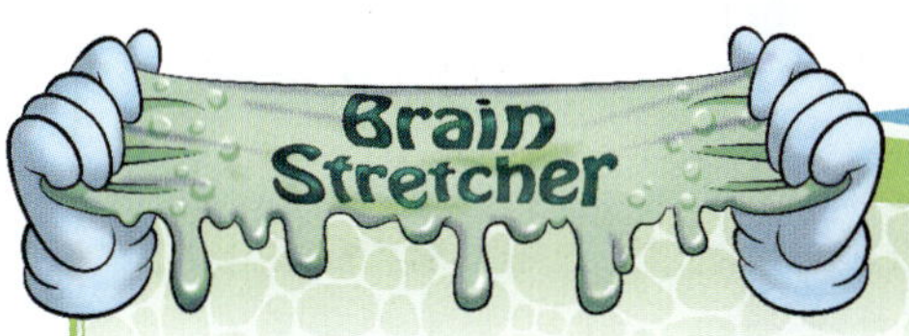

The Seven-Day Week: There is no orbit of planets or heavenly bodies that give us the seven-day week. It was given to us directly by God. God created our world in only six days, and on the seventh day He rested. Like the world He created, God created the Sabbath, the seventh day, for our benefit and His glory. Today, we set aside Sunday because that was the day Christ arose from the dead. Sunday is our day of rest that is holy to God. As God rested after He created the world, so we rest from some of the things which we do for the other six days of the week to allow more time to worship Him in prayer and praise.

Try This! Discover why winter is cold.

Materials needed:
- ✓ flashlight
- ✓ globe
- ✓ piece of cardboard with a one-inch-square hole cut in it
- ✓ small, removable sticker

1. Use the removable sticker to mark the position of your city on the globe.
2. Darken the room by turning out the lights and closing the window coverings.
3. Find your city on the globe. Turn the globe as shown for winter and summer, and shine the flashlight through the cardboard.

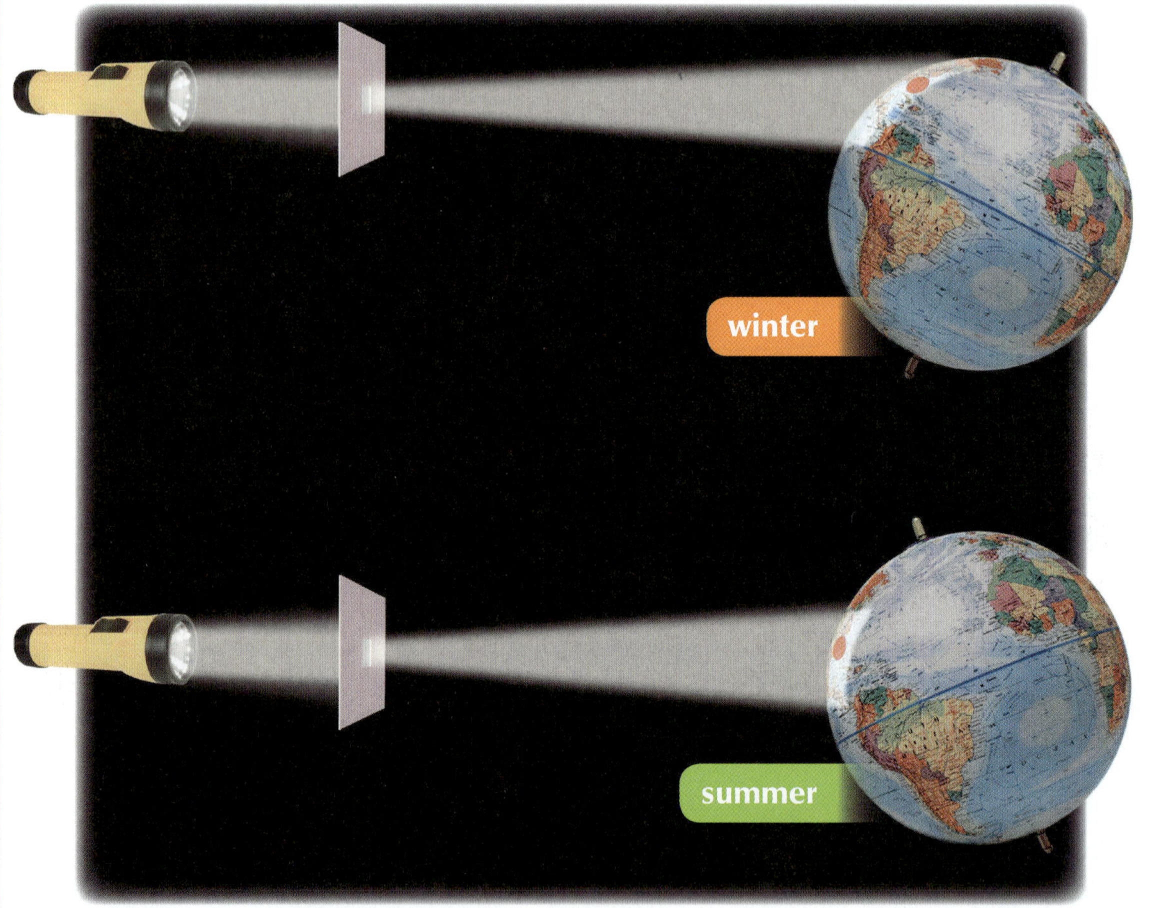

Notice the tilt of the earth on its axis. What shape does the square beam of light make on the globe in winter? Is the shape different in the summer position? Do you see that winter is cooler because the same amount of light is spread over a larger surface?

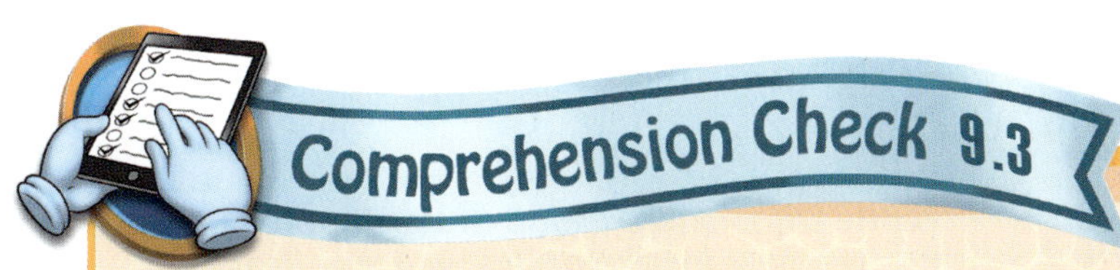

True/False: *If the statement is true, write* true. *If the statement is false, replace the underlined word(s) with a word or phrase that will make the statement true. Do not write* false *in any blank.*

____________ 1. The earth rotates on its axis every 364¼ days.

____________ 2. A long time ago, people could determine the day of the month by looking at the moon.

Multiple Choice: *Circle the correct answer.*

3. The earth spins, or __?__, on its axis.
   a. drops b. revolves c. rotates

4. The earth __?__ around the sun in a continual path.
   a. revolves b. rotates c. tilts

5. The earth __?__ on its axis at an angle that does not change.
   a. revolves b. sits c. tilts

Identify: *Label the pictures, using the words* rotate *or* revolve.

6. ____________ 7. ____________ 8. ____________

Think and Conclude.

9. The earth's ____________________ gives us our day and night.
10. The earth's ____________________ gives us our year.
11. The earth's ____________________ gives us our seasons.

Big Dipper: the seven stars appearing as a long-handled cup, which are part of *the Great Bear*

Little Dipper: the group of stars appearing as a small cup, also called *the Little Bear*

9.4 Constellations

For many years, people have noticed the constellations in the night sky. Do you know how to recognize some of the constellations? As you learn where to look for these star pictures, give praise to the Creator Who always sees you and is thinking of you.

> *Praise ye Him, sun and moon: praise Him, all ye stars of light.*
>
> *Psalm 148:3*

The Big Dipper

One of the most familiar groups of stars seen in the Northern Hemisphere is the **Big Dipper**. *It is made up of seven stars which are part of the constellation called the Great Bear.* The Big Dipper got its name because its stars are in the shape of an old-fashioned, long-handled cup, or dipper. Locating the Big Dipper will help you to find other stars.

The Great Bear

The last star in the handle of the *Big Dipper is the nose of the Great Bear*. The other stars in the handle form the back of the bear's head, and the bowl of the Dipper is the bear's saddle. This huge bear of the northern skies looks very much like a polar bear, as it should, since it is near the North Pole. It can be seen best from February through June.

The Little Dipper

The **Little Dipper** looks like a smaller, fainter version of the Big Dipper. It is most easily seen on a clear night. The Little Dipper has another name. Can you guess what it might be? *The Little Dipper is also called the Little Bear.* Both the Big and Little Dippers are visible all year long.

Leo the Lion

Not far from the Great Bear is another constellation, *Leo the Lion*. You can see Leo in March, April, and May. First, locate the Big Dipper's cup. Next, trace a straight line from the two stars closest to the handle through the front leg of the Great Bear to two very bright stars. The brighter star, Regulus [rĕg′yə·ləs], is Leo's front paw. Another bright star, Denebola [də·nĕb′ə·lə], is at the tip of Leo's tail.

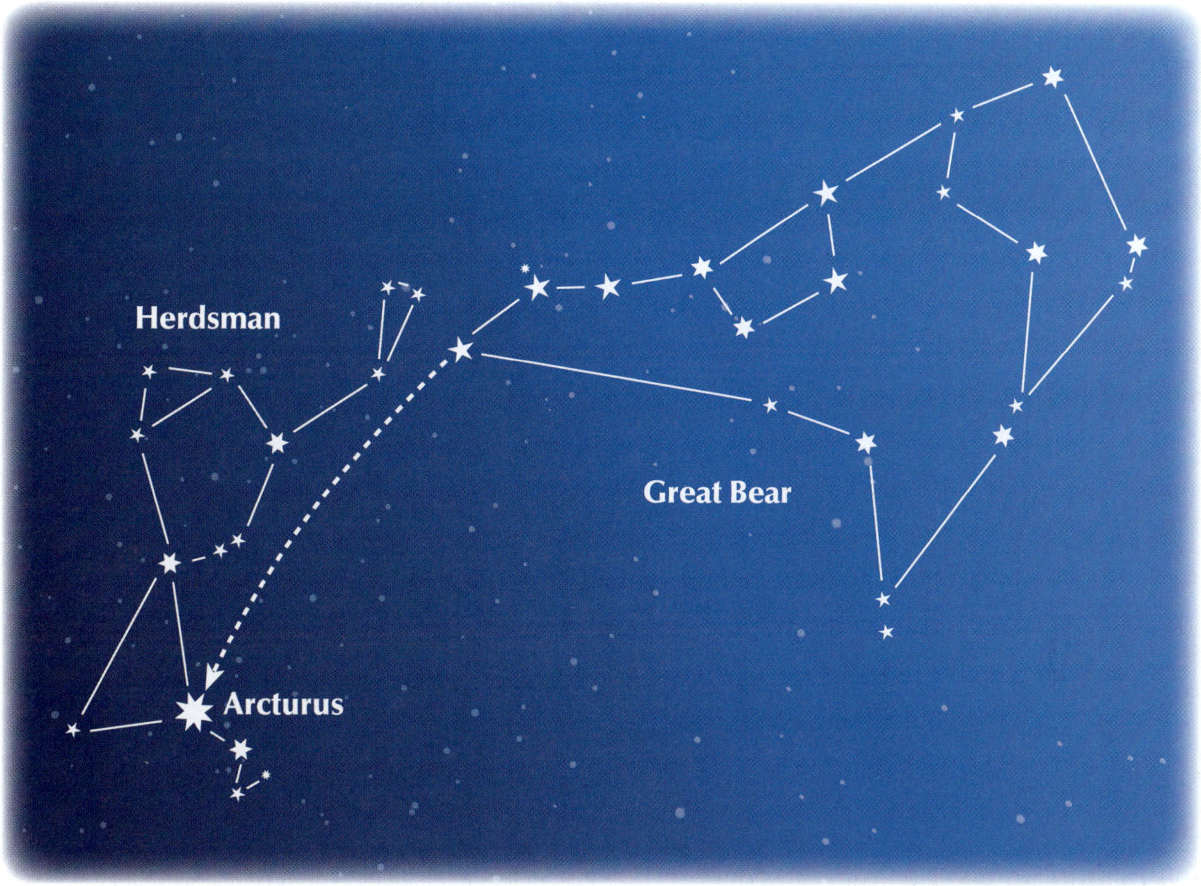

The Herdsman

The Great Bear is looking at the horn of *the Herdsman*. What job do you think a herdsman has? A herdsman is a person who tends a herd of animals.

The Herdsman has a pointed hat, and he looks as if he is sitting on a grassy bank blowing his horn. You can see the Herdsman best from April through August. A curved line drawn from the end of the Big Dipper's handle points to Arcturus [ärk·to͝or′əs], a magnificent star in the Herdsman. Arcturus is a bright orange star—one of the four brightest stars in our sky.

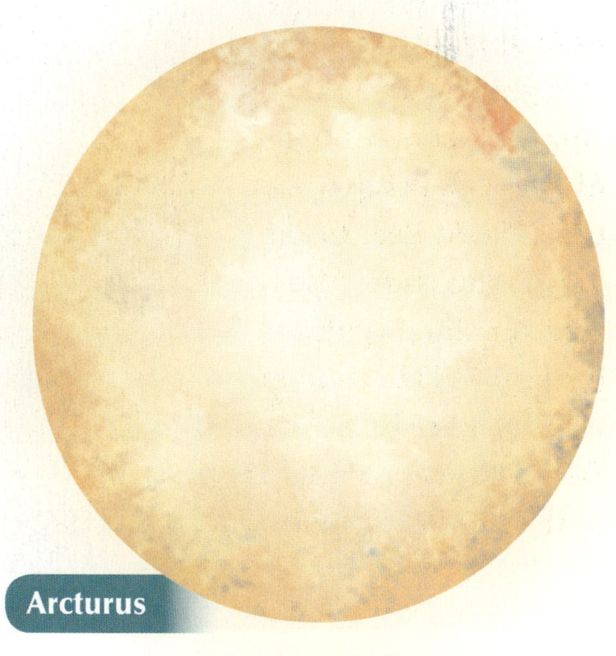

Connect the stars.

1. Can you find the Big Dipper, Great Bear, Leo the Lion, and the Herdsman in the picture?
2. Connect the stars below to show the Big Dipper, and then complete the Great Bear and the rest of the constellations.

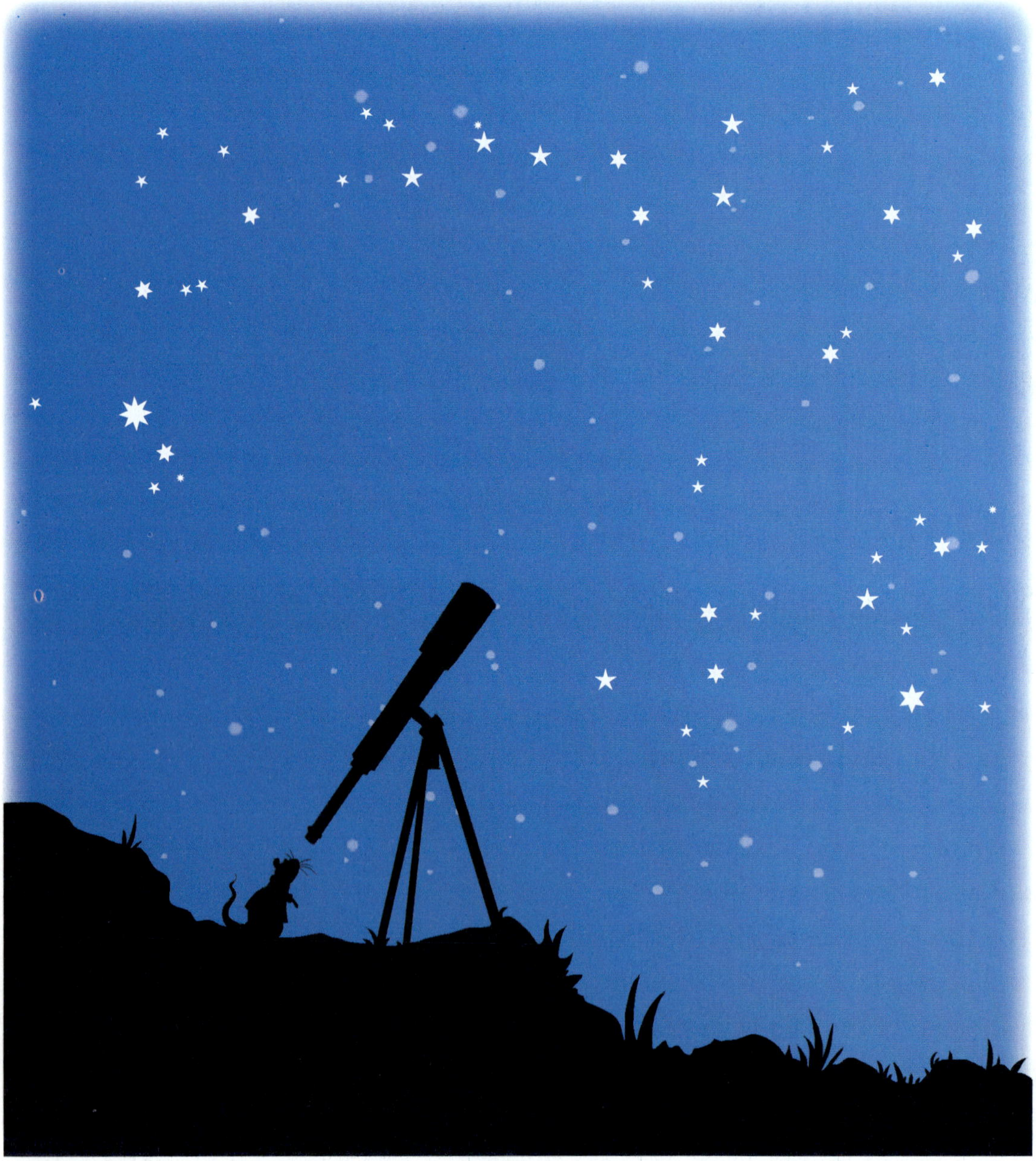

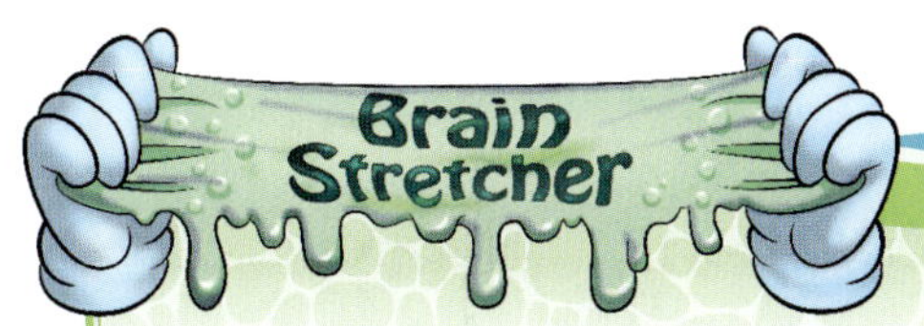

Visibility of Stars: Have you ever wondered why we see the constellations and the other stars only at night? During the daytime, the earth's atmosphere scatters the light from the sun. This scattering makes the sky appear blue. It also makes the sky so bright that we cannot see the stars. If you were on the moon where there is no atmosphere, you could see the stars all day long.

Try This! Discover why stars shine more brightly at night.

Materials needed:
- ✓ white and black construction paper
- ✓ pin
- ✓ window (or other light source)

1. Make several pinholes in a piece of white construction paper and in a piece of black construction paper.
2. Hold each paper up to the light in the room or up to the window.

Which paper makes the light shining through the holes more noticeable? How does this demonstrate what happens to the starlight during the daylight hours?

Comprehension Check 9.4

Identify: *Label each constellation.*

1. ______________________

2. ______________________

3. ______________________

4. ______________________

5. ______________________

GPS: Global Positioning System

sextant and astrolabe: instruments used in navigation

horizon: the line where the earth seems to meet the sky

North Star: the star consistently located almost directly above the earth's axis at the North Pole, also called Polaris or the Pole Star

Southern Cross: four stars forming the shape of a cross with the long crossbar pointing to the South Pole

9.5 Navigation and the Stars

Has your family ever been lost during a road trip or vacation? Perhaps you have watched your family find directions by using **GPS**, or *the Global Positioning System*. GPS uses satellites to help you know exactly where you are on the planet. It helps you navigate, or plan a route using your location, to get from where you are to the place you want to be.

Navigational Instruments

How did people navigate before cellphones, satellites, and GPS were invented? You might remember some early tools, such as the lodestone and compass. A *compass* points north and can help you find the right way to go, but to plan a route *from* your location, you need to *know* your location. Since the 1700s, sailors have used a sextant to find their way. *A* **sextant** *is an instrument used in navigation*. It can measure the angle between two objects. For example, if a sailor were to look through the scope of a sextant, he could find the angle between the **horizon**, *the line where the earth seems to meet the sky*, and the North Star. If he knew the time and date, he could then plan his course by knowing his position on Earth. A sextant can also be used to find the angle between the sun or moon and the horizon.

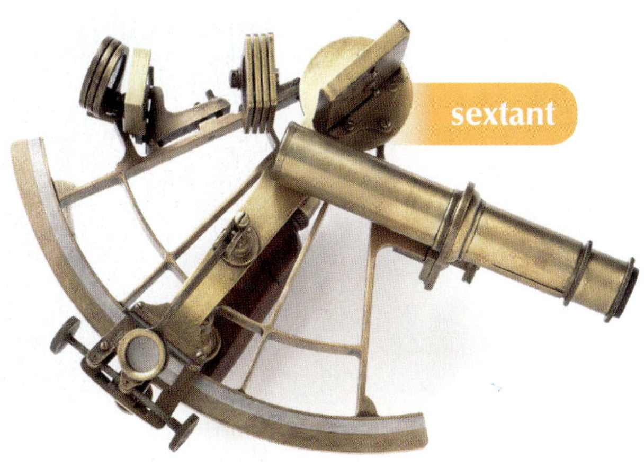
sextant

Knowing where the heavenly bodies in the sky could be seen at certain times of year helped early explorers to navigate. Since creation, the stars in the sky have always seemed to move in the same predictable pattern across our sky. Do you remember what is actually moving to cause this to happen? The stars themselves do not move across our sky; it is the earth that is rotating on its axis as it revolves around the sun, causing the stars to *appear* to move.

Before the sextant was invented, people used *another navigational instrument* called an **astrolabe** to calculate the location of heavenly bodies or to find one's location on Earth. An astrolabe has a map of the constellations and movable parts that help with navigation.

Since you know how to find the Big Dipper and Little Dipper, you can find the North Star like the early navigators and explorers did.

The North Star

The **North Star** is a bright star that always seems to be in the same place. *It is almost directly above the earth's axis at the North Pole*. The North Star is at the end of the "handle" of the Little Dipper. The Little Dipper seems to rotate around the North Star as if it were being swung by its handle. Because the earth rotates on its axis, all other stars seem to rotate around the North Star. This star's position above the North Pole gives it its other names: *Polaris* and *the Pole Star*. For centuries, the North Star has been used for navigation. Explorers on land and sea have used it to find directions, and so can you! *When you face the North Star, you are facing north*, south is behind you, east is to your right, and west is to your left. You can also find the North Star by finding the Big Dipper. On the side of the cup of the Big Dipper are two very bright stars. They are called *the Pointer Stars* because they point in a straight line to the North Star.

The Southern Cross

Fewer constellations can be seen from the Southern Hemisphere than from the Northern Hemisphere. One of the most famous is the smallest constellation of all, the **Southern Cross**. *Its four most brilliant stars form the shape of a cross with the long crossbar pointing to the South Pole*. What a beautiful picture for God to place in the southern sky!

Orion

During winter in the Northern Hemisphere, one of the most spectacular constellations in the sky is **Orion**, *the hunter*. This star group, which is mentioned three times in the Bible (Job 9:9; 38:31; Amos 5:8), has more bright stars than any other constellation. You can easily recognize Orion by finding *the three brilliant stars* that are aligned side-by-side, *forming his belt*. Orion appears to be holding a shield and raising a club high above his head. A sword hangs from Orion's belt. You may not be able to see Orion's sword nor his head without binoculars unless you live in a place where the sky is very dark at night, but you will certainly be able to see his belt and his two bright stars.

> *Seek Him that maketh the seven stars and Orion.* *Amos 5:8*

Give the correct answer.

1. What does GPS stand for? ____________________
2. Name an instrument used in navigation. ____________________

   ____
3. What is another name for the North Star? ____________________

   ____
4. What direction are you facing when you face the North Star? ________

   ____
5. What two stars in the Big Dipper point to the North Star? ________

   ____
6. What constellation points to the South Pole? ____________________

   ____
7. What part of Orion, the hunter, is easy to recognize and locate? ______

   ____

Try This! Construct a star viewer.

Materials needed:
- ✓ large shoebox
- ✓ black construction paper
- ✓ pencil
- ✓ tape
- ✓ flashlight

1. Make a constellation by drawing dots on black construction paper for each star's location. Push the tip of a pencil through the dots on the paper to make small holes.
2. Cut an opening in one end of the shoebox. Tape your constellation over the open end.
3. Place the flashlight inside the shoebox, pointing it toward the open end.
4. In a darkened room, turn on the flashlight to see your star pictures on the wall or ceiling.

9.6 The Sun: The Greater Light

The Light That Rules the Day

The greater light that God made is the sun. On the fourth day of Creation, God made the sun to shine, and it has been shining ever since.

> *And God made two great lights; the greater light to rule the day, and the lesser light to rule the night: He made the stars also.* Genesis 1:16

The sun is about 93 million miles from the earth. It is a star similar to the other stars you see at night. Because the sun is the closest star to the earth, it is the most important star to us. Its closeness to the earth makes it appear to be a very large star, but compared to some stars, *it is only average-sized*. Some stars are called supergiant stars. Betelgeuse [bēt′l·jo͞oz′], for example, is a red supergiant star and is much larger than our sun.

When we compare the sun to the earth, the sun is colossal in size. If we had a large ball the size of the sun, we could drop more than one million balls the size of the earth into it. Though the sun is one million times the earth in volume, the sun is not one million times heavier than the earth. The earth is made of solids, liquids, and gases, but *the sun is made entirely of hot, glowing gases*. Since gases are less dense than solids or liquids, it would take only about 333,000 Earths to equal the *weight* of the sun.

Since the sun sends out powerful rays that can permanently damage your eyes, *you should avoid looking directly at it.* It is also unsafe to use binoculars or an ordinary telescope to look at the sun. These instruments increase the power of the sun's harmful rays and, therefore, increase the damage that the rays can do, even causing blindness. When scientists study the sun, they use telescopes that filter out the sun's harmful rays and enlarge the sun's image.

The Sun's Energy

God has designed the sun to *give off a steady supply of energy as it burns*. Scientists think that the sun produces its energy by atomic reactions. This reaction probably happens as the atoms in the sun's center join together to produce energy. Day after day, the sun gives off almost exactly the same amount of *light* and *heat energy*. Many other stars do not give off steady amounts of energy. We can be thankful that our sun does not behave as those stars do. If the sun were not dependable, the earth would suffer very harsh changes in climate.

Only a very tiny amount of the sun's energy ever reaches the earth. Most energy is lost in space because the sun releases its energy in all directions. The sun supplies our planet with two types of energy. Do you remember what they are? Even though most of the sun's energy does not travel to Earth, the energy that does reach the earth is very powerful—powerful enough to supply our need for heat and light.

Heat

The sun is so hot that nothing can exist there without changing into a hot gas. Knowing this, do you think that scientists would want to explore the sun in a spacecraft? A spacecraft could not go near the sun without being changed into a glowing, hot gas.

Light

If we could drive by car at sixty miles per hour, without ever stopping, it would take us about 177 years to travel the 93 million miles from the earth to the sun. Fortunately, light travels much faster than we can. Light travels from the sun to the earth at the speed of 186,000 miles per second. It would take a ray of light about one second to travel around the earth seven times. The sunlight you see as you look out the window began to travel to the earth just over eight minutes ago.

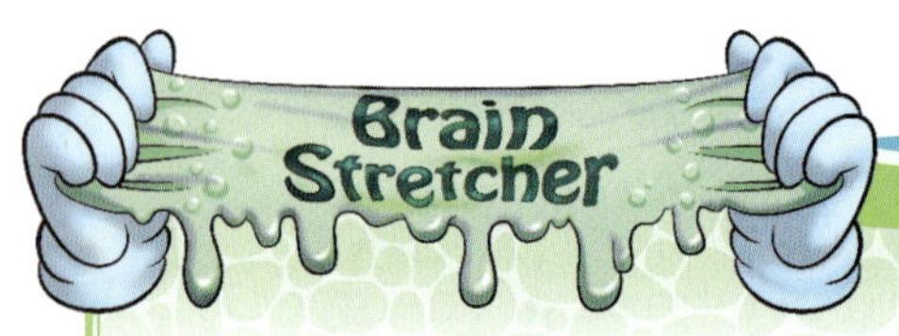

Star Colors: Have you ever gathered around a campfire with friends or family? You may have noticed that not every part of the fire is the same color. At the base of the flames the fire is blue; then the flame changes color from white to yellow to orange to red. Which part of a flame do you think is the hottest? The blue part of the flame is where the fire is hottest. Scientists can tell which stars are the hottest by looking at their color. Blue stars are some of the hottest stars, while red stars are the coldest. Our sun is a yellow star. What does that tell you about its surface temperature?

The Sun's Gravity

As you know, gravity is the force, or pull, that holds objects to the earth. The sun also has gravity. *Because the sun is much larger than the earth, it has a much stronger pull of gravity.* The sun's gravity keeps the planets in orbit around the sun. Without the sun's gravity, the earth and the other planets would move in straight lines because of inertia, and would not curve to circle the sun.

Brain Stretcher

The Sun's Orbit: Did you know that the sun travels through space at a speed of about twelve miles per second? Remember, the sun has an orbital path in the Milky Way galaxy. As it moves, it carries the earth and the other seven planets with it. The sun also rotates, or spins like a top, completing its rotation in twenty-four days. Before astronomers knew about the sun's movement through the sky, the Bible described it.

> *His going forth is from the end of the Heaven, and His circuit unto the ends of it: and there is nothing hid from the heat thereof.*
> *Psalm 19:6*

A long time ago, some people used this verse to try to prove that the Bible wasn't true, because they believed that the sun stood still. How wrong they were! The sun and the planets move together in perfect harmony. They were created and placed in their courses by the wise Designer of the universe. God's plan for the universe works well!

Comprehension Check 9.6

True/False: *If the statement is true, write* true. *If the statement is false, replace the underlined word(s) with a word or phrase that will make the statement true. Do not write* false *in any blank.*

_______________ 1. Compared to other stars, the sun is very large.

_______________ 2. The sun is made of hot, glowing liquids.

_______________ 3. Never look directly at the sun.

_______________ 4. The sun gives off a steady supply of energy as it burns.

_______________ 5. The sun gives light and sound energy.

_______________ 6. The sun has more gravity than the earth.

moon: a heavenly body that orbits a planet

luminous object: an object that gives off its own visible light

nonluminous object: an object that does not give off its own visible light

illuminated object: an object that is seen as the result of reflected light

satellite: an object that orbits a planet or star

lunar cycle: moon cycle

The Moon: The Lesser Light

Our Nearest Neighbor

When God created the sun to rule the day, He also created a lesser light to rule the night. We call that light the moon. *A* **moon** *is a heavenly body that orbits a planet.*

The earth's moon seems to be the largest heavenly object that we can see in the sky, but it is actually quite small compared to the sun and the earth. The width of the earth is 4 times that of the moon, and the width of the sun is about 400 times that of the moon. You may ask, "Why does the moon look about the same size as the sun?" It is because the moon is much closer to us than the sun. Remember, the sun is about 93 million miles away from us. The moon is only 240 thousand miles away. But as seen from the earth, both the moon and the sun seem to be about the same size.

Moonlight

On some nights, the moon is so bright that it seems to light up the whole sky. It even casts shadows. Yet the moon does not produce any light of its own. The light that we see from the moon comes from the sun. Since the moon reflects sunlight, it *appears* to be shining.

Some objects can give off their own light. The sun and the other stars are **luminous objects** because *they give off their own visible light. Any object that does not give off its own visible light* is a **nonluminous object**. The moon is nonluminous.

How can we see nonluminous objects, such as the moon, since they do not give off any visible light of their own? *We see all nonluminous things because of reflected light.* Reflected light is bounced back in the direction from which it came. Light travels from its source and comes in contact with nonluminous objects. Some nonluminous objects can transmit light; others block light. Do you remember how we describe an object that blocks light? Opaque objects may absorb, or soak up, the light; or they may reflect, or throw back, the light. The reflected light bounces back to your eyes, and you are able to see the object. Smooth things such as mirrors, jewels, water, glass, and metals are good reflectors of light. *Objects that are seen as the result of reflected light* are called **illuminated objects**.

Suppose you go for a walk in the woods at night. You bring your flashlight along as a source of light. As light from your flashlight shines on the trees and rocks around you, you are able to see these objects because they reflect the light from your flashlight back to you.

The moon is an opaque object that receives rays of light from the sun. Some of these rays are reflected back from the surface of the moon. This is the moonlight, or illuminated light, that we see. The earth and the other planets also reflect light from the sun. When astronauts travel into space and look back at the earth, they can see the earth shining from its reflected light.

The Moon's Movement

The moon is the *earth's only natural satellite.* A **satellite** *is any object that orbits a planet or star*. Today there are also manmade satellites that orbit the earth. Some of them gather data about the earth, such as weather or ocean conditions. Others help transmit information, such as radio waves or cellphone calls. Some can even take pictures of the earth or work with GPS. The moon is a natural satellite created by God. It takes the moon about thirty days to revolve, or travel, around the earth.

Besides revolving around the earth, the moon also rotates on its axis. It takes almost the same amount of time for the moon to rotate once on its axis as it does for the moon to revolve once around the earth. For this reason, there is a side of the moon that we never see. Spacecrafts that have photographed the far side of the moon have shown us that it looks very much like the side we can see.

The Lunar Cycle

The moon is shaped like a sphere. It appears to change shape because we see different amounts of the sunlit section of the moon as it revolves around the earth. When the moon is directly between the sun and the earth, we cannot see it because the sunlit side is facing away from the earth. *This position in which we cannot see the moon* is called the **new moon**. Though we do not see the moon in this phase, it is called a new moon because it is the *first phase of the moon cycle*, or **lunar cycle**. As the moon changes position, we see more and more of the sunlit side. The second phase, where we *only see a small part of the moon* is called a **crescent moon**. Each position gives the moon a different phase.

The Moon's Phases

crescent moon

new moon

crescent moon

quarter moon

quarter moon

gibbous moon

full moon

gibbous moon

Sometimes, you can *see only half of the moon*. We call this phase a **quarter moon**. You may have noticed a phase in the lunar cycle in which the moon was *almost a circle* but not quite complete. This is a gibbous moon. A **gibbous moon** happens *between a quarter moon and a full moon*.

When the earth is directly between the sun and the moon, we see the **full moon**. Each night after we see a full moon, we gradually see less and less of the moon. *The phases of the lunar cycle are always in this pattern: new, crescent, quarter, gibbous, full, gibbous, quarter, crescent, new.* The lunar cycle is complete when the moon is between the earth and the sun again, and there is another new moon. Remember, it takes about thirty days, or about one month, for a new moon to become a new moon again.

Matching: *Write the letter of the correct answer in the blank.*

_____ 1. a heavenly body that orbits a planet

_____ 2. an object that gives off its own visible light energy

_____ 3. an object that does not make its own visible light

_____ 4. an object that can be seen because of reflected light

_____ 5. any object that orbits a planet or star

A. illuminated
B. luminous
C. moon
D. nonluminous
E. satellite

Think and Predict.

Because the lunar cycle is a pattern, we can look at the moon in the night sky and predict its next phase. Draw and label a picture of the missing phases of the lunar cycle.

NASA (National Aeronautics and Space Administration): the flight and space science agency of the U.S. government

space probe: an unmanned spacecraft that collects pictures and data in space

space shuttle: the first type of rocket-launched spacecraft designed to be reused

space station: a satellite in which astronauts can live for months at a time

9.8 Space Discovery

When you observe a full moon at night, do you see any clues about its surface? If you look at the moon through a pair of binoculars, you will be able to see some of its features, such as high mountains, deep craters, and flat plains.

While observing the earth's moon using a spyglass in 1610, Galileo saw these surface features, too. He also noticed something interesting about Jupiter. He saw three heavenly bodies that looked like stars. Their arrangement made him very curious. He continued to observe these "stars" until he discovered something very surprising. Their positions had changed, and he could see a fourth heavenly body! They were behaving like the earth's moon as they orbited planet Jupiter. Until this time, all the philosophers believed that every heavenly body revolved around the earth. The discovery of Jupiter's moons was probably the first step in learning that the planets of our solar system do not revolve around the earth but around the sun instead.

Jupiter's moons

Exploring the Moon

American space exploration officially began in 1958 when the United States established its own flight and space science agency. You may know it as **NASA**, or the *National Aeronautics and Space Administration.* Its first project was to find a way to safely send a human into space. The Soviet Union, now called Russia, had already accomplished much in space discovery. Many scientists, engineers, mathematicians, and pilots worked with NASA on several projects and inventions to send the first American into space—Alan Shepard. Later, John Glenn became the first American to orbit Earth. NASA's next goal was to put the first human on the moon. No country had accomplished this.

Apollo 11 moon landing

In July 1969, the *Apollo 11 spacecraft landed American astronauts Neil Armstrong and Buzz Aldrin on the moon.* Armstrong and Aldrin were the first men to walk on the moon. The Apollo program used space rockets that could leave the earth's atmosphere and enter into space. A *space rocket enables a spacecraft to go into outer space* because of its propulsion engine. If you remember how some sea creatures use jet propulsion, you will understand a little bit about how rocket force works.

Several minutes after a spacecraft is launched, the rocket part detaches to remove the extra weight. After the mission is complete, astronauts can return home in the space-capsule part of the spacecraft. The space capsule protects them from being burned when passing back through the atmosphere. A parachute guides them down safely for a water landing called a splashdown.

What was it like for the astronauts on the moon? *There is no atmosphere on the moon.* Therefore, there is no weather and no air to be stirred up. The earth's sky is a pretty blue because of the atmosphere. The moon's sky is black because it has no molecules of air around it. Without water on the moon, there are no plants or animals since they need air and water to survive.

With a lack of atmosphere on the moon, sound waves cannot travel as they do on Earth. This is why astronauts use radio waves for communication in space. Do you remember Robert Boyle's experiment with air and sound? You could not hear any sounds on the moon because there are no air molecules to carry the vibrations. Sound waves cannot travel through empty space.

Temperatures on the moon are either extremely high or extremely low because the moon has no atmosphere. Do you remember how the earth's atmosphere protects us from the rays of the sun and helps to keep us warm at night? Astronauts who have traveled to the moon wore special clothing to protect themselves from the moon's extreme temperatures.

You will remember that since the moon is smaller and lighter than the earth, it has *less gravity.* The moon's gravity is about $\frac{1}{6}$ of the earth's gravity. If you weighed sixty pounds on the earth, you would weigh ten pounds on the moon. Because the force of gravity is less, you would also be able to jump much higher on the moon than you can on the earth. The moon has just enough gravity to pull the ocean causing the earth's tides.

The surface of the moon is dry and rocky. Part of the Apollo 11 mission was to collect soil and rock samples from the moon. Geologists later examined the samples and found that some rocks were made of a combination of minerals similar to Earth rocks. They named one of the moon rocks ***armalcolite***, after the three Apollo 11 astronauts—Neil Armstrong, Buzz Aldrin, and Michael Collins. Can you find parts of their names in the word *armalcolite*?

The Apollo 11 crew: Neil Armstrong, Michael Collins, and Buzz Aldrin

armalcolite

Apollo 11 orbit

photos of Jupiter and Galileo's 4 moons taken by *Voyager 1*

Exploring the Solar System and Beyond

After the lunar landing, other space missions were sent out to learn more about the solar system. Many of these missions were unmanned. This means that there were no astronauts on board. The spacecraft and its equipment were controlled from Earth. In 1977, the space probes *Voyager 1* and *Voyager 2* set out to explore the planets in the solar system. **Space probes** *are unmanned spacecraft that collect pictures and data in space*. *Voyager 1* and *2* were able to take photographs and gather data about the planets and other heavenly bodies. It took twelve years for *Voyager 2* to reach the last planet we know about—Neptune. Both probes continued to explore the solar system, each in its own direction. It is believed that *Voyager 1* reached *interstellar space* in 2012. This means that *Voyager 1* was the *first spacecraft* to officially leave the solar system and the energy and magnetic field of the sun. *Voyager 2* left our solar system in 2018. Both probes will continue to send data back to Earth as long as their power supply lasts.

The 1980s marked the beginning of the era of the space shuttle. The **space shuttle** was *the first type of rocket-launched spacecraft designed to be reused*. It could be launched into space because of its rocket power, but it was able to return to Earth and launch again for other space missions. Sally Ride, the first American woman in space, flew on the *Challenger* space shuttle in 1983. The same year, the first African-American astronaut, Guion Bluford, traveled into space aboard the *Challenger*. Americans were gaining confidence and excitement in their space program. Christa McAuliffe, a school teacher, was part of the 1986 *Challenger* mission. Sadly, less than two minutes after liftoff, the *Challenger* broke apart. America mourned the lives of the seven men and women on board who died. Through the *Challenger* disaster, NASA learned how to make space exploration safer. In 1990, the space shuttle *Discovery* took the Hubble Telescope into orbit. The Hubble Telescope takes pictures of heavenly bodies and sends them to Earth using radio waves. In 2011, the space shuttle program came to an end after many successful missions.

space shuttle

space station

Scientists are now able to live and work in space. Over fifteen countries send astronauts to live and work together in the International Space Station. *A* **space station** *is a satellite in which astronauts can live for months at a time*. Up to six astronauts can live there at once, some staying in space for more than a year at a time. Sometimes, one part of the group is replaced by another group. If you look up into the night sky on certain evenings, you might be able to see the International Space Station. It will look like a very bright, moving star because of the sunlight it reflects.

Comprehension Check 9.8

True/False: *If the statement is true, write* true. *If the statement is false, replace the underlined word(s) with a word or phrase that will make the statement true. Do not write false in any blank.*

__________ 1. <u>NASA</u> is the flight and space science agency of the U.S. government.

__________ 2. <u>Challenger</u> took the first people to walk on the moon.

__________ 3. A space rocket <u>can</u> go into outer space.

__________ 4. The moon <u>does</u> have an atmosphere like Earth

__________ 5. On the moon, you would weigh <u>more</u> than you would on Earth.

__________ 6. <u>*Voyager 1*</u> was the first spacecraft to reach interstellar space.

__________ 7. A space shuttle <u>cannot</u> be reused for another space mission.

__________ 8. A space station is a <u>rocket</u> on which astronauts can live.

Think and Predict.

9. Do you think playing basketball on the moon would be easier or more difficult? Why or why not? ____________________

10. How is the International Space Station an illuminated object in the night sky? ____________________

May 30, 2020, marked a historic event as NASA and SpaceX partnered to send Americans back into space. The SpaceX Dragon capsule, powered by the reusable Falcon 9 rocket, successfully launched from Kennedy Space Center. Bob Behnken and Doug Hurley flew inside the capsule to the International Space Station. The Falcon 9 rocket returned to Earth and landed on a barge in the Atlantic Ocean.

photo of Omega Nebula taken by Hubble telescope

The Origin of the Universe

Think of the earth, our home, and of all the things—living and nonliving—that are on it. Then think of the sun and the moon, the lights that rule the day and the night, and the innumerable stars in the millions of galaxies. Where did all these things come from? God is the wise Creator who gave us the written record of Creation in His Word, the Bible. He also gives scientists the ability to observe and study His creation. In science class, the scientific method has helped us understand more about the things God made.

> *Through faith we understand that the worlds were framed by the word of God, so that things which are seen were not made of things which do appear.* *Hebrews 11:3*

Genesis 1:1 says, "In the beginning God created the heaven and the earth." Before the creation of the universe, there was only God. God made matter and energy out of nothing. He spoke, and it was so. You have learned many things about matter and energy this year. Can you name some examples? God made the waters, and His Spirit "moved upon the face of the waters" and brought into existence the wonderful world that we know today.

On the first Creation day, God created light and then divided the light from darkness, making day and night. What special properties of light can you recall? Why do we have day and night?

On the second day, God created the firmament, which He called heaven. How does our atmosphere make life on our planet possible? Do you remember what it is made of?

God gathered the waters together to make the dry land appear on the third day. What can you remember about the ocean? What do we call the large land masses that rise up out of the sea? Our planet is filled with breathtaking landforms and useful resources for us to enjoy.

On the same day, God also commanded the earth to bring forth in abundance the wonderful variety of plants we enjoy today, each with the ability to reproduce new plants of the same kind. What are the main ways plants can reproduce? In creating the plants, God was preparing the earth to be man's home. Can you explain why plants are especially important to all life on Earth? The Bible tells us that God twice declared His work to be good on the third day—after separating the seas and the land and after He created plants.

The sun, moon, stars, and other heavenly bodies appeared on the fourth day. God made them to divide day from

night, to give light to the earth, and to be for signs, seasons, days, and years. How many heavenly bodies can you name? Again, God saw that His work was good.

On the fifth day, God created the birds and all the animals that live in water. What water vertebrates can you name? Can you think of any water invertebrates? What unique ability did God give to most birds? He also created each bird and each water animal with the ability to reproduce after its own kind, and He commanded them to fill the earth and the seas. At the end of the fifth day, God again pronounced it good.

The sixth day was the last day of Creation. God made the land-dwelling animals, from the largest dinosaur to the smallest insect. For His final act of creation, God made man, the only creature He made in His image (Genesis 1:24–31). God made man's body from the dust of the ground and breathed into him the breath of life and man became a living soul. All the living things upon the earth were made to benefit man; it was part of God's wise design for man to rule over and care for the rest of creation. At the end of the sixth day, when God surveyed all that He had made, He saw that it was very good.

On the seventh day, God rested, or ceased, from His work, because it was complete. He blessed the seventh day and set it apart as a special day of rest.

Throughout history, there have been people, even scientists, who have thought up their own stories of how the universe came to be. Many myths, hypotheses, and theories of the earth's origin have been developed and discarded, yet no one but the Designer and Creator of the universe is qualified to tell us how our magnificent universe really came into existence.

By the end of July 1969, the American space team would make a landing on the moon, but before making an actual landing they wanted to be certain everything was working perfectly. NASA sent up three astronauts to orbit the moon without landing, to check all of the equipment, and then to return to Earth. It was almost Christmastime in December of 1968 before

the Apollo 8 mission was launched. Everything went well. Millions of people all over the world were watching their televisions as the astronauts orbited the moon.

As Christmas Day approached, people all over the world heard the astronauts read from the Bible: "In the beginning God created the heaven and the earth." They took turns reading the first ten verses of Genesis. "And God called the dry land Earth; and the gathering together of the waters called He Seas: and God saw that it was good." And then they finished with, "Merry Christmas, and God bless all of you—all of you on the good Earth."

Throughout history, from the age of Adam to the age of space exploration, people who have really wanted to know how our great universe came to exist have found the answer by looking to the Creator. Have you "seen" God in your study of science this year? Ask God to help you see His glory as you continue to understand His world.

Then shall ye call upon Me, and ye shall go and pray unto Me, and I will hearken unto you. And ye shall seek Me, and find Me, when ye shall search for Me with all your heart. Jeremiah 29:12–13

Chapter 9 Concepts Review

1 Basic Astronomy Concepts 9.1–9.3

A. Remember

Matching: *Draw lines to match the description with the correct term.*

1. group of stars that form a picture
2. huge star system
3. star and all the heavenly objects that orbit it
4. heavenly body that orbits a star

galaxy

solar system

planet

constellation

Identify: *Draw a picture of the solar system. Label the planets; draw a star next to each outer planet.*

B. Think Like a Scientist

Short Answer: *Write the correct answer in the blank.*

1. What determines our year? ______________________________

____

2. What determines our day and night? ______________________________

____

3. What determines our seasons? ______________________________

____

4. What determines our months? ______________________________

____

C. Fun with Terms

Word Scramble: *Unscramble the words; then write the correct term in the blank.*

1. **ixas**—an imaginary line through the center of the earth from the North Pole to the South Pole ______________________________

2. **rnine tlneaps**—the four planets closest to the sun having solid surfaces ______________________________

3. **lMyik ayW**—the galaxy our sun and solar system belong to ______________________________

4. **rteuo tnapsel**—the four planets farthest from the sun, made mostly of gases ______________________________

5. **vloever**—to orbit a central point ______________________________

6. **otaetr**—to spin ______________________________

7. **iudsaln**—an instrument using the earth's position to tell time ______________________________

Sun, Moon, and Star Concepts 9.4–9.8

A. Remember

Matching: *Write the letter of the correct answer in the blank.*

______1. Big Dipper	______3. Little Dipper	______5. Herdsman
______2. Orion	______4. Southern Cross	______6. Leo the Lion

B. Think Like a Scientist

Identify: *Label each phase of the moon; then circle the lunar phase in which none of the moon is illuminated and draw a box around the lunar phases in which half the moon is illuminated.*

C. Write

Essay: *How could knowing the positions and names of the constellations help you if you were ever lost in the woods?*

____

____

____

____

____

____

____

____

____

____

____

____

D. Think Like a Scientist

Short Answer: *Write the correct answer in the blank.*

1. How does the moon shine even though it is a nonluminous object? ________

   ____

2. Is the sun the largest star in the sky? ______________________________

3. In what way is the sun dependable? ________________________________

   ____

4. Why is the sun able to give off light and heat energy? _______________

   ____

5. Why does the sun have a stronger pull of gravity than the earth? ________

   ____

E. Fun with Terms

Matching: *Write the letter of the correct answer in the blank.*

______ 1. also called Polaris or the Pole Star

______ 2. part of the Great Bear

______ 3. an object that gives off its own visible light

______ 4. a satellite in which an astronaut can live for months at a time

______ 5. flight and space agency of the U.S. government

______ 6. line where the earth seems to meet with the sky

______ 7. an unmanned spacecraft that collects pictures and data in space

______ 8. also called the Little Bear

______ 9. orbits a planet or star

______ 10. the phases of the moon

A. Big Dipper
B. horizon
C. Little Dipper
D. luminous
E. lunar cycle
F. NASA
G. North Star
H. satellite
I. space probe
J. space station
K. transparent

F. Write

Essay: *What did you learn from science that helped you appreciate God as the Master Designer? Explain your answer.*

____

____

____

____

____

____

____

____

____

____

Challenge: *Explain how your knowledge of weather patterns and navigational instruments would help you fly a plane from New York, New York, to London, England? What might make you change your route?*

Credits

Photo credits are listed from top to bottom, left to right on each page. Abbreviations used for sources are as follows: AA-Animals Animals/Earth Scenes; AGE-age fotostock America, Inc.; AL-Alamy Stock Images; DP-Depositphotos.com; GI-Getty Images; LoC-Library of Congress; ShSk-Shutterstock.com; ScSo-Science Source; WmC-Wikimedia Commons. If all photos on a page are from the same source, the source is given only once.

Chapter 9 Constellations. Drawings from *The Stars: A New Way To See Them*, by H.A. Rey. Copyright © 1952, 1962, 1967, 1970, 1975, 1976 by H.A. Rey. Reprinted by permission of Houghton Mifflin Harcourt Publishing Company. All rights reserved.

Cover-song_mi/GI; Cover-uschools/GI, altedart/GI, GordZam/GI, Steve Debenport/GI, cookelma/GI, Damocean/GI, FlamingPumpkin/GI, Steve Debenport/GI, Wavebreakmedia/GI, BrianEKushner/GI, tarasov_vi/GI, pedrosala, sololos/GI, 3DSculptor/GI, hairballusa/GI; i-uschools/GI, altedart/GI, GordZam/GI, Steve Debenport/GI, cookelma/GI, Damocean/GI, FlamingPumpkin/GI; ii-Kenneth Canning/GI, xxmmxx/GI, uschools/GI, Steve G Schmeissner/ScSo, Bet_Noire/GI; iii-uschools/GI, evdayan/GI; iv-StrahilDimitrov/GI; v-GlobalP/GI, amirage/GI, texturis/GI, fcafotodigital/IS, Pannonia/GI; vi-MaximFesenko/GI, SKapl/GI, NanoStockk/GI; vii-rekemp/GI, jonya/GI, Floortje/GI; viii-NOAA/NASA, Nataniil/GI; x-pyotr021/GI; images used throughout-releon8211/GI, ChubarovY/GI, Vect0r0vich/GI, ayagiz/GI, song_mi/GI, uschools/GI; 1-pyotr021/GI, evdayan/GI; 2-Georgijevic/GI, archy13/GI, JohnnyGreig/GI; 3-archy13/GI, pagadesign/GI, Tatomm/GI, bombuscreative/GI; 4-WmC; 5-gorodenkoff/GI, AlexRaths/GI, shironosov/GI, gorodenkoff/GI; 7-Dean Mirchell/GI, Arthur K Hale/GI, AndreasKermann/GI; 8-sbayram/GI, underworld111/GI, borchee/GI, Sarah8000/GI, Azure-Dragon/GI, Roman Samokhin/GI, BamBamImages/GI, LazingBee/GI; 9-pepifoto/GI, DrPAS/GI, MaximFesenko/GI; 10-Joshua Moore/GI, dionisvero/GI, Gregory_Dubus/GI; 13-monkeybusinessimages/GI, deepblue4you/GI; 15-LeventKonuk/GI; 19-Delmaine Donson/GI; 20-stayorgo/GI, weerapatkiatdumrong/GI, kourafas5/GI; 22-pedrosala/GI, tarasov_vi/GI, Image Source/GI, mipan/GI, Giambra/GI, MicroStockHub/GI; 23-Rost-9D/GI, Bet_Noire/GI, jack0m/GI, Pongasn68/GI, JMrocek/GI, alex-mit/GI(2); 24-francisblack/GI; 26-BlackJack3D/GI; 30-sitox/GI, Magnascan/GI, Jonathan Tieh/unsplash Photos, Imagvixen/GI, haryigit/GI, K_Thalhofer/GI; 31-DNY; 32-Magnascan/GI(water); 33-ansonsaw/GI, malerapaso/GI, Matt_Brown/GI; 34-Steve Debenport/GI, ismailciydem/GI, evgenyatamanenko/GI, Syldavia/GI, Rachel_Web_Design/GI, Matt_Brown/GI, artisteer/GI, Alter_photo/GI, stockcam/GI, Marilyn Nieves/GI; 35-xxmmxx/GI, Boonchuay1970/GI, Vladimiroquai/GI, Aleaimage/GI, Floortje/GI, pidjoe/GI, juliannafunk/GI, masterzphotois/GI; 36-Art Wager/GI; 40-marrio31/GI, standret/GI; 47-zaricm/GI, malerapaso/GI; 48-gaspr13/GI, fstop123/IS, fstop123/IS, guvendemir/GI, jpgfactory/GI, MicroStockHub/GI, borchee/GI, haryigit/GI, perets/GI, Paul Bradbury/GI, Toxitz/GI; 50-WmC; 51-RichVintage/GI, Hraun/GI, D3Damon/GI; 53-borsheim/GI, Zyabich/ShSk; 55-iLexx/GI (E. spark); 56-haryigit/GI, alikemalkarasu/GI, DNY59/GI; 58-nycshooter/GI; 59-dgstudiodg/GI, pagadesign/GI, alikemalkarasu/GI, joji/GI, Floortje/GI; 60-jeka1984/GI, themacx/GI, Dmytro Yashchuk/GI, ValeriyLebedev/GI, Anja W./GI, Warchi/GI; 61-jeka1984/GI, yupiyan/GI, ivkuzmin/GI, Alex Frood/GI, mammuth/GI, CoreyFord/GI, atese/GI, tdub_video/GI; 62-guvendemir/GI; 63-Milkos/GI, Milkos/GI, LightFieldStudios/GI, Bank215/GI; 64-bikec/GI, damircudic/GI, themacx/GI; 65-PeopleImages/GI, sihuo0860371/GI; 67-Pat_Hastings/GI; 68-sitox/GI; 69-Stepan_Bormotov/GI (balloons); 70-sitox/GI; 71-sunnyunai/GI, Toxitz/GI; 76-Gearstd/GI; 77-LuminaStock/GI; 79-LoC; 80-kevinjeon00/GI, turn_stock_photographer/GI, thumb/GI, JMWScout/GI, dem10/GI; 81-malerapaso/GI, Barcin/GI; 84-D3Damon/GI, haryigit/GI; 86-PeopleImages/GI, damircudic/GI, themacx/GI, sihuo0860371/GI, Anja W./GI, Pat_Hastings/GI; 87-Matt_Brown/GI, FGirgun/GI, pikepicture/GI, Alter_photo/GI, artisteer/GI, stockcam/GI; 88-mbbirdy/IS, StrahilDimitrov/GI, valerieann/GI; 89-Ideas_Studio/GI; 90-AlexandrMoroz/GI, dani3315/GI, anigoweb/GI; 91-BongkarnThanyakij/GI, Choreograph/GI, Arisa Thebanchornchai/GI, Gladiathor/GI; 92-Lorado/GI, FatCamera/GI, RichVintage/GI, Yobro10/GI, omgimages/GI; 94-AlbertPego/GI, cookelma/GI, DNY59/GI, DrPas/GI, johan63/GI, GlobalP/GI, WDnet/GI, Sergii Petruk/GI, Vladimiroquai/GI, wirOman/GI, TuiPhotoengineer/GI; 95-SeventyFour/GI; 97-skynesher/GI; 98-Bill Sykes/GI, anouchka/GI, santypan/GI, Vladimiroquai/GI, Vladimiroquai/GI, Callipso/GI, mauribo/GI, umdash9/GI, Henrik5000/GI, rbv/GI, Ostill/GI; 100-Rawpixel/GI, benedek/GI; 102-ttsz/GI; 103-ilbusca/GI, cmannphoto/GI; 104-rusm/IS, PeterHermesFurian/GI; 105-DNY59/GI, Dezein/GI; 106-PhonlamaiPhoto/GI, Henrik5000/GI, monticelllo/GI, GibsonPictures/GI; 107-Marianna Lishchenco/GI, dsafanda/IS, Kesu01/IS; 110-Vitoriano Jr/IS, fergregory/IS, KevinHyde/IS; 113-BanksPhotos/GI, fstop123/IS; 114-Joaquin Ossorio-Castillo/GI, Tanarch/GI, eurobanks/GI, DavidGraham86/GI; 115-stevecoleimages/IS, David Cope/GI, nemoris/IS, Milkos/GI, LightFieldStudios/GI; 116-gregepperson/IS, CatLane/GI; 117-ttsz/IS, ZargonDesign/GI; 119-Joaquin Ossorio-Castillo/GI, eurobanks/GI, Tanarch/GI; 120-bpablo/GI, BrianAJackson/GI, Vladimiroquai/GI, Nerthuz/GI; 122-stevecoleimages/IS, gabort71/GI, arnphoto/GI, Magdevski/GI, tomch/GI, ollo/GI, NNehring/GI, montego666/GI, ivanmollov/GI, masterzphotois/GI; 124-Pgiam/IS; 125-Pgiam/IS, skydie/IS; 126-Ron_Thomas/IS; 127-narcisa/IS, july7th/IS, Duncan_Andison/IS(2), dancestrokes/IS, Alexandrum79/IS, DrPAS/IS; 128-SolStock/GI; 129-vavit/GI(sunflower), Vizweskaya/GI(oak); 130-Khadi Ganiev/GI, aluxum/GI, godadex/IS; 131-keira01/GI, podfoto/IS, Alter_photo/IS; 132-pawel.gaul/IS; 133-danleap/IS; 134-joakimbkk/IS, meltonmedia/IS, joakimbkk/IS; 135-DNY59/IS, Eerik/GI, meltonmedia/IS, nickkurzenko/GI, SondraP/GI, Craig Chanowski/GI, Melinda Fawver/Dreamstime, K. Wagner-Blickwinkel/AGE, Bob Gibbons/AL, Arco Images GmbH/AL, Arco Images GmbH/AL, blickwinkel/AL, DNY59/GI, lightasafeather/GI, nameinfame/GI, Photo_HamsterMan/GI; 136-leventina/IS, ithinksky/IS, ithinksky/IS; 137-kickimages/IS; 138-bgfoto/IS, SerrNovik/IS; 139-csp_charlesgibson/Fotosearch, Christina-J-Hauri/GI, aimintang/GI, epantha/GI, Duncan_Andison/GI, katatonia82/GI, anna1311/GI, undefined undefined/GI, Roman Samokhin/GI, Hirurg/GI, YinYang/GI, unpict/GI, DNY59/GI, dionisvero/GI; 140-DNY59/GI, petrovval/GI, weisschr/GI, SelenaRus/GI, bkkm/GI, xxmmxx/GI, scisettialflo/GI, tiler84/GI, Art Directors & TRIPP/AL; 141-MeePoohyaphoto/IS, tiler84/IS, Denyshutter/IS, Adisak Rungjaruchai/Dreamstime, fcafotodigital/IS, repinanatoly/IS, joloei/IS; 142-CampPhoto/GI; 143-borchee/GI; 144-OGphoto/GI, RugliG/IS; 148-vav63/GI, ligora/GI, domnicky/GI; 149-Celiena/GI, ph2212/GI, Pilat666/GI, KaboradaM/GI; 150-Monia33/GI, belchonock/DP; 151-Jag_cz/GI, Beata Haliw/GI, yurybosin/GI; 152-Sergey Shcherbakov/GI, PatrikStedrak/GI, TriggerPhoto/GI, BanarTABS/GI; 153-kolesnikovserg/GI, naotoshinkai/GI, drakuliren/GI, alexmak72427/GI, kiorio/GI; 154-ZimaNady_kigd/GI; 156-chengyuzheng/GI (beans), Pannonia/GI, Sezeryadigar/GI, BWFolsom/GI; 157-ampol/GI; 158-tiler84/GI, Peterfactors/GI, amirage/GI, texturis/GI, Garsya/GI, skydie/IS, maksime/GI, Hirurg/GI, malerapaso/GI; 161-songdech17/GI, Barcin/GI, Anest/GI; 162-Ivan Marjanovic/GI, LegART/GI, serjedi/GI; 163-Artem_Egorov/GI; 167-hudiemm/GI, mashuk/GI, ElenaSeychelles/GI, Everett Collection Inc/AL, NNhering/GI; 168-Vizweskaya/GI (oak); 171-Soldt/GI (flower); 172-Stephanie_Zieber/GI, Freder/GI; 173-amriphoto/GI, Jennifer McCallum/GI, kramikel/GI, cookelma/GI, MarkGomez/GI; 174-Redders49/GI, Hung_Chung_Chih/GI, lisas212/GI, Spiderplay/GI, cmannphoto/GI; 175-stanley45/GI, SteveByland/GI, Derrick Alderman/AL, MikeLane45/GI, mauribo/GI; 176-JoannaTkaczuk/ShSk, Piter1977/GI, AlexRaths/GI, blickwinkel/AL, Hailshadow/GI, Tanya_F/GI, blickwinkel/AL; 177-GlobalP/GI, ra-photos/GI; 178-GlobalP/GI, reptiles4all/GI, phasinphoto/GI, Heather LaVelle/ShSk, Zoran Kolundzija/GI, eyfoto/GI, Bob_Eastman/GI, JasonOndreicka/GI; 179-bennymarty/GI, nattanan726/GI; 180-eyfoto/GI, JasonOndreicka/GI; 182-Abdesign/GI, Joseph Beck/GI, Kyle Bedell/GI, Carol Hamilton/GI; 183-webguzs/GI, blisken/GI, MriyaWildlife/GI, yanta/GI, Backyard-Photography/GI; 185-OGphoto/GI, FlamingPumpkin/GI, ca2hill/GI, BirdImages/GI, Mirko Vickovic/GI, William Fergsuon © California Academy of Sciences; 186-GK Hart/Vikki Hart/GI, geoffsp/GI, LifeJourneys/GI, JFLCapture/GI; 188-Dean_Fikar/GI, MikeLane45/GI, The Birds of North America https://birdsna.org, maintained by The Cornell Lab of Ornithology, Tempau/GI, Farahat/GI, Bassador/GI; 189-FrankHildebrand/GI, DieterMeyri/GI, epantha/GI, Shackleford-Photography/GI, Kaphoto/GI, Kenneth Canning/GI; 190-gqxue/GI; 191-assistantua/GI; 192-KenCanning/GI, eurotravel/GI; 194-LuCaAr/GI, LuCaAr/GI; 195-sduben/GI; 196-OGphoto/GI; 197-mirceax/GI, KathyKafka/GI; 198-outtakes/GI, alukich/GI; 199-mbolina/GI, Zwilling330/GI; 200-MichaelStubblefield/GI, NNehring/GI, SteveByland/GI, alukich/GI, Shackleford-Photography/GI, suefeldberg/GI; 201-AlesVeluscek/GI, leksele/GI, Wildnerdpix/GI, Vito_Elefante/GI; 202-RomoloTavani/GI, Trifonenko/GI, MikeCardUK/GI, KeithSzafranski/GI; 203-Utopia_BB/GI, Mirko Vickovic/GI; 204-thepalmer/GI, GlobalP/GI, 4FR/GI; 205-Jonathan Ross/GI; 208-KeithSzafranski/GI, SteveByland/GI, John_Wijsman/GI, mattbrownecouk/GI, sbayram/GI; 209-LYounghhs/GI; 211-HannamariaH/GI; 212-TacioPhilip/GI; 213-ConstantinCornel/GI; 214-igreen_images/GI; 215-Andyworks/GI, Martina_L/GI, imagebroker/AL; 216-hdere/GI; 218-Sergey Lisitsyn/GI, JodiJacobson/GI, Georgette Douwma/Minden Pictures; 219-adamkaz/GI, joebelanger/GI, HuxleyMedia/GI; 220-StreetFlash/GI, Sionme/GI, joebelanger/GI; 221-TiktaAlik/GI, stanley45/GI, yothinpi/GI, timquade/GI, sommail/GI; 223-Antagain/GI; 224-nechaev-kon/GI, Chinnasorn Pangcharoen/GI, ConstantinCornel/GI (crane fly), macroworld/GI; 225-ConstantinCornel/GI, ElsvanderGun/GI, ConstantinCornel/GI, Linas Toleikis/GI; 226-Patricia Head/AA, Corel, arlindo71/GI, coopeder1/GI; 227-Antagain/GI, abadonian/GI, paulrommer/GI, Steve G Schmeissner/ScSo, Antagain/GI; 229-Antagain/GI; 230-Mathisa_s/GI, changaim/GI; 231-GlobalP/GI, PK6289/GI, A & J Visage/AL; 232-Paul_Cooper/GI, ElementalImaging/GI, Topaz777/GI, Martina_L/GI; 233-SusanWoodImages/GI, Seregraff/GI, Dr. John Brackenbury/ScSo, sbonk/GI; 234-William Fergsuon © California Academy of Sciences, sebastianosecondi/GI; 235-GlobalP/GI, prajit48/ShSk, Antagain/GI, blickwinkel/AL; 236-scol22/GI, Mantonature/GI, DarrenMower/GI, Yitao/GI, William Fergsuon © California Academy of Sciences; 237-ithinksky/GI, buslig22/GI; 238-Satoshi Kuribayashi/Minden Pictures, kasira5698/GI, Tom McHugh/ScSo; 239-Betty4240/GI, Harry Rogers/ScSo,

Kagenmi/GI, Kagenmi/GI, William Fergsuon © California Academy of Sciences; 240-Marianne Campoiongo/GI, Stockbyte; 241-Hajakely/GI, Ignatiev/GI, Antagain/GI; 242-Adisak Mitrprayoon/GI, Noppharat05018977/GI; 244-Inventori/GI, Antagain/GI; 245-John Abbott/Minden Pictures, BanksPhotos/GI; 246-znm/GI; 247-WmC; 248-Richard Becker/AL, WmC, Koldunov/GI; 251-NightAndDayImages/GI, Andyworks/GI, michaelmill/GI, PatricioHidalgoP/GI, NoDerog/GI, igreen_images/GI, wwing/GI, Hailshadow/GI; 252-SandraMatic/GI, IdealPhoto30/GI, PeskyMonkey/GI, no_limit_pictures/GI, brytta/GI, Stephanie Sigafoos/GI, dszc/GI, dinadesign/GI; 254-GlobalP/GI; 255-prajit48/ShSk, Antagain/GI, blickwinkel/AL, GlobalP/GI; 256-rusm/GI, Stefonlinton/GI; 257-ValerijaP/GI; 258-joshschutz/GI, t_kimura/IS, janniwet/IS; 259-Hansen, M., DeFries, J.R.G. Townshed, and R. Sohlberg (1998) UMD Global Land Cover Classification, 1 Kilometer, 1.0, Department of Geography, University of Maryland, College Park, Maryland, 1981-1994., Evgeny_D/IS; 262-Maor Winetrob/GI, Byrdyak/GI, brittak/GI; 263-RichLindie/GI; 265-ArgentiHewitt/GI, SVproduction/GI, rotyoung/GI, Ferenc Cegledi/GI, Daijianyou/GI, FotoMaximum/GI, Gregory_Dubus/GI; 266-kavram/GI, FrankHildebrand/GI; 267-AlinaMD/GI, AlinaMD/GI, Eileen Kumpf/GI, SteveByland/GI, Byronsdad/GI, Martin-Kubik/GI, arlindo71/GI; 268-Jeremy Christensen/GI, CreativeNature_nl/GI, jlmcloughlin/GI; 269-Nerthuz/GI, troutnut/GI, GlobalP/GI, Dennis Wegewijs/GI, XiaImages/GI, paisan191/GI, Chinnasorn Pangcharoen/GI, VitalyEdush/GI, SKapl/GI, RiverNorthPhotography/GI, LazyFocus/GI, quickshooting/GI; 270-Werhane/GI; 271-superjoseph/GI; 272-AlinaMD/GI(2), Matt_Gibson/GI, kjekol/GI, tacojim/IS; 273-Stefonlinton/GI; 274-donald_gruener/GI, KenCanning/GI, dennisvdw/GI; 276-ekolara/GI; 278-SashaFoxWalters/GI, ktatarka/GI; 280-Lubo Ivanko/GI; 281-1001slide/GI, photoBlueIce/GI, WLDavies/GI, assalve/GI; 282-znm/GI, pierivb/GI, 1001slide/GI; 283-StuPorts/GI, Dennis Wegewijs/GI, anankkml/GI; 284-AlexmarPhoto/GI; 285-DimaBerkut/GI, Yerbolat Shadrakhov/GI, Ocs_12/GI, christophe_cerisier/GI, Binty/DP, GlobalP/GI, PatrickPoendl/GI; 286-James Moore/GI, Dantesattic/GI; 287-VSanandhakrishna/GI, Sara Edwards/GI, fischerscoop/GI, Pavliha/GI, loki1982/GI; 288-tzooka/GI, jamesdvdsn/GI, Pamela_Lynn/GI, PeopleImages/GI, kojihirano/GI, PetlinDmitry/GI; 289-Anna Bliokh/GI; 291-RomoloTavani/GI, VichoT/GI; 293-Goddard_Photography/GI, redtea/GI, keewizane/GI, Tenedos/GI, pilipenkoD/GI; 294-Lazareva/GI, Gannet77/GI; 297-Zdenek Machacke/unsplash Photos, pilipenkoD/GI, Redders49/GI, joebelanger/GI, wrangel/GI, eco2drew/GI, shalunts/GI, forogradieMG/GI, spyder24/GI; 298-Shur_ca/GI, GlobalP/GI; 300-Michael Ver Sprill/GI, Yarinca/GI; 301-julichka/GI, chengyuzheng/GI, roc8jas/GI, bonishphotography/GI, ziquiu/GI, emer1940/GI; 302-romrodinka/GI; 303-M-Sat Ltd/ScSo, jaminwell/GI; 304-Boogich/GI, icholakov/GI, 6381380/GI; 305-John Pontier/AA, romrodinka/GI, TasfotoNL/GI, TriggerPhoto/GI; 306-19Thunvar/GI, Mark Kostich/GI, HHelene/GI, CreativeNature_nl/GI, Dan Martin/GI; 307-Mark Kostich/GI, milehightraveler/GI, Donyanedomam/GI, traveler1116/GI, 33karen33/GI, Saddako/GI, anankkml/GI; 309-Heiko Kiera/ShSk; 310-KelcieMarie/GI, jomoba/GI, oshun/GI; 311-apomares/GI; 312-Boogich/GI; 313-Maartje van Caspel/GI; 316-ekolara/GI; 320-enderbirer/GI; 321-wawritto/GI; 322-aeduard/GI, Meinzahn/GI; 323-Saturated/GI, hsvrs/GI, benedek/GI; 324-Sean Pavone/GI, LoC; 326-Floortje/GI; 330-janrysavy/GI; 331-sreenath_k/GI, Duncan Rawlinson Photography; 333-Explorer/ScSo, NASA; 335-Ron and Patty Thomas/GI, CHBD/GI; 337-Dafinchi/GI, halock/GI, Coldmoon_photo/GI; 338-SunChan/GI, SunChan/GI, BlackJack3D/GI; 339-Aviator70/IS, A-Basler/IS, korionov/GI, Khadi Ganiev/GI; 341-Khadi Ganiev/GI; 342-NeilBradfield/GI; 344-NanoStockk/GI, dszc/GI; 345-kosher-pixels/IS; 346-andersen_oystein/GI; 347-CampPhoto/GI (ditch); 350-hepatus/GI, joshblake/GI; 351-Anna Usova/GI; 352-rekemp/GI, JohnGollop/GI, MarcelC/GI, bjphotographs/GI, MahirAtes/GI, Guy Corbishley/AL; 353-Estellez/GI, Photon-Photos/GI, KrimKate/GI, lissart/GI, Tycson1/GI, rekemp/GI; 354-KrimKate/GI, jeffwqc/GI; 355-Tycson1/GI, malerapaso/GI, Gary Ombler/GI, noppadon_sangpeam/GI; 356-Erlend Haarberg/NG Image Collection; 357-ideabug/GI, traveler1116/GI, mmac72/GI, JoyTasa/GI; 359-Rixipix/GI, Juan Carlos Munoz/GI, John Cancalosi/GI; 360-loonger/GI, GoodLifeStudio/GI, prill/GI, martinspurny/GI, Richard Greiner/EyeEm/GI, Zens photo/GI; 361-paulrommer/GI, hidesy/GI, Universal Images/SuperStock, Atypeek/GI, Judith Dzierzawa/GI; 362-ca2hill/GI, dottedhippo/GI, Stocktrek Images, Inc. /AL; 363-Joe Iacuzzo, Joe Iacuzzo, fotoVoyager/GI; 364-Ghedoghedo/WmC; 366-baranozdemir/GI; 367-zorandimzr/GI, adamkaz/GI; 368-ArtMarie/GI; 369-Anton Shaparenko/GI, fstop123/GI, Steve Debenport/GI, jonya/GI, PeopleImages/GI, cuhrig/GI, AlexRaths/GI, AndreasWeber/GI, Believe_In_Me/GI; 370-Mimadeo/GI; 371-Dennis Hallinan/AL, ScSo/USGS/GI; 372-nathanphoto/GI; 373-Samson1976/GI, photobykrumm/GI; 374-Charles Schug/GI, borchee/GI; 377-Khadi Ganiev/GI, NeilBradfield/GI, alle12/GI, malo85/IS; 381-Stichelbaut Benoit/AGE; 382-JSC/NASA; 384-4kodiak/GI; 385-Mark Garlick/ScSo, NOAA/NASA; 386-sbayram/GI, Viorika/GI, kertlis/GI, GarysFRP/GI, mdesigner125/GI, clintspencer/GI; 387-Mark Garlick/ScSo; 388-Jodi Jacobson/GI; 390-mdesigner125/GI, kertlis/GI, sbayram/GI, Nataniil/GI; 392-spooh/GI; 393-focusstock/GI; 394-bjdlzx/GI; 395-primeimages/IS; 400-pavel_klimenko/GI, loops7/GI, primeimages/GI, Edward Hattersley/AL, woottigon/GI, gsagi/GI; 402-primeimages/GI, mdesigner125/GI, BanksPhotos/IS, Dzmitrock87/GI; 404-peangdao/GI; 405-anilyanik/GI; 406-pavel_klimenko/GI; 407-ValentynVolkov/GI, Aleaimage/GI; 408-AHDesignConcepts/GI, Robin_Hoood/GI, c1a1p1c1o1m1/GI, sbayram/GI; 410-vencavolrab/GI, DuchesseArt/GI; 411-anilyanik/GI; 412-BanksPhotos/IS; 413-Jurkos/GI; 414-Elen11/GI; 415-CIMSS/NOAA; 418-NOAA; 420-JAXA/SSAI/Hal Price/NASA, Placebo365/GI, FEMA; 421-DenisTangneyJr/GI, shaunl/GI; 422-AFP Contributor/GI; 423-sushi7688/GI, Claudiad/GI, Mario Tama/GI, JacobH/GI; 425-onurdongel/GI, rekemp/GI, imamember/GI; 426-gsagi/GI, matteusus/GI, tomislz/GI, Niclasbo/GI; 427-SpiffyJ/GI; 428-JL-art/Adobestock; 431-genekrebs/GI, WmC, Photos.com/GI; 435-JL-art/Adobestock; 436-/JPL/USGS/NASA, maciek905/GI; 437-bjdlzx/GI, Neutronman/GI, knickohr/GI, ManuelHuss/GI; 439-JPL/USGS/NASA, JPL/STScI/NASA, JPL/U of AZ/NASA, NASA/JPL, JPL/USGS/NASA, AlexLMX/GI, NSSDC/NASA, JPL/MSSS/NASA; 440-NSSDC/NASA, JPL/MSSS/NASA, JPL/U of AZ/NASA, JPL/USGS/NASA, JPL/STScI/NASA, JPL/NASA; 442-rwarnick/IS, JPL/USGS/NASA, NSSDC/NASA, JPL/MSSS/NASA, JPL/U of AZ/NASA, JPL/STScI/NASA, JPL/NASA, JPL/USGS/NASA; 444-VladimirB/GI; 445-SiberianArt/GI, JPL/USGS/NASA, NSSDC/NASA, JPL/NASA, JPL/USGS/NASA, JPL/MSSS/NASA, JPL/U of AZ/NASA, JPL/STScI/NASA; 446-Corel, bjdlzx/GI, Floortje/GI; 447-Corel, hadzi3/GI, janrysavy/IS, NASA, AlexLMX/GI; 448-Corel, 449-Corel, hadzi3/GI, janrysavy/IS, NASA; 450-kathykonkle/GI, Corel; 451-454, 456-Corel; 457-HadelProductions/GI; 458-Corel, Vac1/GI, jodiecoston/GI; 459-460-Corel; 464-vav63/GI, NASA Goddard; 465-JPL-Caltech/NASA; 467-ricardoreitmeyer/GI; 468-bjdlzx/GI, NASA(3); 469-bjdlzx/GI, NASA(3); 470-NASA; 471-ARC/NASA; 472-KSC/NASA; 473-MSFC/NASA, Thomas Witzke/WmC, MSFC/NASA; 474-JPL/NASA, JPL/DLR/NASA, JPL-Caltech/NASA; 475-KSC/NASA, MSFC/NASA; 476 Tony Gray and Tim Powers/NASA; 477-JPL/NASA, MSFC/NASA; 478-Stiggdriver/GI, Ron_Thomas/IS, narcisa/IS; 479-MSFC/NASA, hadzi3/GI, NASA, mevans/GI, FGorgun/GI, AlizadaStudios/GI, jarenwicklund/GI; 480-JSC/NASA; 484-Corel, NASA (7); 503-GordZam/GI, joakimbkk/IS, dionisvero/GI, BirdImages/GI, vavit/GI, OGphoto/IS; 504-bryndin/GI, danleap/IS, Vizweskaya/GI, joakimbkk/IS, dionisvero/GI, Jessica Kirk/GI, Michael Burrell/GI, TomFawls/GI, naotoshinkai/GI, McPhoto/Diez /AL; 505-filmfoto/GI, ArminStautBerlin/GI, Renae Soleil Photography/Stockimo/AL, Lokibaho/GI, Mazirama/GI, Azure-Dragon/GI, Khrizmo/GI; 506-lucky-photographer/GI, Didier Descouens/WmC, ideeone/GI, Theodore Trimmer/ShSk, Natthakan Jommanee/EyeEm/GI, Zoonar GmbH/AL, Arterra Picture Library/AL; 507-TatyanaMishchenko/GI, fotoVoyager/GI, epantha/GI, aimintang/GI, iquacu/GI, ThereseMcK/GI, weisschr/GI; 508-TatjanaVer/ShSk, Bob Gibbons/AL, harrison macintosh/GI, Melinda Fawver/Dreamstime, magicflute002/GI, Christina-J-Hauri/GI; 509-xxmmxx/GI, scisettialfio/GI, ViliamM/GI, joakimbkk/IS, dionisvero/GI, visualica.com/GI, blickwinkel/AL, studio2013/GI, traveler1116/GI, kertlis/GI; 510-pelicankate/GI, Skapie777/GI, ByMPhotos/GI, VisionsbyAtlee/GI, meltonmedia/IS, nickkurzenko/GI, Bob Gibbons/AL, kazakovmaksim/GI, VvoeVale/GI; 511-Joe_Potato/GI, heibaihui/GI, Samo Trebizan/GI, Photo_HamsterMan/GI, nameinfame/GI, Roxana Rachman/GI, Dorling Kindersley ltd/AL, Murphy_Shewchuk/GI; 512-magicflute002/GI, Scisetti Alfio/ShSk; 513-Vidok/GI, t_kimura/GI, Vizweskaya/GI, AntiMartina/GI, undefined undefined/GI, Roman Samokhin/GI, Hirurg/GI, Xsandra/GI; 514-RoseMaryBush/GI, Amenohi/GI, Bob Hilscher/GI, hongquang09/GI, DHuss/GI, Kathryn8/GI, magicflute002/GI; 515-ozgurdonmaz/GI, terex/GI, tr3gi/GI, Ekaterina Aleshinskaya/GI, unpict/GI, Lemanieh/GI, lepas2004/GI, Vidok/GI; 516-aga7ta/GI, amygdala_imagery/GI, l3ankadi/GI, _Vilor/GI(2), Martin Wahlborg/GI; 517-Ladislav Kubeš/GI, Azure-Dragon/GI, Pilat666/GI, Avalon_Studio/GI, Alexey Sorokin/GI, aimintang/GI, epantha/GI; 518-Adventure_Photo/GI, Nadezhda_Nesterova/GI, Alexandrum79/GI, firina/GI, Jill_InspiredByDesign/GI, Chushkin/GI; 519-Marina Indova/GI, Adventure_Photo/GI, skibreck/GI, Savushkin/GI, MRaust/GI; 520-epantha/GI, wingmar/GI, wingmar/GI, anouchka/GI, digihelion/GI; 521-honbliss/GI, Zoya2222/GI, csp_charlesgibson/Fotosearch, Christina-J-Hauri/GI, loonger/GI, Amy_Lv/GI, HaraldBiebel/GI; 522-shelma1/GI, kjohansen/GI, Anghi/GI, adrianciurea69/GI; 523-caoyu36/GI, UliU/GI, epantha/GI, Gleti/GI, kellymarken/GI; 524-daverhead/GI, jimbo3904/DP, Maasik/GI, dndavis/GI, hsvrs/GI; 525-ulchik74/DP, portgrimes/GI, domnitsky/ShSk, Jay Pierstorff/GI; 526-petarlackovic/GI, John Warburton-Lee Photography/AL, Brilt/GI, Sieboldianus/GI, Antonel/GI; 527-PamWalker68/GI, Lezh/GI, Jack Glisson/AL, Karel Bock/GI, Joshua Moore/GI; 528-Vasyl Rohan/GI, idp garden collection/AL, seven75/GI, Kenpei/WmC (CC BY-SA 3.0); 529-mtreasure/GI, jonnysek/GI, BigJoker/GI; 530-Thomas Lenne/AL, stanley45/GI, SteveByland/GI, 33karen33/GI, BirdImages/GI; 531-outtakes/GI, Steve_byland/DP, htrnr/GI; 532-BrianEKushner/GI, Silfox/GI, MichaelStubblefield/GI; 533-mirceax/GI, KenCanning/GI, Nadanka/GI; 534-Abdesign/GI, tmarko/GI, hstiver/GI; 535-merlinpf/GI, MichaelStubblefield/GI, creighton359/GI; 536-KenCanning/GI, KathyKafka/GI, Stefonlinton/GI; 537-rpbirdman/GI, Frank Fichtmüller/GI, CreativeNature_nl/GI; 538-paulacobleigh/GI, ca2hill/GI, OGphoto/GI; 539-twildlife/GI, photographybyJHWilliams/GI, matthewo2000/GI; 540-CarolinaBirdman/GI; 541-ConstantinCornel/GI, Gutinov/GI, Savushkin/GI, smuay/GI; 542-N-sky/GI, sommail/GI, André De Kesel/GI, Arco Images GmbH/AL; 543-JeffGoulden/GI, tony mills/AL, Mantonature/GI; 544-Garry Gay/AL, vencav/DP, hekakoskinen/GI; 545-macropixel/DP, happyfoto/GI, coopeder1/GI, KarelGallas/GI; 546-Joesboy/GI, CC:BY-SA 3.0/WmC, badwiser/DP, Young Swee Ming/ShSk; 547-Natural History Library/AL, taden1/DP, lucidwaters/DP; 548-teptong/DP, Simon002/GI, photosbyjimn/GI; 549-ePhotocorp/GI, happyfoto/GI, smuayc/DP, HelloRF Zcool/ShSk, GlobalP/GI; 550-Eremeychuk Leonidy/AL, Denis Crawford/AL, OGphoto/IS; 551-OGphoto/GI, buslig22/GI, impr2003/GI; 552-mjf795/GI

Glossary

📖 Location in Terms box indicated by bold page number.

abdomen: last body part of an insect; contains the heart and stomach (pp. **222**, 227)

abyss: the deepest, darkest zone of the ocean (pp. **295**, 297)

abyssal plain: the valley-like floor of the ocean between continents (pp. **295**, 296)

acorn: the fruit of an oak tree (pp. **138**, 139)

air mass: a large body of air that has the same temperature and humidity (pp. **424**, 428)

algae: non-flowering green plants without roots, stems, or leaves (alga, *sing.*) (p. **160**)

altitude: the height above sea level (pp. **256**, 257)

amplitude: the measure of a wave's strength, measured from its height to its resting position (pp. **60**, 62)

annual rings: layers of wood in the trunk; each ring represents one year of the tree's life (pp. **132**, 134)

antennae: "feelers," or sense organs, on the heads of some invertebrates (antenna, *plural*) (pp. **212**, 213)

apex predator: a predator at the top of its food chain with no natural enemies (pp. **267**, 268)

arthropod: an invertebrate having jointed legs (pp. **218**, 221)

atmosphere: the air surrounding the earth (pp. **381**, 382)

atom: a particle that makes up a molecule (pp. **26**, 27)

attract: to pull together (pp. **50**, 52)

axis: an imaginary line through the center of the earth from the North Pole to the South Pole (p. **443**)

bacteria: single-celled organisms (bacterium, *sing.*) (pp. **160**, 162)

balanced forces: opposing forces that prevent work from being done (pp. **87**, 90)

bark: the protective layer on the outside of the trunk of a tree (pp. **132**, 133)

barometer: weather instrument meteorologists use to detect air pressure changes (pp. **388**, 390)

bedrock: the solid rock beneath the soil; part of the earth's crust (pp. **337**, 339)

Big Dipper: the seven stars appearing as a long-handled cup, which are part of *the Great Bear* (p. **450**)

biodiversity: variety and amount of life within an ecosystem (pp. **273**, 274)

blizzard: a severe snowstorm causing colder temperatures, strong winds, and blowing snow (pp. **419**, 421)

boiling point: the temperature at which a liquid rapidly changes to a gas (pp. **39**, 40)

broadleaf tree: a tree having flat, broad leaves and producing seeds (p. **138**)

browser: an animal that feeds on trees and shrubs (pp. **280**, 282)

burrow: to tunnel (pp. **280**, 281)

cable: bundled wires, covered in plastic or another insulator (pp. **54**, 55)

camouflage: the ability to blend in with surroundings (pp. **236**, 239)

carnivore: an animal that eats meat (pp. **267**, 268)

center of gravity: the point of balance in an object that represents the effects of gravity on all its particles (pp. **109**, 110)

chlorophyll: the green matter in plants which traps sunlight energy; gives a plant the ability to produce its own food (pp. **125**, 128)

climate: the weather conditions an area receives over time (pp. **256**, **381**, 382)

closed circuit: an uninterrupted path of electricity; allows an electrical current to flow (pp. **54**, 56)

cold-blooded: having a body temperature that changes with the environmental temperature (pp. **172**, 176)

collision: two objects bumping into each other, causing energy transfers (pp. **97**, 98)

colony: a group of animals living together (p. **218**)

color spectrum: the series of colors created when white light is separated (pp. **67**, 68)

complete metamorphosis: four stages of growth in insects—egg, larva, pupa, adult (p. **230**)

composite flower: flowers made of a combination of many smaller flowers (p. **148**)

compound eyes: large eyes that are made of many small eyes (pp. **222**, 224)

conclusion: a decision based on evidence (pp. **1**, 4)

condensation: the process by which water vapor changes back into very tiny droplets of liquid water (pp. **405**, 406)

conductor: a material that can easily transfer (electrical) energy (p. **54**)

conifer tree: a tree that produces seeds in cones (pp. **132**, 135)

constellation: a group of stars that forms a picture (p. **436**)

consumer: an organism that depends on producers for food energy (pp. **125**, 128)

continental shelf: the underwater land along the coast of a continent (pp. **295**, 296)

continents: the largest land masses that rise out of the oceans (pp. **330**, 332)

contract: to get smaller (pp. **39**, 40)

controlled variables: the parts of an experiment that stay the same (pp. **12**, 14)

converted energy: one kind of energy changed to another kind of energy (pp. **46**, 47)

core: the innermost part of the earth (pp. **326**, 329)

crest: the highest point of a wave (pp. **60**, 62)

crown: the leaves and branches of a tree (pp. **132**, 133)

crust: a solid layer of rock beneath the soil; made of two kinds—oceanic crust and continental crust (pp. **326**, 328)

crustal plates: large areas of the earth's crust (pp. **330**, 332)

crystal: a solid made of atoms that are always arranged in a precise, repeating pattern (p. **337**)

current electricity: the flow of electrons along a path (p. **54**)

cyclone: a storm that begins to rotate around a low-pressure area (pp. **412**, 414)

data: a collection of facts (pp. **1**, 4)

deciduous tree: a tree that loses its leaves and is bare for a part of the year (p. **138**)

decomposer: an organism that breaks down dead things to help the soil (pp. **160**, 162)

delta: a triangular landform at the mouth of a river made from deposited soil (pp. **302**, 303)

dense: packed closely together (p. **31**)

density: how tightly packed molecules are in an object (pp. **34**, 36)

dependent variable: the part of an experiment that is measured to test results (pp. **12**, 14)

desert: a dry land with little plant growth (p. **284**)

disk flower: flowers that form a disk at the center of a composite flower (p. **148**)

dissolve: to break down into smaller pieces until it can no longer be seen (pp. **42**, 43)

dormant: alive but resting (pp. **155**, 157)

down: small, fluffy feathers (pp. **190**, 191)

drought: a prolonged period of dryness (pp. **419**, 422)

earthquake: a trembling or shaking in the earth's crust caused by plate activity (pp. **330**, 333)

ecosystem: a community of organisms, such as plants and animals, interacting with their physical environment (p. **261**)

electromagnet: a device that generates electromagnetic force (pp. **101**, 102)

electron: a negative particle of an atom (pp. **50**, 51)

elm: a tree with toothed leaves that produces small, flat seeds encased in paper-like wings (pp. **138**, 140)

embryo: the living part of the seed that becomes the new plant (p. **155**)

energy: the ability to do work (p. **46**)

epicenter: where earthquake damage is often greatest; located above the focus (pp. **346**, 347)

equator: the imaginary line around the earth equally distant from the poles (p. **326**)

erosion: the loss of soil by water or wind (p. **342**)

estuary: a body of water where the ocean tide meets a river or stream, causing salt water to mix with fresh water (pp. **302**, 303)

evaporation: the process by which liquid water becomes water vapor (pp. **405**, 406)

evergreen tree: a tree that keeps its leaves throughout the year (pp. **132**, 135)

evolution: the unproven theory that animals slowly turned into other animals over millions of years instead of being created by God (p. **181**)

exoskeleton: an outside skeleton (p. **212**)

expand: to get larger (pp. **39**, 40)

experiment: a procedure designed to answer a question and to test a hypothesis (pp. **1**, 4)

extinct animal: an animal that no longer exists (pp. **201**, 202)

fault: a break between two moving plates (pp. **330**, 333)

ferns and mosses: non-flowering green plants which grow from spores instead of seeds (p. **160**)

flooding: an overflow of water (pp. **419**, 422)

focus: the place where plates move inside the earth's crust; where the energy of an earthquake comes from (pp. **346**, 347)

fog: a cloud at ground level (pp. **399**, 402)

food chain: the transfer of energy from one living thing to another for survival (p. **267**)

force: the push or pull on an object (pp. **87**, 88)

forecast: a weather prediction (pp. **388**, 390)

fossil: the remains or impression of a living thing found in sedimentary rock that has been hardened (pp. **358**, 359)

fossil fuel: a natural fuel, such as coal, oil, or natural gas; made from remains of living organisms (p. **365**)

foundation: the base of a building (p. **113**)

frame: the skeleton of a building, giving it its shape and support (p. **113**)

freezing point: the point at which a liquid changes to a solid (pp. **39**, 40)

friction: a force that works against motion along a surface (pp. **87**, 89)

front: the boundary between two different air masses (pp. **424**, 428)

fruit: a plant structure that holds and protects seeds (pp. **138**, 139)

fuel: a material that is burned to release energy (pp. **46**, 47)

fulcrum: the part in a lever that holds and balances weight (pp. **113**, 117)

fungi: organisms that produce spores instead of seeds, such as mushrooms, lichens, molds, or yeasts (fungus, *sing.*) (pp. **160**, 162)

galaxy: a huge star system (pp. **436**, 437)

gas: matter that can fill a space but escapes easily (pp. **31**, 32)

gem: a rare and valuable mineral (pp. **337**, 338)

generator: a machine that converts mechanical energy into electrical energy (p. **54**)

geologist: a scientist who studies the earth (pp. **321**, 322)

geology: the study of the earth and its structure (p. **321**)

germinate: to sprout (pp. **155**, 157)

gills: organs that filter dissolved oxygen from water for breathing (pp. **172**, 176)

gizzard: the part of a bird's stomach used for grinding food (pp. **190**, 192)

glacier: a large, moving body of ice on land (pp. **290**, 291)

global winds: bands of wind that usually flow in the same predictable direction at certain latitudes (pp. **393**, 395)

glucose: the simple sugar made by green plants (pp. **125**, 128)

GPS: Global Positioning System (p. **457**)

graph: a diagram that shows data (p. **17**)

grassland: a large, flat, open area of grasses (p. **280**)

gravity: the attracting force between two objects (such as the pull of the earth to itself) (pp. **109**, 110)

grazer: an animal that feeds on grasses (pp. **280**, 282)

greenhouse effect: the atmosphere's ability to keep heat from easily escaping into space (pp. **393**, 394)

groundwater: water found beneath the earth's surface that supplies wells and springs (pp. **330**, 331)

habitat: the natural home of an animal or plant (pp. **181, 256**)

hemisphere: half of a sphere (p. **326**)

herbivore: an animal that eats only plants (pp. **267**, 268)

hibernation: a deep winter sleep to conserve energy (pp. **273**, 278)

horizon: the line where the earth seems to meet the sky (p. **457**)

host: the animal or plant on which a parasite feeds (pp. **230**, 233)

humus: the soft organic material in soil made from the decayed remains of living things (pp. **125**, 130, **337**)

hurricane: the most severe type of tropical cyclone; begins over a warm ocean (pp. **412**, 414)

hygrometer: weather instrument that measures the relative humidity of air (pp. **399**, 400)

hypothesis: a sensible guess; a reasonable prediction (pp. **1**, 4)

Ice Age: a period of cold temperatures bringing ice and snow after the Flood (pp. **290**, 291)

iceberg: a large piece of ice that has broken from a glacier (pp. **290**, 291)

ice cap: a year-round covering of ice and snow over polar land (p. **290**)

ice shelf: frozen, floating seawater connected to land (pp. **290**, 291)

igneous rock: rocks that can be formed from hot magma or lava (p. **354**)

illuminated object: an object that is seen as the result of reflected light (pp. **466**, 467)

incomplete metamorphosis: three stages of growth in insects—egg, nymph, adult (p. **230**)

incubate: to keep an egg warm until it hatches (pp. **201**, 202)

independent variable: the part of an experiment that is changed to test the hypothesis (pp. **12**, 14)

inertia: the property of matter that makes it resist a change in motion (p. **93**)

inner planets: the four planets closest to the sun, having solid surfaces—Mercury, Venus, Earth, and Mars (p. **439**)

instinct: a God-given ability or behavior that is inherited rather than learned (pp. **172**, 177)

insulator: a material that cannot easily transfer energy (pp. **54**, 55)

interval: the amount from one line to the next on a graph (pp. **17**, 18)

invertebrate: an animal without a backbone (pp. **172**, 173)

jet stream: a band of wind high above the earth that flows from west to east around the globe (pp. **393**, 396)

jet propulsion: a special way some invertebrates have been engineered to maneuver in water (pp. **212**, 215)

kinetic energy: moving, working energy (p. **46**)

laboratory: a place where scientists work (p. **20**)

landform: a natural formation of rock or soil on the earth's surface (pp. **330**, 334)

lava: melted rock that flows from a volcano (pp. **346**, 349)

laws of motion: Newton's discoveries about how motion always works (p. **93**)

leaf litter: layer of dead and decaying leaves on the forest floor (pp. **125**, 130)

lift: the force that keeps a bird or airplane in the air (p. **190**)

lightning: electricity that travels from one cloud to another or between a cloud and the ground (pp. **412**, 413)

liquid: matter that cannot keep its shape, but can fill the shape of a container (pp. **31**, 32)

Little Dipper: the group of stars appearing as a small cup, also called *the Little Bear* (pp. **450**, 452)

loam: a mixture of sand, silt, and clay (pp. **337**, 338)

lubricant: a substance that minimizes friction (pp. **105**, 107)

luminous: giving off light energy (p. **60**)

luminous object: an object that gives off its own visible light (p. **466**)

lunar cycle: moon cycle (pp. **466**, 468)

lungs: organs that use oxygen from air for breathing (pp. **172**, 174)

machine: a device that makes work easier (pp. **105**, 106)

magma: hot, liquid rock within the earth (pp. **326**, 329)

magnetism: the attracting and repelling force created by a magnetic field (p. **101**)

mantle: the layer of the earth beneath the crust (pp. **326**, 329)

maple: a tree with hand-shaped leaves that produces winged fruits (pp. **138**, 140)

mass: the amount of matter in an object (pp. **34**, 35)

matter: a physical substance that has weight and takes up space (p. **26**)

melting point: the temperature at which a solid changes into a liquid (p. **39**)

metamorphic rock: igneous or sedimentary rock that seems to have been changed by heat or pressure (pp. **354**, 357)

metamorphosis: a change in form (p. **230**)

meteorologist: a scientist who studies the weather (pp. **388**, 390)

metric system: a universal decimal measuring system based on the meter and the kilogram (pp. **20**, 21)

midnight zone: the dark depth of the ocean after the twilight zone (pp. **295**, 297)

migration: the movement from one place to another with the change of seasons (pp. **273**, 277)

Milky Way: the galaxy our sun and solar system belong to (pp. **436**, 437)

mimicry: one animal looking like another animal for protection (pp. **236**, 240)

mineral: a substance found naturally in the earth that has a crystal-like structure (p. **337**)

mixture: two or more different kinds of matter combined together (p. **42**)

molecule: a tiny particle of matter (pp. **26**, 27)

molting: the shedding of an exoskeleton, skin, or other body covering (pp. **212**, 213)

monsoon: a seasonal wind that can bring heavy rain to some places in southern Asia (pp. **419**, 422)

moon: a heavenly body that orbits a planet (p. **466**)

motion: a change in position (p. **87**)

mouth: the place where a river flows into the ocean (or other body of water) (pp. **302**, 303)

NASA (National Aeronautics and Space Administration): the flight and space science agency of the U.S. government (pp. **471**, 472)

natural resource: a material found in or on the earth which is helpful to people (pp. **365**, 368)

needleleaf tree: a tree with narrow leaves, or needles (pp. **132**, 135)

niche: the special job of an organism in its environment (p. **261**)

nocturnal: to be active during the night (pp. **284**, 288)

nonluminous object: an object that does not give off its own visible light (p. **466**)

nonrenewable energy source: an energy source which cannot be used again, such as fossil fuels (pp. **365**, 366)

North Star: the star consistently located almost directly above the earth's axis at the North Pole, also called Polaris or the Pole Star (pp. **457**, 459)

nucleus: the positive center of an atom (pp. **50**, 51)

oak: a tree that produces acorns (pp. **138**, 139)

oasis: fertile place in the desert where there is water (pp. **284**, 285)

ocean basin: the surface of the earth covered by an ocean (pp. **295**, 296)

oceans: the earth's largest bodies of water (p. **330**)

omnivore: an animal that eats both plants and meat (pp. **267**, 268)

opaque: allowing no light to pass through (p. **64**)

open circuit: an interrupted path of electricity; does not allow electrical current to flow (pp. **54**, 56)

organism: a living thing (pp. **125**, 128)

outer planets: the four planets farthest from the sun, made mostly of gases—Jupiter, Saturn, Uranus, Neptune (pp. **439**, 440)

palm: a branchless tree or shrub; does not have annual rings or bark (pp. **138**, 140)

parasite: an organism that attaches itself to another organism and feeds on it (pp. **230**, 233)

permafrost: permanently frozen subsoil (p. **273**)

petal: the colorful part of many flowers that attracts pollinators (p. **142**)

photosynthesis: the food-making process in green plants (pp. **125**, 128)

physical property: a property that can be observed or measured (p. **34**)

physical world: the world we live in and interact with, made of matter (p. **26**)

phytoplankton: single-celled algae; able to use the sun's energy for photosynthesis (pp. **290**, 293)

pistil: the vase-shaped center part of the flower that makes undeveloped seeds (p. **142**)

pitch: the highness or lowness of sound (pp. **75**, 77)

planet: a heavenly body that orbits a star (p. **439**)

plankton: microscopic, floating water plants and animals (pp. **290**, 293)

point: a location on a line graph that represents data (pp. **17**, 18)

polarity: having two opposite sides that work differently (p. **101**)

pollen: yellow dust on flowers made by the stamens (p. **142**)

pollination: the moving of pollen from a stamen to the pistil of the flower (pp. **142**, 144)

potential energy: stored, waiting energy (p. **46**)

precipitation: any form of water which falls from the sky to the earth (pp. **405**, 407)

predator: an animal that hunts other animals for food (pp. **181**, 183)

pressure: the measurement of the force exerted against an object (pp. **388**, 389)

prey: the animal a predator hunts for food (pp. **181**, 183)

primary root: the first root (pp. **155**, 158)

prism: a transparent, solid figure used to separate white light into colors (pp. **67**, 68)

producer: an organism that is able to produce its own food by using sunlight energy (pp. **125**, 128)

property: a way to describe or sort matter according to its characteristics (p. **31**)

range: continental area (pp. **181**, 188)

ray flower: flowers that radiate from the disk of a composite flower (p. **148**)

reflect: to throw back light, changing its direction (pp. **64**, 65)

refraction: the bending of light rays as they enter a new material (p. **67**)

relative humidity: a measure of the amount of water vapor in the air (pp. **399**, 400)

renewable energy source: an energy source which cannot be used up, such as wind, water, and solar energy (pp. **365**, 366)

repel: to push apart (pp. **50**, 52)

revolve: to orbit a central point (pp. **443**, 444)

roots: the part that anchors the tree and draws water and minerals from the soil (pp. **132**, 133)

rotate: to spin (p. **443**)

salinity: saltiness (pp. **256**, 258)

satellite: an object that orbits a planet or star (pp. **466**, 467)

savanna: a tropical grassland (pp. **280**, 281)

scavenger: an animal that eats other animals that have already died (pp. **280**, 283)

scientific method: orderly, step-by-step process in which scientists work (pp. **1**, 4)

sector: a part of a circle graph that represents a certain fraction or part of the whole (pp. **17**, 18)

sediment: matter that settles to the bottom or sides of a body of water (pp. **342**, 344)

sedimentary rock: rocks formed from sediment that has been hardened in a cementing process (pp. **354**, 355)

seed coat: the covering that protects the embryo until it is time for the seed to sprout (pp. **155**, 156)

segment: a body section (pp. **212**, 214)

sepal: leaf-like structure that surrounds the base of a flower's petals; protects the flower's bud (p. **142**)

setae: tiny sharp projections on an earthworm (pp. **212**, 214)

sextant and astrolabe: instruments used in navigation (pp. **457**, 458)

shadow: a dark area or shape cast by an opaque object (p. **64**)

shoot: a new sprouting plant (pp. **155**, 158)

soil horizons: the layers contained in most soil—humus, topsoil, subsoil, and bedrock (pp. **337**, 339)

solar system: a star and all the heavenly objects that orbit it (p. **439**)

solid: matter that can keep its shape (p. **31**)

solute: the part of a solution that is dissolved (pp. **42**, 43)

solution: a mixture in which one kind of substance is dissolved into another (pp. **42**, 43)

solvent: the part of a solution that dissolves another substance (pp. **42**, 43)

Southern Cross: four stars forming the shape of a cross with the long crossbar pointing to the South Pole (pp. **457**, 460)

space probe: an unmanned spacecraft that collects pictures and data in space (pp. **471**, 474)

space shuttle: the first type of rocket-launched spacecraft designed to be reused (pp. **471**, 475)

space station: a satellite in which astronauts can live for months at a time (pp. **471**, 475)

species: different types of animals (p. **181**)

sperm cells: cells in the pollen produced by the stamen (pp. **142**, 144)

spiracles: holes insects use for breathing (pp. **222**, 227)

spore: a single cell that is able to reproduce some organisms (pp. **160**, 161)

stamen: the center part of the flower that makes pollen (p. **142**)

static electricity: a buildup of electrical charge (p. **50**)

stored food: nourishment for the young plant until it can make food on its own (pp. **155**, 156)

storm surge: the rise of ocean water surrounding a hurricane (pp. **412**, 416)

streamlined: designed to move easily through water or air (pp. **172**, 176, **190**)

sundial: an instrument using the earth's position to tell time (pp. **443**, 444)

sunlight zone: the ocean depth with the most light; has the most ocean life because of photosynthesis (pp. **295**, 297)

symbiotic relationship: two different organisms that live closely together and benefit from each other (pp. **218**, 220)

technology: science put to practical use (pp. **1**, 2)

thermal energy: heat energy (pp. **26**, 29)

thorax: the middle body part of an insect where wings and legs are attached (pp. **222**, 226)

thunderhead: a thick, dark storm cloud (p. **412**)

thunderstorm: a storm that brings heavy rain, strong winds, and lightning (p. **412**)

tornado: a cyclone that develops over hot land (p. **419**)

tornado warning: notification that a tornado has been spotted (p. **419**)

tornado watch: notification that conditions are right for tornado formation (p. **419**)

transferred energy: energy moved from one place to another (pp. **46**, 48)

transitional form: a term evolutionists use to describe a "missing link" between two kinds of animals to show how one kind of animal could have turned into another kind of animal (pp. **358**, 361)

translucent: allowing light through but scattering it so objects cannot be seen clearly (pp. **64**, 65)

transmit: to allow waves to pass through (p. **64**)

transparent: allowing light to pass straight through (pp. **64**, 65)

tree line: the altitude at which trees do not grow (p. **273**)

tremor: a minor earthquake (pp. **346**, 347)

trenches: the deepest places of the ocean floor (pp. **295**, 296)

trend: data on a graph that can help predict future data (pp. **17**, 18)

trough: the lowest point of a wave (pp. **60**, 62)

trunk: the stem of a tree (pp. **132**, 133)

tsunami: a set of giant ocean waves often caused by underwater earthquakes or volcanic eruptions (pp. **346**, 347)

tundra: treeless land because of permafrost (p. **273**)

twilight zone: the ocean depth with only blue light; does not allow photosynthesis (pp. **295**, 297)

unbalanced forces: a greater force that overcomes opposing forces, resulting in work (pp. **87**, 91)

variables: the parts of an experiment that are controlled (pp. **12**, 14)

vent: the opening of a volcano where magma, gases, hot ash, and rock can erupt (pp. **346**, 349)

vertebrate: an animal with a backbone (pp. **172**, 173)

vibrate: to move rapidly back and forth (p. **71**)

volcano: a place in the earth's crust where magma can erupt as lava (pp. **330**, 333)

volume (in matter): the amount of space matter takes up (pp. **34**, 35)

volume (in sound): the loudness or softness of sound (pp. **75**, 76)

warm-blooded: having a body temperature that stays the same temperature no matter the environmental temperature (pp. **172**, 173)

water cycle: the movement of water from the earth to the air and back again (p. **405**)

water runoff: precipitation that eventually trickles into streams (pp. **405**, 408)

wavelength: the distance from one crest to the next crest (pp. **67**, 68)

weather: the condition of the air (pp. **381**, 382)

weathering: the breaking down of rock by the forces of nature (p. **342**)

weather phenomenon: a weather event caused by specific conditions (p. **412**)

weed: a plant that grows where it is not wanted (pp. **148**, 149)

weight: the force of gravity on an object (pp. **109**, 110)

wetland: land covered with water for at least part of the year (pp. **302**, 303)

work: movement resulting from force (pp. **87**, 88)

zooplankton: small water animals (pp. **290**, 293)

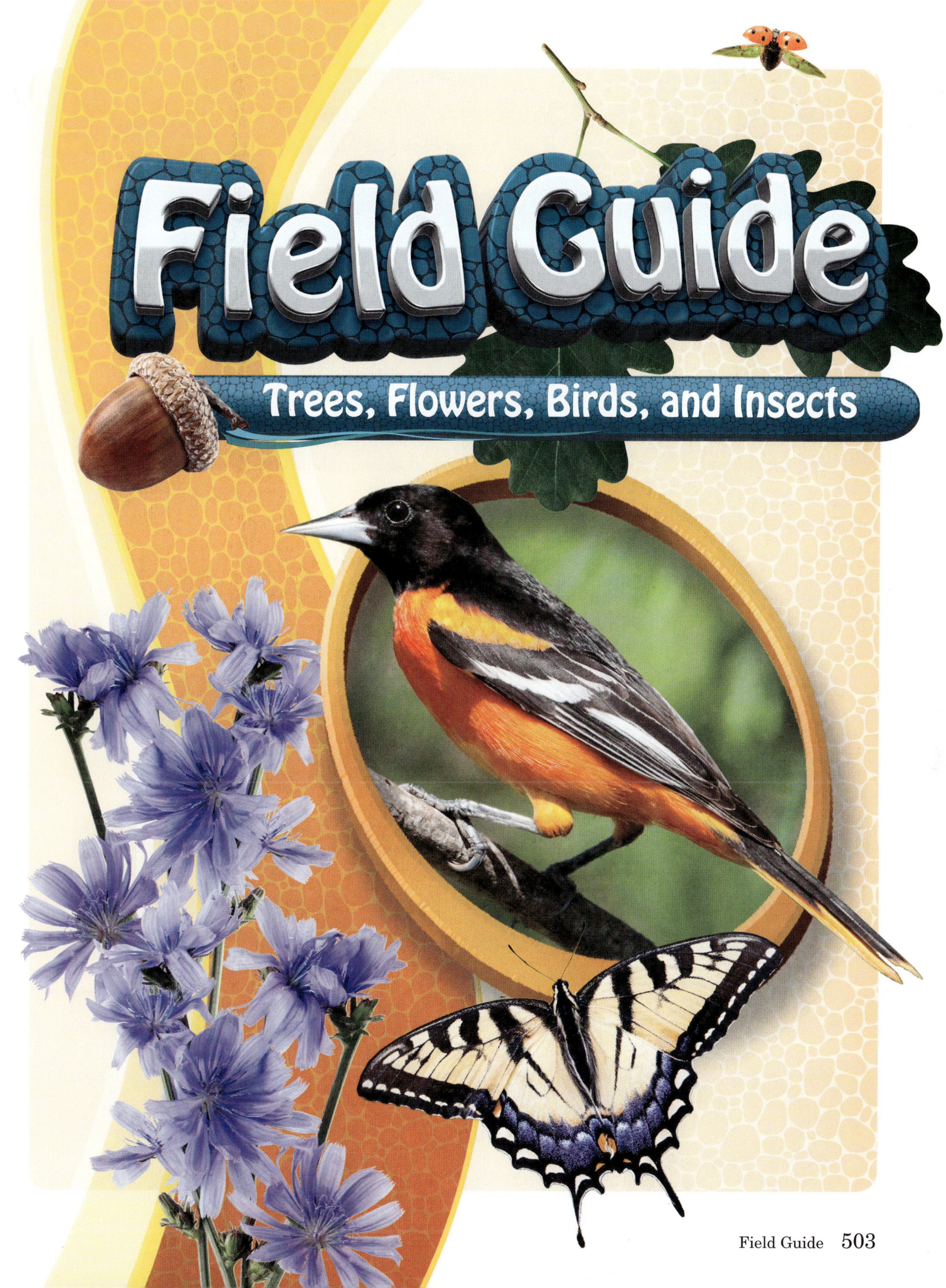
Field Guide
Trees, Flowers, Birds, and Insects

Trees

Physical Characteristics

Interesting Facts

American Holly

Broadleaf

Evergreen

Forms flowers and fruit

State tree of Delaware

Bald Cypress

Needlelike leaves

Deciduous

Conifer

State tree of Louisiana

Birch

Broadleaf

Deciduous with white bark

Forms winged seeds

State tree of New Hampshire

Provincial tree of Saskatchewan and Quebec

Blue Palo Verde

Broadleaf

Deciduous

Forms flowers and seed pods

State tree of Arizona

Buckeye

Broadleaf

Deciduous

Forms flowers and fruit

State tree of Ohio

California Redwood / Giant Sequoia

Needleleaf

Evergreen

Conifer

State tree of California

Candlenut (Kukui)

Broadleaf

Deciduous

Forms flowers and fruit

State tree of Hawaii

Cedar

Needleleaf

Evergreen

Conifer

Provincial tree of British Columbia

Cottonwood

Broadleaf

Deciduous

Forms flowers and cottony seeds

State tree of Kansas and Nebraska

Dogwood

Broadleaf

Deciduous

Forms flowers and fruit

State tree of Missouri and Virginia

Elm

Broadleaf

Deciduous

Forms fruits which are small, flat seeds encased in paper-like wings

State tree of Massachusetts and North Dakota

Fir

- Needleleaf
- Evergreen
- Conifer
- State tree of Oregon
- Provincial tree of New Brunswick, Yukon Territory

Hemlock

- Needleleaf
- Evergreen
- Conifer
- State tree of Pennsylvania and Washington

Magnolia

- Broadleaf
- Deciduous
- Forms large flowers and fruit
- State tree of Mississippi

Maple

- Broadleaf
- Deciduous
- Leaf resembles shape of a hand
- Forms winged fruits encased in paper-like wings
- State tree of New York, Rhode Island, Vermont, West Virginia, Wisconsin

Oak

- Broadleaf
- Deciduous
- Forms flowers and fruit
- State tree of Connecticut, Georgia, Illinois, Iowa, Maryland, New Jersey, symbol of District of Columbia
- Provincial tree of Prince Edward Island

Palm

- Palm
- Evergreen
- Forms fronds and fruit
- State tree of Florida, South Carolina

Pecan

- Broadleaf
- Deciduous
- Forms flowers and fruit
- State tree of Texas

Pine

- Needleleaf
- Evergreen
- Conifer
- State tree of Alabama, Arkansas, Idaho, Maine, Michigan, Minnesota, Montana, Nevada, New Mexico, North Carolina
- Provincial tree of Ontario

Quaking Aspen

- Broadleaf
- Deciduous
- Forms flowers and cottony seeds
- State tree of Utah

Redbud

Broadleaf

Deciduous

Forms flowers and seed pods

State tree of Oklahoma

Spruce

Needleleaf

Evergreen

Conifer

State tree of Alaska, Colorado, South Dakota

Provincial tree of Manitoba, Newfoundland, Labrador

Tamarack (Eastern Larch)

Needleleaf

Deciduous

Conifer

Provincial tree of Northwest Territories

Write two entries for the Field Guide. Include a drawing or picture of each tree.

Tulip Poplar

- Broadleaf
- Deciduous
- Forms flowers and fruit
- State tree of Indiana, Kentucky, Tennessee

Flowers

- Physical Characteristics
- Habitat
- Interesting Facts

Apple Blossom

- A tree flower; five white or pink fragrant petals on each flower

  Fruit ripens in autumn
- Grows in temperate climates worldwide
- State flower of Arkansas, Michigan

Aster

- Composite

  Daisy-like flower with yellow disk flowers and white, blue, pink, red, or purple ray flowers
- Grows in open areas in North America except northern and eastern Canada
- Approximately 250 kinds of asters

Bitterroot

- A low-growing, leafless flower that ranges in color from white to purple
- Grows in western United States and Canada, blooming in early spring
- Edible roots

State flower of Montana

Black-Eyed Susan

- Composite, member of the aster family

A yellow ray flower with dark brown or purplish disk flowers on a rounded or cone-shaped center

- Grows in open prairie areas throughout North America except northern Canada and Alaska
- Once used by Native Americans to treat infections

State flower of Maryland

Bluebonnet

- Blue, pea-like flowers found in clusters of up to fifty on one stem
- Grows in open fields and along roadsides; petals resemble a woman's sunbonnet
- Also called buffalo clover or wolf flower

State flower of Texas

Camellia

- Evergreen shrub with red, pink, white, or variegated petals
- Grows best in cool, shady areas with well-drained soil
- Leaves used to make tea

State flower of Alabama

Carnation

- Fringed, double flower may be white, yellow, pink, red, or purple with a spicy-sweet scent and a slender stem with swollen joints
- Grows in flower gardens in many parts of the world

Flower petals are edible.

State flower of Ohio

Chicory

- Composite

Light purple flowers on a hairy stem that grows three to five feet tall

- Found in fields and along roadsides throughout North America except northern Canada
- Roots sometimes ground to make a coffee-like beverage; can be eaten raw in a salad

Chrysanthemum

- Composite, member of the aster family

Bushy plant that grows in a variety of shapes; flowers can be white, yellow, pink, orange, gold, purple, or red

- Grows best in temperate climates worldwide
- Stems and leaves give off a strong odor when touched

Clover

- Red, pink, or white flowers with three-part leaves
- Grows in open areas, along roadsides, and in lawns throughout most temperate climates worldwide
- Bees collect its pollen to use in honey production; clover used for livestock feed

  Red clover: State flower of Vermont

Columbine

- Bi-colored flower with blue outer and white inner petals
- Grows in Canada and United States as far south as Florida and Texas
- State flower of Colorado

Cosmos

- Composite, member of the aster family
- Tall plant with thin stems and delicate leaves; white, orange, pink, or red blossoms
- Grows well in hot climates; needs little water

  Easy to grow; attracts many butterflies

Dahlia

- Composite, member of the aster family

  Flowers are of various colors and may be shaped like daisies or pompoms
- Grows in moist, temperate climates
- Some varieties have blooms up to twelve inches in diameter.

Daisy

- Composite, member of the aster family

  Yellow disk flowers form center; white ray flowers radiate from center
- Grows well in open areas throughout North America except northern Canada
- Also called ox-eye daisy, white daisy, marguerite, or whiteweed

Dandelion

- Composite with ray flowers but no disk flowers

  One yellow bloom that opens and closes each day; hollow, milky stem, and leaves that can be used in salads
- Grows well in grasslands and open areas throughout North America; spreads quickly in yards
- Blossom dries into a puffball that carries the seeds on wind currents

Dogwood

- A tree flower

  Small, greenish white flowers surrounded by four to six large pink or white bracts, or special leaves
- Grows well in fertile soil in woods, thickets, and bogs in southern Canada and most of the United States
- State flower of North Carolina and Virginia

  Provincial flower of British Columbia

Fireweed

- Two- to six-foot-tall plant with a spike of light pink to red flowers
- Grows in most of North America except for very hot regions
- One of the first flowers to grow in an area which has been burned by a forest fire

Provincial flower of Yukon Territory

Forget-Me-Not

- Flowers grow in clusters; usually blue, but may be pink or white; light green, small, hairy leaves; height varies from four to twenty-four inches
- Grows in eastern United States to southeastern Canada and the Pacific coast and Alaska
- Prefers a temperate climate

State flower of Alaska

Goldenrod

- Composite, member of the aster family

Tall plant with tough stem and hundreds of tiny, yellow flower heads growing in clusters along the stem

- Grows in North American woodlands, swamps, mountains, and fields from the Great Plains east to the Atlantic Ocean
- Produces large amounts of pollen and nectar; attracts wide variety of insects

State flower of Kentucky and Nebraska

Hibiscus

- Large, colorful flowers; may be red, pink, yellow, or orange with noticeably long stamens
- Shrub which grows well in gardens in the tropics and subtropics
- Hibiscus tea made from flower and leaves

State flower of Hawaii

Indian Paintbrush

- Tall prairie flower with red and green blooms
- Grows up to three feet; found in western areas of North America
- Attaches to a host plant and takes water and nutrients from it

State flower of Wyoming

Iris

- Six petals—three drooping, and three forming a dome; usually have yellow "beard" in the center
- Grows well in many climates; reaches a height of four feet; sword-like leaves
- Blooms resemble orchids.

State flower of Tennessee

Blue flag iris: Provincial flower of Quebec

Ironweed

- Composite, member of the aster family
- Dark purple flowers on a stem up to ten feet tall
- Grows on prairies and grasslands in central and eastern North America
- Name comes from its strong, tough stem

Lady's Slipper

- Usually white, pink, yellow, or magenta; grows from six to sixteen inches tall
- Found in woods, thickets, bogs throughout eastern North America except southeast
- Wild orchid; sometimes called the moccasin flower because of its shape
- State flower of Minnesota
- Provincial flower of Prince Edward Island

Lilac

- Shrub or small tree; grows up to ten feet; oval or heart-shaped leaves; tiny white, yellow, deep red, or purple flowers found in clusters up to ten inches long
- Grows well in brushy areas of southern Canada and throughout the United States
- Oil used in cosmetics and cleaning products
- State flower of New Hampshire

Lily

- One or two short-lived flowers; long, narrow leaves
- Found in fields and meadows of southern-central Canada, central and eastern United States

Some varieties are edible.

State flower of Utah

Provincial flower of Saskatchewan

Magnolia

- A tree flower with small to large pink or white petals
- Grows in southern areas of the United States
- Seeds are eaten by rodents and large birds.

State flower of Louisiana and Mississippi

Marigold

- Composite, member of the aster family

Yellow or orange flowers with overlapping petals

- Grows well in warm climates and bright sun
- Edible flower used in salads or for decoration; strong scent

Mountain Avens

- Member of the rose family with short, woody stems and small, white flowers; toothed, leathery leaves
- Grows in open, high-altitude, rocky areas in the northwestern United States and Canada
- Provincial flower of Northwest Territories

Mountain Laurel

- A flowering evergreen shrub

  Fragrant, star-shaped clusters of flowers which bloom in spring and summer
- Grows in shady areas of the eastern United States
- State flower of Connecticut and Pennsylvania

Orange Blossom

- A fragrant tree flower

  Waxy, white blossom with five petals
- Grows in central and southern Florida; native to Southeast Asia
- Produces oranges

  State flower of Florida

Oregon Grape

- A flowering evergreen shrub

  Blooms in early summer; small, yellow clusters of flowers produce a blue berry that ripens in autumn
- Grows in the Pacific coastal areas
- State flower of Oregon

Pasqueflower

- Cup-shaped purple flowers; two- to ten-inch stalk, leaves, and seeds covered with fine hairs
- Grows in prairies and open slopes of Alaska, Canada, Midwestern United States, Mexico
- Also called prairie crocus

  State flower of South Dakota

  Provincial flower of Manitoba

Peach Blossom

- A tree flower

  Pink blooms appear in early spring before tooth-edged leaves
- Grown in orchards or in the wild; produces peaches in late summer
- State flower of Delaware

Peony

- Shrub with large, glossy leaves; large, single or double blooms; white, pink, crimson, or rose-colored
- Native to mostly Europe and Asia, two varieties native to North America
- Flowers close up at night or when sky is overcast

  State flower of Indiana

Pitcher Plant

- Hollow, pitcher-shaped leaves contain sweet nectar that attracts insects; trapped insects become plant's food.
- Grows in bogs and marshes of eastern United States, central and southeastern Canada
- Also known as Indian-cup or Indian-jug

  Provincial flower of Labrador and Newfoundland

Poppy

- Single five-petaled, brilliant yellow, orange, or red flower on a stalk with lacy leaves; grows to twenty-four inches
- Grows wild in grasslands, meadows, and dunes of southwestern Canada, western and southwestern United States, and Mexico
- On Veterans Day or Remembrance Day, the red poppy is worn around the world to remember the soldiers who gave their lives for freedom during World War I.

State flower of California

Purple Saxifrage

- Clusters of small, funnel-shaped purple flowers close to the ground
- Grows in mountainous, arctic climates; plant forms cushion over rocky areas soon after the snow melts
- Name means "rock-breaker"

Provincial flower of Nunavut

Rhododendron

- A flowering shrub that often forms dense thickets; from ten to twenty feet tall; leaves four to twelve inches long with funnel-shaped flowers that can be white, yellow, pink, red, or purple
- Grows in temperate climates on moist mountain slopes and streambanks from New England to Georgia; from British Columbia to California and in Southeast Asia
- State flower of Washington and West Virginia

Rose

- Bushy plant with five-petaled pink, white, red, or yellow flowers and usually five to nine leaflets making up each leaf; normally has thorns on the stem; cultivated varieties often have several layers of sweetly scented petals always in multiples of five
- Grows in gardens and open woods worldwide
- Rose hips are the fruit of the rose and are rich in vitamin C.

  State flower of Georgia, Iowa, New York, North Dakota, Oklahoma; National flower of the United States

  Provincial flower of Alberta

Sagebrush

- Grows up to twelve feet in well-watered areas and produces small white or yellow flowers in late summer
- Grows in desert areas of southwestern United States
- Strong, pungent fragrance when wet

  State flower of Nevada

Saguaro Cactus Blossom

- Waxy, fragrant, white flowers bloom in clusters on cactus plants over fifty feet tall.
- Grows in desert areas of North America; blooms appear in May and June
- State flower of Arizona

Sunflower

- Composite, member of the aster family; produces edible seeds; yellow ray flowers and dark disk flowers grow on a flat center
- Grows in fields and prairies throughout North America
- Follows the sun's path

  Flower may be over one foot wide; stalk may be fifteen feet tall.

  State flower of Kansas

Syringa

- A flowering shrub that grows up to ten feet with clusters of white, fragrant flowers having four to five petals
- Grows in western United States; blooms from May to July
- State flower of Idaho

Tansy

- Composite, member of the aster family with orange-yellow flowers that look like daisies without white ray flowers; fern-like leaves along stem
- Grows along roadsides and open areas throughout United States (except the Southwest) and into much of Canada
- Also called bitter buttons because of strong, bitter odor; also known as golden-buttons

Thistle

- Composite, member of the aster family with cup-like pink or purple flower heads composed of slender disk flowers; spiny stem scattered with jagged, edible leaves; between two and six feet tall
- Grows in open areas throughout North America (except northern Canada)
- Dead bloom replaced by fluffy hairs similar to the dandelion's which carry away the seeds

Trailing Arbutus

- Creeping plant with pink to white trumpet-shaped flowers; five-petaled blossoms grow in clusters; oval, leathery leaves in pairs; stem grows about six inches off the ground
- Grows in sandy or boggy areas in central and southeastern Canada and eastern United States
- Also known as ground laurel or mayflowers

  State flower of Massachusetts

  Provincial flower of Nova Scotia

Trillium

- Common woodland flower with three petals, three sepals, three leaves on a single stem; four to twenty inches tall depending on variety
- Grows in rich woods and thickets in eastern parts of Canada and the United States
- Also called wake-robin

  Provincial flower of Ontario

Violet

- Low-growing plant; dark green heart-shaped leaves; five-petaled flower grows on its own stem
- Found in woods, meadows, lawns, swamps, and open areas worldwide
- Edible; flowers and leaves used to decorate salads and cakes or other desserts

State flower of Illinois, New Jersey, Rhode Island, Wisconsin; Provincial flower of New Brunswick

White Hawthorn

- A flowering shrub that grows up to twenty feet
- Found all over North America
- Produces small, apple-like fruit eaten by animals and used to make jam

State flower of Missouri

White Pinecone and Tassel

- Not a flower, but a cone that produces seeds
- Grows in eastern United States
- Largest conifer in northeastern United States; blooms year-round

State flower of Maine

Yellow Jessamine

- Evergreen, flowering vine with clusters of fragrant, yellow blooms that appear in late winter or early spring
- Grows in southeastern United States into temperate climates of Central America
- Poisonous if eaten

State flower of South Carolina

Yucca

- Member of the lily family that can grow as tall as a tree; greenish-white flowers grow on tall spikes
- Grows well in dry, hot climates
- State flower of New Mexico

Zinnia

- Composite, member of the aster family that produces flowers of every color except blue; flowers grow on stiff, hairy stems with oval leaves
- Native to southern areas of North America; prefers warm climates
- Attracts many butterflies

Write one entry for the Field Guide.
Include a drawing or picture of the flower.

Birds

- Physical Characteristics
- Habitat
- Diet
- Interesting Facts

Atlantic Puffin

- Fishing beak; swimming feet
- Open sea; seacoast to lay single eggs; nests in large colonies along eastern Canada, New England coastal islands, North Atlantic Ocean
- Fish, shellfish, shrimp
- Provincial bird of Newfoundland and Labrador

Baltimore Oriole

- Nectar/insect-eating beak; perching feet
- Eastern North America from southern Canada to northern Mexico in edges of woodlands, trees bordering fields and streams
- Caterpillars, beetles, wild fruit, nectar
- State bird of Maryland

Blue Jay

- Seed-eating beak; perching feet; Steller's jay's head and upper portions are black
- Forests, woodlands, orchards, parks, gardens; blue jay found in eastern North America; Steller's jay found in western North America
- Nuts, fruit, seeds, insects, other birds' nestlings; buries seeds and acorns
- Blue jay: Provincial bird of Prince Edward Island

  Steller's jay: Provincial bird of British Columbia

Bluebird

- Insect-eating beak; perching feet; eastern bluebird has blue head and wings, white belly, reddish breast; mountain bluebird completely blue
- Eastern bluebirds found east of the Rocky Mountains; mountain bluebirds found in Alaska, western Canada, southwestern United States; open country, orchards, farmlands, gardens
- Insects, fruit
- Eastern bluebird: State bird of Missouri and New York

  Mountain bluebird: State bird of Idaho and Nevada

Brown Pelican

- Scooping beak; swimming feet
- Along the Atlantic and Pacific coasts of North and South America in salt bays, beaches, coastal islands
- Fish, small marine invertebrates
- State bird of Louisiana

Brown Thrasher

- Insect-eating beak; perching feet
- East of the Rockies from south central Canada and northern New England southward to Gulf of Mexico
- Mostly insects along with some berries and nuts
- State bird of Georgia

California Gull

- Heavy, insect-eating beak; swimming/wading feet
- Pacific coast (winter) and inland lakes (summer); found from northern prairie provinces of Canada, south to northwestern Wyoming and Utah, and west to northeastern California
- Fish, insects, small rodents, eggs, scavenged foods
- State bird of Utah

California Quail

- Seed-eating beak, running feet
- From California to southern Oregon and northern Nevada; Vancouver, Canada; Hawaii; open countryside with brush and thickets for cover
- Seeds, buds, berries, some insects (especially in winter)
- State bird of California

Cardinal

- Seed-eating beak; perching feet
- Southeastern Canada throughout much of the United States and Mexico; dense shrubs in backyards, parks, forests
- Seeds, fruit, some insects
- State bird of Illinois, Indiana, Kentucky, North Carolina, Ohio, Virginia, West Virginia

Chickadee

- Seed/insect-eating beak; perching feet
- Black-capped chickadee found in deciduous forests, woods, thickets in United States, southern and western Canada
- Insects in spring, summer, and autumn; seeds, berries, other plant matter in winter
- Chickadee: State bird of Maine, Massachusetts

  Black-capped Chickadee: Provincial bird of New Brunswick

Chicken

- Seed/insect-eating beak; scratching feet

  Rhode Island Red is a medium-sized chicken with brilliant red feathers; Blue Hen is a blue strain of the American gamecock.
- Domestic farms and yards
- Seeds, small insects, worms
- Blue Hen Chicken: State bird of Delaware

  Rhode Island Red Chicken: State bird of Rhode Island

Finch

- Seed-eating beak; perching feet
- Woodlands and coniferous forests in southern Canada and the United States in areas with thistle plants, near feeders, in orchards, gardens
- Weed seeds, especially thistle and dandelion, wild fruits, some insects
- Goldfinch: State bird of Iowa, New Jersey, Washington

  Purple Finch: State bird of New Hampshire

Grouse

- Insect-eating beak; running, scratching feet
- Ruffed Grouse found in Alaska, Canada, New England, California, South Dakota, Carolinas; Prairie Sharp-Tailed Grouse found in Canada and north central United States
- Berries and fruits of forest plants in summer; bugs, catkins, and twigs in winter
- Ruffed Grouse: State bird of Pennsylvania

  Prairie Sharp-Tailed Grouse: Provincial bird of Saskatchewan

Gyrfalcon

- Bird-of-prey beak and feet
- Northernmost United States and Canada
- Peregrine falcons, smaller birds and mammals
- Provincial bird of Northwest Territories

Hawaiian Goose (Nene)

- Berry and plant-eating bill; partially swimming feet (webbing not complete)
- Hawaiian Islands
- Fruit, seeds, plants
- State bird of Hawaii

Hermit Thrush

- Seed/insect-eating beak; perching feet
- Forests in southern Alaska, southern Canada, and most of the United States
- Insects, berries, worms, small seeds
- State bird of Vermont

Lark Bunting

- Insect-eating beak; perching feet
- Dry plains and prairies; sagebrush with short grass; western plains, from southern Canada through the western-central United States
- Insects (especially grasshoppers), seeds
- State bird of Colorado

Loon

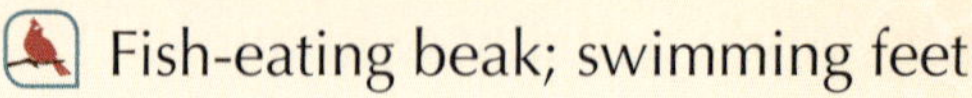

- Fish-eating beak; swimming feet
- Lakes, rivers, coastal areas throughout North America
- Fish, shellfish, frogs, insects
- State bird of Minnesota
  Provincial bird of Ontario

Mockingbird

- Insect-eating beak; perching feet
- Throughout North America; farms, towns, parks, open areas of grasslands; shrubs, lawns, even desert thickets
- Insects, berries, worms
- State bird of Arkansas, Florida, Mississippi, Tennessee, Texas

Osprey

- Bird-of-prey beak and feet
- Rivers and lakes, coastal areas worldwide (except Antarctica)
- Fish
- Provincial bird of Nova Scotia

Owl

- Bird-of-prey beak and feet
- Forests, cliffs, and canyons worldwide (except Antarctica)
- Small mammals, fish, frogs, lizards, small birds
- Great Horned Owl: Provincial bird of Alberta

  Great Gray Owl: Provincial bird of Manitoba

  Snowy Owl: Provincial bird of Quebec

Ptarmigan

- Seed-eating beak; scratching feet
- Arctic tundra and mossy bogs of Alaska and northern Canada
- Plants, berries, leaves; twigs and buds from willow trees
- Willow Ptarmigan: State bird of Alaska

  Rock Ptarmigan: Territorial bird of Nunavut

Raven

- Bird-of-prey beak and feet
- Most of Northern Hemisphere in forests, mountains, rocky coasts, and deserts
- Almost anything: fish, frogs, large insects, small mammals and reptiles, eggs, nestlings; garbage and carrion (dead animals)
- Provincial bird of Yukon Territory

Ring-Necked Pheasant

- Seed/insect-eating beak; running feet
- Brush, pastures, along grassy edges of woods, outskirts of large cities
- Grain, berries, seeds, insects, small animals
- State bird of South Dakota

Roadrunner

- Insect-eating beak; running feet
- Desert of southwestern United States; open, dry places with scattered trees, cacti, and stones
- Grasshoppers, lizards, snakes, mice, small gophers, fruit, seeds
- State bird of New Mexico

Robin

- Insect-eating beak; perching feet
- From Alaska and Canada to southern Mexico
- Wild fruit, earthworms, insects, seeds
- State bird of Connecticut, Michigan, Wisconsin

Scissor-Tailed Flycatcher

- Insect-eating beak; perching feet
- Southern Great Plains, Texas
- Insects, berries, seeds
- State bird of Oklahoma

Western Meadowlark

- Insect-eating beak; perching feet
- Open grassland near marshes or mountain meadows
- Cutworms, grasshoppers, wild grass seed, waste grain
- State bird of Kansas, Montana, Nebraska, North Dakota, Oregon, Wyoming

Wren

- Insect-eating beak; perching feet
- Thickets, brush, gardens, backyards; Cactus wren found in deserts of southwestern United States; Carolina wren found throughout eastern United States
- Insects, spiders, lizards, fruit
- Cactus wren: State bird of Arizona

  Carolina wren: State bird of South Carolina

Yellowhammer Woodpecker (Northern Flicker)

Insect-eating beak; perching feet

Wooded areas, forest edges, fields, marshes

Insects, fruits, seeds

Only North American woodpecker that searches the ground for insects

State bird of Alabama

Write two entries for the Field Guide. Include a drawing or picture of each bird.

Insects

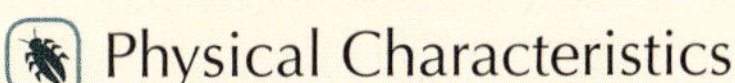

- Physical Characteristics
- Habitat
- Diet
- Interesting Facts

Ant

- Large head with two sets of jaws; oval abdomen joins thorax by small waist; some have a stinger at the end of the abdomen; can carry between ten and fifty times its own weight
- Worldwide; most prefer warm climates; some live in trees, but most live underground; social insects—found under rotting logs or rocks
- Bits of fruit, vegetables, bread crumbs, meat
- Queen ant lays eggs; worker ants care for eggs.

Bed Bug

- Reddish-brown, tiny, flat; unpleasant scent
- Temperate climates worldwide
- Blood from humans and warm-blooded animals
- Sucks blood from host at night; rarely seen during the day

Beetle

- Hard wing coverings for protection; biting-chewing mouth
- Almost every area of the world on land and in fresh water
- Plant parts, other animals, waste materials
- Titan beetle is longest beetle (seven inches).

Booklice

- Tiny, transparent, wingless insects
- Found most often in areas of high humidity and temperature
- Mold and fungi, dead matter, grains
- Often found around old books and papers in libraries or in facilities where paper is stored

Burying Beetle

- One and one-half inches long; shiny, black body with four orange markings on the abdomen and one behind the head; largest carrion beetle; strong flier
- Rhode Island, Oklahoma, Arkansas, Nebraska, South Dakota
- Carrion (animals that have died)
- Looks for small, dead animals, such as birds or mice, to bury; digs a tunnel for itself near the dead animal where female lays eggs; larvae feed on dead animal; endangered species

State insect of Rhode Island

Checkerspot Butterfly

- Brush-footed; black wings with white and orange markings
- From Canada to Great Lakes regions and eastern United States
- Caterpillars feed on composite flowers including black-eyed Susan and sunflowers.
- Named for the first Lord Baltimore because of colors on his shield

  State insect of Maryland

Cricket

- Grasshopper-like insect, usually brown or black; up to two inches long; thin antennae; jointed legs for jumping; "ears" located on its front legs; wings lie flat, one on top of another; wings used for chirping (males)
- Worldwide
- Vegetable refuse, other insects
- Chirps slowly in cool air and more rapidly in warm air

Damselfly

- Similar to the dragonfly, but more slender; usually folds its wings at rest; wing color varies among species; some wingspans reach six inches
- Shallow, freshwater areas throughout temperate climates
- Other insects, even other damselflies
- Usually lays eggs in water

  State insect of Nevada

Dogface Butterfly

- Yellow lower wings; black upper wings with markings resembling a dog's head
- California, from Sierra Nevada foothills to Coast Ranges and south to San Diego
- Caterpillars feed on leafed plants of pea family, alfalfa, and clover.
- The first state insect chosen in the United States

  State insect of California

Dragonfly

- Especially large eyes that can see moving insects eighteen feet away; catches prey in flight; clear wings; flies rapidly forward, backward, and vertically and can change direction instantly
- Near ponds and lakes worldwide
- Other insects
- Rarely walks; legs are used for perching and for grasping prey

  State insect of Alaska, Washington

Earwig

- Has wings but rarely flies; flattened body enables it to crawl into narrow spaces; two pincers at the end of its abdomen for defense and capturing prey
- Worldwide (except in polar regions)
- Flowers, fruit, insects
- Nocturnal

Firefly

- Soft-bodied beetle; flat, brown or black with reddish-orange markings near head; organ on underside of abdomen has special cells that produce flashes of light
- Most tropical and temperate regions
- Some adults do not eat; others eat pollen and nectar.
- All males and some females fly; females without wings are called glowworms.

State insect of Indiana, Pennsylvania, Tennessee

Flea

- Small, flightless insect; parasite; bristled body; carrier of disease
- Worldwide
- Blood of mammals and some birds
- Has strong legs that allow it to jump 200 times its body length

Grasshopper

- Thick, tough exoskeleton; chewing mouthparts; powerful legs that enable it to jump as far as thirty inches; hearing organs on its abdomen; compound eyes
- Grasslands, lowland tropical forests, temperate regions
- Plants
- Migrating grasshoppers are called locusts.

Hairstreak Butterfly

- Hair-like markings on underside of wings; Colorado hairstreak has purple wings with brown underwings; Wyoming green hairstreak has brown upper wings with green underwings
- Worldwide (mostly in North American tropical areas)
- Caterpillars feed on gambel oak and wild buckwheat.
- Fourth graders in Colorado are responsible for their state choosing a state insect.

State insect of Colorado, Wyoming

Honeybee

- Two pairs of wings; stinger; "pollen baskets" on hind legs; chewing-lapping mouth; long, dagger-like tongue
- Worldwide (except Antarctica)
- Nectar and pollen from flowers
- Social insect; lives together in a hive which consists of many small, six-sided wax cells to store honey and bee larvae

State insect of Alabama, Arkansas, Georgia, Kansas, Kentucky, Louisiana, Maine, Mississippi, Missouri, Nebraska, New Jersey, North Carolina, Oklahoma, South Dakota, Tennessee, Texas, Utah, Vermont, West Virginia, Wisconsin

Housefly

- Gray with yellowish abdomen; four lines down thorax; two see-through wings; two thick, short antennae between its eyes; foot pads grip any surface; tastes with its feet
- Tropical and temperate climates; found anywhere people live
- Spoiled food, decaying matter
- Uses antennae to feel changes in the movement of air or to smell rotting meat and garbage

Kamehameha Butterfly (Pulelehua)

- Brown underwings; orange and black upper wings with black and white spots
- Hawaii
- Caterpillars feed on native Hawaiian plant leaves.
- Not found outside of Hawaiian Islands

State insect of Hawaii

Ladybug

- Beetle; small, round red or yellow body with black spots on wing covers; short legs
- Worldwide (except in Arctic regions)
- Other insects, particularly aphids
- Larvae can eat 400 aphids each day.

State insect of Delaware, Massachusetts, New Hampshire, New York, North Dakota, Ohio, Tennessee

Monarch Butterfly

- Large butterfly with wingspan of 3½ to 4 inches; orange and black markings; bitter taste to predators
- North, Central, and South America; Hawaii; Australia; India
- Caterpillars feed on milkweed leaves.
- Migrate yearly from Canada to southern California, Florida, or Mexico for winter; may travel over 1,800 miles each way

State insect of Alabama, Idaho, Illinois, Minnesota, Texas, Vermont, West Virginia

Mosquito

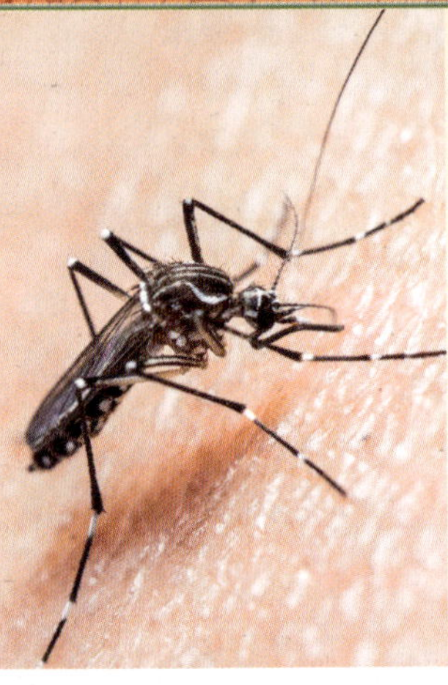

- Long, thin bodies and legs; piercing-sucking mouth; feathery antennae
- Worldwide (except extremely cold areas)
- Male and female eat the juices of plants and flowers; female eats blood to provide protein for eggs.
- Only the female bites; pierces skin with two tubes, one with a chemical to keep blood from clotting and one to suck the blood.

Moth

- Males have feather-like antennae; wings lay flat at rest.
- Worldwide
- Many do not have mouths (do not eat at all); those with mouths drink nectar.
- Most species are nocturnal

Mourning Cloak Butterfly

- Deep red to dark brown wings with bright blue spots and a yellow border; hairy, brush-like front legs
- Throughout North America
- Caterpillars feed on willows, aspen, paper birch leaves.
- Has the longest lifespan of any butterfly—ten months

  State insect of Montana

Praying Mantis

- Carnivorous; large, grasshopper-like insect; large front legs with spines for catching and holding prey
- Temperate and tropical climates worldwide
- Insects, small mice and frogs
- Holds its forelegs in what looks like a position of prayer, ready to grab prey

State insect of Connecticut, South Carolina

Roach

- Flat, oval, usually dark brown or black; most have wings but few fly; active at night
- More common in warm regions of the world
- Living or dead plant or animal material
- Can transmit bacteria to human food sources

Silkworm

- Caterpillars produce strands of silk for cocoon used in silk production; adult silkworm moth lives only a few days
- Raised only in captivity
- Prefers mulberry leaves
- Silkworms have been used to produce silk for thousands of years.

Silverfish

- Wingless; three tail bristles; silvery scales
- Mostly indoors worldwide
- Materials containing starch, such as wallpaper and book bindings
- Nocturnal; usually hides during the day

Termite

- Red, brown, black or whitish in color; short antennae; biting-chewing mouth
- Found worldwide in warm climates; dry wood above ground or moist, rotting wood underground
- Wood, leather, cables; crops such as sugar cane and fruit trees
- Tiny creatures inside intestines of termite help it digest wood.

Tiger Swallowtail Butterfly

- Yellow spots around the edge of each wing; spots of blue and orange on hind wings; "tails" on hind wings; emits musky odor for protection
- Throughout North America; mostly east of the Rocky Mountains
- Caterpillars feed on leaves from woody plants such as wild cherry, willow, birch, and cottonwood.
- One of the most easily recognized butterflies in Eastern United States

  State insect of Arizona, Delaware, Georgia, Oregon, South Carolina, Virginia

Viceroy Butterfly

- Mimics Monarch butterfly, but smaller; wings have extra black line across the middle
- Most of North America
- Caterpillars feed on leaves of willow and poplar trees.
- Does not feed on milkweed or have a bitter taste, but because it resembles the monarch butterfly so closely, birds are afraid to eat it

State insect of Kentucky

Wasp

- Smooth body, stripes around abdomen; females have stinger; two pairs of wings
- Worldwide
- Fruit, dead insects, nectar
- Queen gathers wood fiber from dead trees or posts, chews it well, and forms it into paper sheets for nest with six-sided cells.

White Admiral Butterfly

- Black with wide, white stripes through middle of both wings; blue and red spots
- Southern Canada, northeastern United States
- Caterpillars feed on willow, aspen, poplar, and cherry trees.
- Rarely seen on flowers; prefers rotting fruit

Provincial insect of Quebec

Zebra Longwing Butterfly

- Long, black wings with thin, zebra-like stripes; makes a creaking sound when frightened
- Damp, tropical areas from Texas to Florida, Central America
- Caterpillars feed on leaves of passion flowers.
- Poisonous to predators

State insect of Florida

Write two entries for the Field Guide. Include a drawing or picture of each insect.